BOILERMAN
1 & C

Prepared by

BUREAU OF NAVAL PERSONNEL

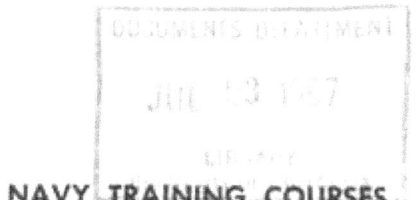

NAVY TRAINING COURSES

NAVPERS 10536–B

UNITED STATES
GOVERNMENT PRINTING OFFICE
WASHINGTON: 1957

PREFACE

This book is a revision of the Boilerman 1 and C training course which was first published in 1950 and was revised in 1953. It is written for enlisted men of the Navy and the Naval Reserve who are preparing for advancement to the rates of Boilerman 1 and Chief Boilerman. Study of this text should be combined with practical experience and with study of the appropriate references listed in chapter 1.

The qualifications for advancement in the Boilerman rating are given in appendix II of this book. Since the examinations for advancement in rating are based upon these qualifications, it is suggested that you refer to them frequently while studying this training course.

As one of the Navy Training Courses, this book has been prepared by the U. S. Navy Training Publications Center for the Bureau of Naval Personnel, with technical assistance from the Bureau of Ships, the Naval Boiler and Turbine Laboratory, and the U. S. Naval School Boiler-men, Philadelphia.

Hi

THE UNITED STATES NAVY

GUARDIAN OF OUR COUNTRY

The United States Navy is responsible for maintaining control of the sea and IS a ready force on watch at home and overseas, capable of strong action to preserve the peace or of instant offensive action to win in war.

It is upon the maintenance of this control that our country's glorious future depends; the United States Navy exists to make it so.

WE SERVE WITH HONOR

Tradition, valor, and victory are the Navy's heritage from the past. To these may be added dedication, discipline, and vigilance as the watchwords of the present and the future.

At home or on distant stations we serve with pride, confident in the respect of our country, our shipmates, and our families.

Our responsibilities sober us; our adversities strengthen us.

Service to God and Country is our special privilege. We serve with honor.

THE FUTURE OF THE NAVY

The Navy will always employ new weapons, new techniques, and greater power to protect and defend the United States on the sea, under the sea. and in the air.

Now and in the future, control of the sea gives the United States her greatest advantage for the maintenance of peace and for victory in war.

Mobility, surprise, dispersal, and offensive power are the keynotes of the new Navy The roots of the Navy lie in a strong belief in the future, in continued dedication to our tasks, and in reflection on our heritage from the past.

Never have our opportunities and our responsibilities been greater.

CONTENTS

Chapter Page

1. Requirements for advancement 1
2. Pumps and forced draft blowers 7
3. Auxiliary turbines and accessories 52
4. Fuel oil service equipment 86
5. Boiler fittings and instruments 110

6. Boiler water treatment and feed systems 144

7. Boiler operations 179

8. Boiler efficiency 222

9. Boiler retubing 236

10. Boiler refractories 257

11. Boiler cleaning and maintenance 291

12. Fireroom maintenance 325

13. Navy repair procedures 347

14. The oil king 392

15. Automatic combustion control 430

16. Propulsion turbines 488

17. Engineering casualty control 508

18. Records and reports 536

19. Inspections and trials 562

Appendix

I. Answers to quizzes 597

II. Qualifications for advancement in rating 615

Index 625

CREDITS

All illustrations in this edition of Boilerman 1 and C are official U. S. Navy illustrations, with the exception of the copyrig^hted illustrations designated below.

Source

American Society for

Testing Materials

Bailey Meter Co

Combustion Engineering Co.

De Laval Steam Turbine Co

Leslie Co

United States Naval Institute:

Naval Auxiliary Machinery..

Naval Boilers

Naval Turbines

Westinghouse Electric Corp

Figures

8-1. 15-15.

15-2, 15-3, 15-4, 15-6, 15-7, 15-8, 15-9, 15-10, 15-11, 15-12, 15-13, 15-14, 15-16, 15-17, 15-18, 15-19, 15-20,

15- 21, 15-22. 3-5.

12-4.

2- 8, 4-5, 4-6, 4-7, 5-12. 5-3, 5-11, 6-3.

3- 1, 3-2, 3-3, 3-4, 3-6(A), 3-7,

16- 4. 2-12.

AaiVE DUTY ADVANCEMENT REQUIREMENTS

'^Rocommondotion of potty offlcort, ofDcort ond opprovol by commonding offlcor roquirod for oil odvancomontt. #Soo duty not roquirod of oviotion rotingt; CT, CM, JO, MA and CS rofingt; mon in LS or L6; TARS or onlittod womon.

INACTIVE DUTY ADVANCEMENT REQUIREMENTS

Rocommondotion of potty ofUcori, offlcors ond opprovol by comlnonding officer roquirod for oil odvoncomontt. ^Activo duty poriodt moy bo tubttitutod for drills ond troining duty.

BOILERMAN 1 & C

READING LIST

NAVY TRAINING COURSES

Boilerman S & 2, NavPers 10536«C Basic Hand Tool Skills, NavPers 10085 (metal working skills only) Blueprint Reading and Sketching, NavPers 10077-A (Chapters 2, 3,4,6, 10, 11)

OTHER PUBLICATIONS

Engineering, Operation and Maintenance, NavPers 10813 Bureau of Ships Manual, Chapters 41; 48; 51; 53; 55; 56; 88, Sect. Ill

USAFI TEXTS

U. S. Armed Forces Institute (USAFI) courses for additional reading and study are available through your Information and Education Officer.* A partial list of those courses applicable to your rate follows:

Number Title

Correspondence

CB 799 Potver Plant Engineering

Self-Teaching

MB 799 Power Plant Engineering

♦"Members of the United States Armed Forces Reserve Components, when on active duty, are eligible to enroll for, USAFI courses, services, and materials if the orders calling them to active duty specify a period of 120 days or more, or if they have been on active duty for a period of 120 days or more, regardless of the time specified in the active duty orders."

CHAPTER

1

REQUIREMENTS FOR ADVANCEMENT

The current qualifications for advancement in the Boiler-man rating require that the applicant be able to operate all types of marine boilers and fireroom machinery; transfer, test, and take inventory of fuels and water; and maintain and repair boilers, pumps, and associated machinery.

In order to know more specifically what is required for advancement to First Class or Chief Boilerman, you should make a careful study of the requirements for Boilerman given in the Mamuil of Qualifications for Advancement in Rating, NavPers 18068 (Revised). The portion of this Mamuil which deals with Boilerman qualifications is given in appendix II of this training course. As you study these qualifications, remember that they are the minimum requirements for each rate.

If you doubt your ability to meet any of the requirements for advancement, study appropriate training courses, manufacturers' instruction books, applicable chapters of the Bureau of Ships Manuxd, and other reference material. In addition to reading and studying, you should check yourself out on equipment, to make sure that you can meet the practical factor requirements.

MILITARY REQUIREMENTS

As you advance in the Boilerman rating, your military duties will become correspondingly more important. You will have an increased responsibility for leading and instructing other men, for showing them what to do and how to do it, and for checking their work.

This training course deals primarily with information related to your professional (technical) duties, and does not attempt any detailed consideration of your military duties. The basic information necessary to meet the military requirements of each rate is given in General Training Course for Petty Officers, NavPers 10055. The General Training Course for Petty Officers also lists additional references to be consulted for further information on military qualifications.

PROFESSIONAL REQUIREMENTS

As a petty officer, you must be a technical specialist in the subject matter of your rating. In order to meet the professional (technical) requirements for advancement to BTl or BTC, you will have to acquire new skills and new knowledge. You will have to know more about trouble shooting, maintenance, and repair of all fireroom equipment, and you will have to assume responsibility for the work performed by your men.

As a BTl or BTC, you will be responsible for compiling the data required for engineering records, reports, and material histories; and for making routine inspections, tests, and reports on fireroom machinery and equipment.

You may be responsible to the division officer for the proper setting and standing of all watches during your duty period. You may be required to post the daily watch list in the fireroom, and you may be responsible for instructing and training watch standers in their duties.

The importance of standing watch cannot be overemphasized. No man should be assigned to stand watch at any station until the division officer considers him to be completely qualified for that watch. It's your job to see that each man is assigned according to his ability and experience. A man in need of training should be assigned to work with an experienced man who can give him the necessary instructions. As a general rule, men should be rotated from one station to another so that each man can become qualified to stand watch on every station in the fireroom.

As a duty BTC, you will be responsible for seeing that the daily fireroom routine is carried out in accordance with all standing orders, and that all special orders are properly executed. You should be in the fireroom whenever a boiler is being lighted off or secured. It is your job to see that the boilers are steamed and auxiliary machinery is operated in the most economical way consistent with the ship's operating requirements. You must make sure that all entries in the fireroom log are accurate. You must be sure that the fireroom operating data board shows the correct steam pressure, oil temperature, fuel oil tank in use, standby fuel oil tank, feed water tank in use, standby feed water tank, and all other pertinent information. You must make such inspections of the fireroom as are necessary to ensure the proper and efficient operation of the fireroom plant.

You will be responsible for instructing and training your men in fireroom casualty control, so that emergencies can be handled as rapidly and efficiently as possible. You must assume a great deal of responsibility in the matter of safety. You must make sure that safety precautions are posted at each piece of equipment, and that your men observe all safety precautions relating to the operation and maintenance of fireroom machinery. Safety precautions are covered in the applicable chapters of BuShips Manical, in some manufacturers* instruction books, and in United States Navy Safety Precautions, OPNAV 34 PI.

As a BTl or BTC, you will have the over-all responsibility for many repair jobs. You should learn to handle repair work in a systematic manner, making use of all available information and maintaining appropriate records of the work accomplished. Special consideration should be given to the following items:

1. Before attempting to repair any machinery unit, assemble all pertinent blueprints, drawings, and dimensional data.

2. Use the manufacturers' instruction books. Detailed information concerning operation, maintenance and repair is given in the instruction book which is generally supplied by the manufacturer with ea^h machinery unit.

3. Consult the appropriate chapter or chapters of BuShips Manual for authoritative information and guidance on the operation, maintenance, and repair of all machinery under the cognizance of BuShips.

4. Consult the ship's allowance list to determine the accessories, special tools, and repair parts which are carried on board for the unit.

5. Check the appropriate machinery history card before doing any repair work on a unit. Make the proper notations on the machinery history card AFTER the repair job is completed.

As you advance in the Boilerman rating, it will become increasingly important for you to understand the duties and responsibilities of men in related ratings. You must know what work is done and what equipment is used by the Electrician's Mate, the Machinist's Mate, the Pipe Fitter, the Machinery Repairman, the Metalsmith, the Damage Controlman, and personnel of other Engineering and Hull group ratings. Although it is true that many repair jobs can be properly handled within your own division, some jobs require skills and equipment found only in other divisions. You must learn to work with the rest of the Engineering Department, and to utilize the skills and technical knowledge of other ratings when necessary.

SOURCES OF INFORMATION

One of the most valuable things you can learn about any subject is how to find out more about it. You should know where to look for accurate, up-to-date, authoritative information concerning all fireroom machinery and

equipment. The following publications may serve as a guide to the type of technical source material which you should know about:

Basic Hand Tool Skills, NavPers 10085 Basic Machines, NavPers 10624

Blueprint Reading and Sketching, NavPers 10077-A

Boilerman 3 and 2, NavPers 10535-C

Bureau of Ships Bulletin of Information (Published

through April 1952. Incorporated in Bureau of Ships

Journal, May 1952.) Bureau of Ships Journal (published monthly) Bureau of Ships Manual

Engineering, Operation and Maintenarice, NavPers

10813

Mathematics, NavPers 10069-A and 10070-A

The manufacturers* instruction books which are furnished with most machinery units are valuable sources of information on operation, maintenance, and repair. You should also become familiar with machinery handbooks, machinists' handbooks, and marine engineering handbooks, so that you will know how to go about looking for information in them.

You may find it useful to consult the Navy training courses prepared for other Group VII (Engineering and Hull) ratings. Reference to these training courses will add to your knowledge of the duties of other men in the Engineering Department.

When consulting any reference material, check to be sure that you are using the most recent edition available to you. The Navy training courses are revised at frequent intervals, to bring the material up-to-date; and new courses are added to the list from time to time. The current List of Training Publications, NavPers 10061 (revised), gives the titles and NavPers numbers of available Navy training courses. This list is revised about twice a year; therefore, be sure to consult the most recent edition. Other publications which you should check for information on available training courses include the Naval Training Bulletin, All Hands, and The Naval Reservist. A list of available enlisted correspondence courses is given in the Catalog of Enlisted Correspondence Courses. A pamphlet entitled Training Courses and Publications for General Service Ratings, NavPers 10052 (revised) contains a list of material to be studied by enlisted personnel seeking advancement in rating. This publication is revised from time to time: therefore, be sure to consult the most recent edition.

The BuShips Manual is constantly under revision, and all copies should be kept up-to-date. When using the Manual, or any other publication which is kept current by means of changes, be sure that you have a copy in which all changes have been made. It will not do you much good to study cancelled or obsolete information.

CHAPTER

PUMPS AND FORCED DRAFT BLOWERS

As a BT1 or BTC, you will need to have a considerable knowledge of fireroom auxiliary machinery. In this chapter we will take up pumps and forced draft blowers. The next chapter will deal with the auxiliary turbines which are used to drive the pumps and blowers; and with turbine accessories such as reduction gears, flexible couplings, and governors.

For the most part, the discussion of fireroom auxiliaries in this training course will deal with trouble shooting, maintenance, and repair. Information on operating principles, construction, and procedures for starting, running, and securing this machinery is given in Boiler-man 3 and 2, NavPers 10535-C.

RECIPROCATING PUMPS

Although reciprocating pumps were once widely used for a variety of services, their use on combatant vessels is now generally restricted to emergency feed pumps, fire and bilge pumps, and fuel oil tank stripping and bilge pumps. On auxiliary vessels, reciprocating pumps are still used for a number of services, including auxiliary feed, standby fuel oil service, fuel oil transfer, auxiliary circulating and condensate, fire and bilge, ballast, lube oil transfer, and cargo stripping.

Reciprocating pumps are relatively simple and economical to operate; and they are, for the most part, quite

reliable. They are, however, subject to certain operating difficulties; and, like all machinery units, they require a certain amount of maintenance and repair.

Operating Troubles

If a reciprocating pump fails to start, go through the whole starting procedure again, to make sure that everything has been done correctly. If the pump still fails to start, check the pressure in the steam cylinder by opening the top and bottom steam cylinder drains, one at a time, and seeing how much steam blows out each drain. Secure the throttle.

If you have low steam pressure (15 to 20 psi) at both ends of the steam cylinder, the steam blowing from the drains is exhaust steam which has backed into the pump from the exhaust line. This indicates that there is something wrong with the steam supply; perhaps there is a closed valve, or a valve in which the disk has become detached from the stem.

If high-pressure steam blows from both ends of the steam cylinder, the trouble is probably in the exhaust line. Again, it may be a closed or defective valve.

If you have a high steam pressure at one end of the steam cylinder and low steam pressure at the other end, and if the pump does not move even though the piston is in a position which should allow it to move, check the rod packing glands at the steam end and at the water end. If someone has set up on these glands too tightly, the pump may be locked so that it cannot move. If the packing glands are not at fault, the plunger or the piston is probably frozen.

If the pump has moved but will not run, check to see whether the discharge line is closed. You can check this by opening the pump vents, to see if there is excessive pressure on the discharge side.

If you have high steam pressure at one end of the steam cylinder and low steam pressure at the other, and if the high steam pressure is keeping the piston at the end of a

PISTOI

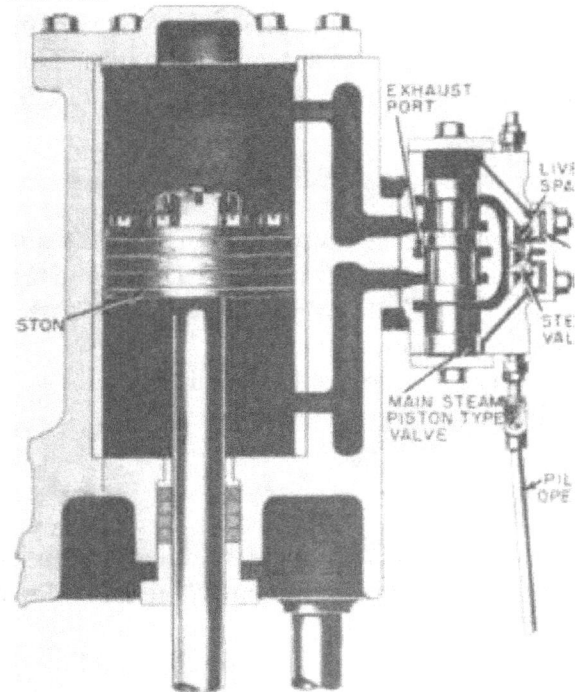

LIVE STEAM SPACE
STEAM PORT FOR PILOT VALVE
STEAM PILOT VALVE
PILOT VALVE OPERATING ROD

Figure 2-1.—Pitton-fype valve gear for reciprocating pump.

stroke, the trouble is probably in the steam valve chest. Disconnect the pilot valve operating rod from the valve-operating assembly. (See figs. 2-1 and 2-2.) Do NOT change the adjustment of the tappet collars. Open the exhaust, suction, and discharge valves; and then crack the throttle. Work the pilot valve by hand. It may be necessary to loosen the packing so that you can work the valve. If this does not correct the trouble, remove the steam valve chest cover and examine the main valve to see if it has overridden or if it has stuck.

If the pump still cannot be made to start, a complete overhaul of the steam end will probably be necessary.

If a reciprocating pump loses suction, it is likely to be irregular in operation, or to race without an appreciable increase in discharge pressure. Loss of suction may be caused by obstructions in the suction line, loss of suction head, the presence of air in the system, and other conditions.

Obstructions in the suction line frequently cause operating trouble in fire and bilge pumps. Be sure that the suction lines are clear and that all valves in the line are open. Clean the suction line strainer and the bilge strainer.

STEAM PISTON ROD
CROSSHEAD ARM

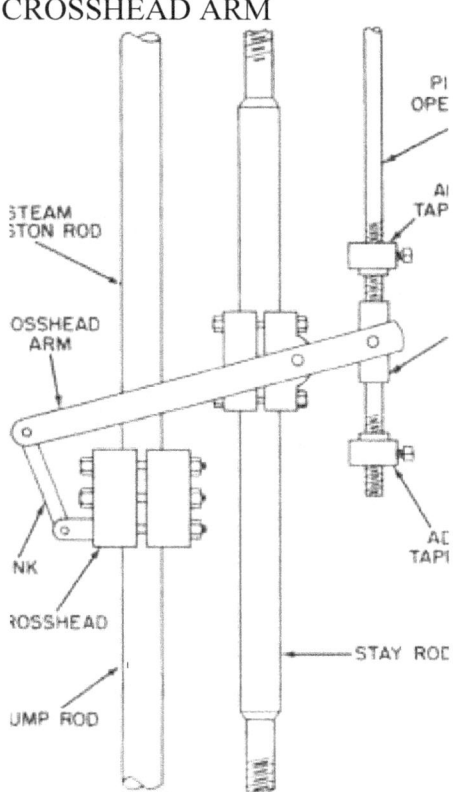

PILOT VALVE OPERATING ROD
ADJUSTABLE TAPPET COLLAR
MOVING TAPPET
LINK
CROSSHEAD
PUMP ROD
ADJUSTABLE TAPPET COLLAR
STAY ROD

Figur* 2-2.—Valve-operoling aitembly.

Loss of suction head may cause a hot liquid to vaporize and the pump to become vapor-bound. The emergency feed pump is particularly subject to this trouble. If the loss of suction head is caused by low discharge pressure from the feed booster pump, the booster pump should be speeded up immediately. If the emergency feed pump does not regain suction at once, and if a standby pump is not available, you will have to cool the pump rapidly. Shift to cold suction (reserve feed tank). Open the vents on the liquid-end valve chest. Cool the pump by turning a hose on it, or by pouring buckets of cold water over it. Continue the cooling measures until you see a steady flow of water coming from the vents. Continue to use cold suction until the head pressure is restored.

The presence of air in the system may cause a reciprocating pump to become air-bound and thus lose suction. The remedy for this condition is to open the aircocks and vents on the liquid-end valve chest, and to leave them open until water flows out.

Although reciprocating pumps are generally considered to be self-priming, pumps having a high suction lift may require priming before they will take suction. The fire and bilge pump can usually be primed from the sea by opening the sea suction valve for a short time.

Loss OF DISCHARGE PRESSURE and LOSS OF CAPACITY are closely related operating troubles: They may be caused by various conditions, including:

1. Low steam pressure
2. High exhaust back pressure
3. Insufficient speed of the pump
4. Excessive suction lift
5. Air leakage into the pump
6. Entrained vapors in the pumped liquid
7. Excessive friction resulting from improperly packed stuflfing boxes or from binding of the packing in the liquid end.
8. Excessive wear of packing on the pump plunger.

If a reciprocating pump races without an appreciable increase in discharge pressure, the trouble may be caused by leaky, broken, or stuck valves in the water end; leaky valves in the suction line; a leaky piston; or other conditions. If a pump which has been running properly suddenly loses discharge pressure on one stroke, the trouble is very likely to be a broken valve in the water end. Whenever a reciprocating pump loses discharge pressure or capacity, or whenever it begins to race without increasing the discharge pressure, it should be stopped as soon as possible so that the trouble may be found and corrected.

Pounding in the water end is most likely to be caused by improper adjustment of the steam cushioning valves; a pump plunger which is slightly loose on the rod; a rod which has come loose at the crosshead; loose water chest valves or a loose valve plate; or loose zinc plates. Water hammer or ram effect in the suction piping may cause pounding in the water end of a pump which has a considerable suction lift; this pounding will usually stop when the pump is run at a slower speed. Pounding in the liquid end may also be caused by insufficient charging of the air chamber (if installed). If pounding occurs in the liquid end of a pump which is not fitted with an air chamber, the installation of a snifter valve may take care of the trouble. However, snifter valves or air chambers should never be installed on- the emergency feed pump, since these devices would tend to draw air into the feed water.

Groaning in the water end is most often caused by packing that is too tight. Other causes include misalignment of the pump, a broken or damaged follower plate, and other broken parts. The pump should be stopped at once and the trouble should be found and corrected. Continued operation of a pump which has any of these defects may cause a scored cylinder or even more serious damage.

Knocking in the steam end may be caused by too long a stroke, by the presence of water in the steam cylinder, by loose piston rings or a loose piston assembly, by the piston being loose on the rod, or by some difficulty in the piston-type valve gear. The pump should be stopped at once and the trouble should be found and corrected.

Groaning in the steam end may indicate that the packing is too tight, that the piston or a piston ring is broken, that rust has formed in the cylinder, or that the pump is out of alignment. The pump should be stopped so that the difficulty may be found and corrected.

Sometimes a reciprocating pump will stick at the end of a stroke, or will stop frequently, even when the throttle valve is opened the proper amount. This kind of ERRATIC operation is generally caused by defects in the steam end of the pump. Some of the most common causes of trouble are:

1. Sticking of the main valve. If the main valve sticks, check to see whether scale, rust, or other foreign matter has become lodged between the piston and the cylinder wall (or between the flat D-type slide valve and its seating surface). Shreds of worn packing from the lower end of the steam cylinder are sometimes brought into the steam cylinder and carried to the main valve, causing the main valve to stick. Poorly fitted piston rings can also cause a main valve to stick.

2. Steam leakage past flat-faced (D-type) slide VALVES. Excessive wear of either the pilot (auxiliary) slide valve or the main slide valve may allow steam leakage sufficient to cause erratic operation of the pump. The remedy for this condition is to grind in or face off the slide valve and its seating surfaces until a smooth, tight fit is achieved.

3. Steam leakage past piston valves. Worn or poorly fitted piston rings are the most likely cause of steam leakage past piston-type valves. Renewing or refitting the rings should take care of the trouble. However, it may be necessary to rebore the valve chest cylinder and to fit an oversize piston valve.

4. Steam leakage past the steam piston. Excessive leakage can usually be remedied by renewing the steam piston rings. It may be necessary to re-bore the cylinder and to fit an oversize steam piston.

5- Blocking of passages in the valve chest. Small ports and passages in the valve chest may become clogged with scale, and thus prevent the pump from operating properly. This trouble is most likely to occur on new vessels, as a result of failure to blow all scale from the steam lines before the pump is connected.

Erratic operation of a reciprocating pump may also be caused by worn bushings and pins in the valve-operating assembly.

Maintenance of Reciprocating Pumps

In order to maintain reciprocating pumps in good operating condition, the following tests and inspections should be made:

Daily: Jack over all idle pumps by hand.

Weekly: Move all pumps by steam.

Quarterly : Inspect the liquid end valves, valve stems, and springs. Inspect the steam valve gear for wear. Check the settings of the relief valves.

The pins of the valve-operating assembly must be kept well oiled at all times. However, you should not allow anyone to lubricate the internal parts of the steam end or of the liquid end of the pump. A slight gland leak-off is sufficient to lubricate the pump rod.

The length of stroke should be adjusted so that the piston will travel a little beyond the counterbore. The length of stroke is adjusted by changing the setting of the tappet collars on the pilot valve operating rod. Detailed information concerning the adjustment of stroke is given in chapter 47 of BuShips Manvxil and in the appropriate manufacturers' instruction books.

As the packing wears, you will have to set up on the glands to prevent or correct excessive leakage. Be very careful to set up evenly on the two sides, so that the gland will not become tilted. If the gland is not set up evenly, the rod may become scored or the gland itself may be broken. If a gland continues to leak after the nuts have been given a few turns, do not attempt to stop the leakage by making the gland very tight, as this would probably cause the packing to score the rod. Excessive leakage cannot be corrected until the pump is shut down and the cause of the leakage is determined.

As a matter of routine maintenance, you should check the fit of the valve chest to the steam cylinder. The joining surfaces are subject to serious steam cutting, but this may be minimized by maintaining a good metal-to-metal fit at the joint.

The valves in the liquid end should be cleaned whenever necessary; and they should be kept absolutely tight. Satisfactory and economical operation of a reciprocating pump cannot be achieved unless these valves are tight.

Salt water pumps require special maintenance measures because of the danger of corrosion. The internal parts of the liquid end should be examined, scaled, and wire-brushed every six months. If zincs are fitted, they should be inspected once a month and replaced when necessary.

Repair of Reciprocating Pumps

Before you begin to repair a pump, you should assemble all drawings, blueprints, and other pertinent data. You should have the complete history of the pump that you are repairing, so that you will know what work has been done, when it was done, what kind of trouble has been encountered with this particular pump, and so forth. Remember, also, that the machinery history must be kept up-to-date; always make the proper notations on the machinery history card after you have completed a repair job.

Since most reciprocating pump troubles are due to defects in the liquid end, a pump overhaul should always
begin with the liquid end. Do not disassemble the steam end until after the liquid end has been examined and repaired.

Scored water cylinder liners can usually be smoothed by stoning. They should not be rebored or renewed unless the scoring is extensive enough to cause excessive leakage and rapid wear of the pump plunger.

After a water cylinder has been rebored, you may find that the piston and follower plate are too small. If it is not possible to obtain a new piston and follower plate of the proper size, one of the following procedures may be used to decrease the clearance:

1. Machine the piston down to a considerably smaller size, and thread it; make a tight-fitting ring with the same thread, and screw it on the piston. Treat the follower plate in the same way.

2. Build up the piston and the follower plate by spraying on metal; or by oxyacetylene or electric welding. Machine the piston and the follower to fit the cylinder.

Breakage of follower plates and studs may be caused by misalignment; or by the studs or their nuts working loose. Whenever a piston is removed for examination or repair, you should check to be sure that the studs and nuts are tight. A corrosion-resistant locking wire should be used to hold the studs and nuts in place.

Worn or broken water chest valves often cause unsatisfactory operation of the pump. You can test for tightness of the water chest valves by closing the discharge valve and cautiously opening the throttle valve. If the water chest valves are tight, the pump will stall.

Scored or pitted valve seats and disks should be ground in. Broken or worn valve cages, stems, springs, and binding screws should be renewed. See that the valve-spring tension is great enough to ensure the quick closing of the valve, but not so great that the valve cannot be lifted easily by hand. The springs should be well secured by split pins.

When removing or replacing force-fit (taper-fit) or screwed valve seats, be careful that you do not warp the valve plate. Taper-fit valve seats should be forced into place by means of a jack. The jack rests on the end of a reseater which, in turn, rests on the face of the valve seat. Screwed valve seats should always be screwed in with white lead. If this is not done, you will find it almost impossible to remove the seats after they have been in place for a while.

Figures 2-3 shows an arrangement of valves in the water chest. Note that in this arrangement the valves seat on valve plates which are made separate from the water chest casing. Some difl^culty may be found in maintaining a tight joint between the valve plate and the water chest casing. If the joint is not tight, the shoulder on the casing should be ground in by hand. The valve plate itself may be refaced on a lathe. In some cases, you will find that a copper gasket installed at the shoulder of the joint will effectively seal the joint and prevent leakage.

Figure 2-4 shows a type of water chest valve commonly used on fire and bilge pumps. Figue 2-5 shows a valve for use on emergency feed pumps.

Corroded snifter valves are sometimes the cause of faulty pump operation. The snifter valve may be a stop-check valve, or a plain check valve. Where a check valve is used, a globe stop valve should be installed between the check valve and the pump so that the amount of air can be regulated. Snifter valves are not subject to any great amount of wear in the course of normal operation. However, they may become corroded from water (particularly salt water) leaking onto them; if this occurs, the valves may require grinding in.

Loose water pistons and steam pistons are sometimes a source of trouble in reciprocating pumps. A piston should always be properly refitted after it has been removed from the rod. The tapered part of the rod and the opening in the piston must be free of foreign matter

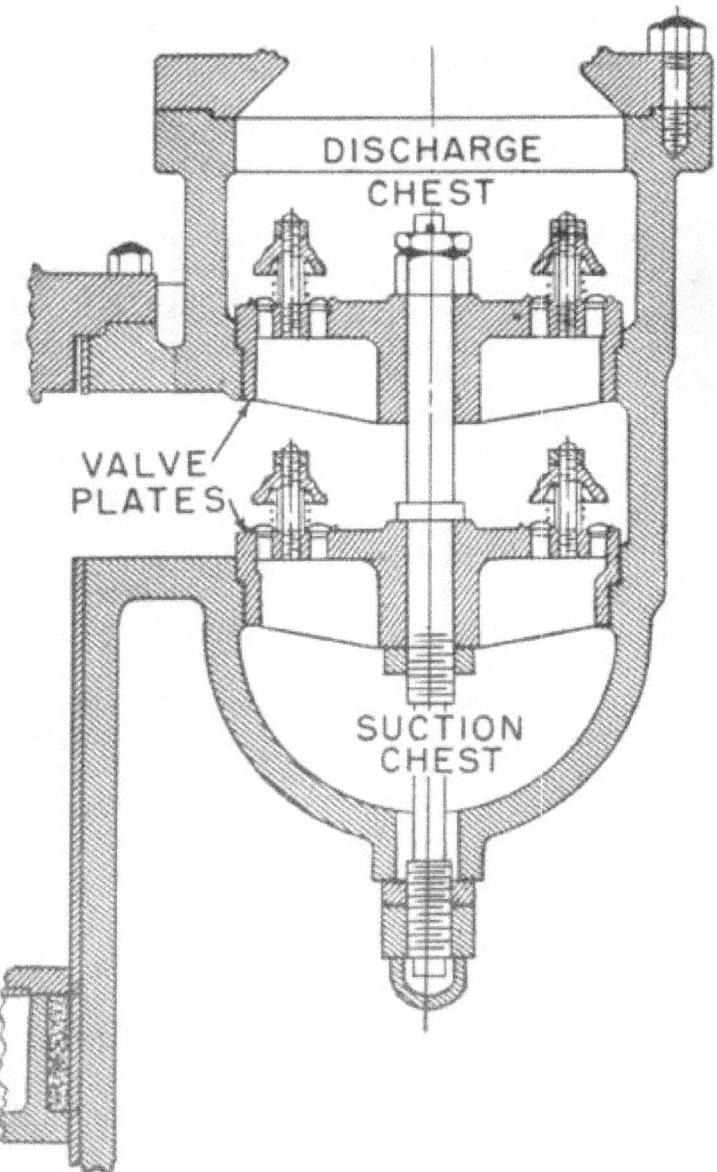

DISCHARGE
CHEST

VALVE
PLATES

SUCTION
CHEST

Figure 2-3.—Water chest arrangement (separate valve plates).

18

Figure 2-4.—Fire ond bilge pump water-chest valve.

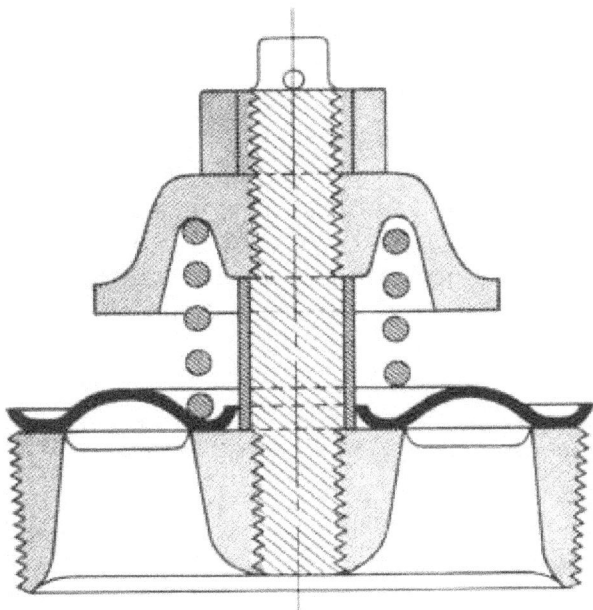

Figure 2-5.—Emergency feed pump water-chest valve.

before the piston is fitted; even a small amount of foreign matter will prevent the piston from giving the proper bearing surface to the tapered part of the rod. The piston should be fitted onto the tapered part of the rod, and the securing nut should be set up handtight. There should be a uniform clearance of %2 ^ Vh of an inch between the piston and the rod shoulder, the size of the clearance depending upon the amount of taper of the rod. After the piston is thus properly placed, the securing nut should be tightened to force the piston tightly against the shoulder of the rod.

Worn steam piston rings are one of the main causes of trouble in the steam end of a reciprocating pump. To fit new rings, you will have to secure the pump plunger and take off the crosshead. Free the packing gland and remove the packing. Remove the top cylinder head, lift the piston, and remove the old rings. Before fitting the new rings, measure the rings to be sure that they are the right size for the cylinder. Spare rings carried aboard ship may be 0.010 inch oversize, since they are intended for use after the cylinder has worn slightly. If the cylinder has not worn to the size of the new rings, the new rings must be machined down to size. The outer surface of the rings should be smoothed with emery cloth or crocus cloth.

When fitting new rings in the steam end, remember that the rings should fit snugly, but without binding, in the piston. Split rings require a small end-gap clearance, to allow for expansion. If the end-gap clearance is not large enough, the rings will buckle and will be likely to score the cylinder when the pump is operated. If the end-gap clearance is not specified in the blueprint, you can compute it by allowing slightly less than 0.001 inch per inch of cylinder diameter. For example, for a 10-inch cylinder, the ring end-gap clearance should be about 0.007 to 0.008 inch.

When split rings are used, steam that enters the spaces between the rings and the piston may become trapped and

build up in pressure. Excessive pressures in these spaces will force the rings out against the cylinder, causing binding and even cutting of the cylinder. If such excessive pressures develop, the situation can be remedied by turning a groove about %2 ii^ch deep and % inch wide around the middle of each ring, and drilling three or four i^-inch holes through the grooved section of the ring. By allowing the pressures to equalize, these holes prevent the build-up of excessive pressures in the spaces between the rings and the piston.

Scored steam cylinders must be rebored, since even relatively small scores allow steam to cut the cylinder walls. If this steam cutting is allowed to continue for any length of time, the steam leakage past the piston will become so excessive as to cause faulty operation of the pump.

The Steam-end valve gear must be kept in good operating condition at all times. Flat steam valves and their seating surfaces should be ground in or faced off so that the valve will make a smooth, tight fit. Auxiliary pistons and piston-type valves are generally fitted with rings, which require renewal from time to time.

Misalignment is one of the most frequent and most serious causes of trouble in reciprocating pumps. Continued operation of a misaligned pump causes scoring of rods and cylinders, breakage of follower plates and bolts, and other damage. Misalignment may be caused by improper installation of the pump or of associated piping; by distortion of bulkheads after installation of the pump; by lack of expansion room at foundations and supports; and by other causes. The alignment of each reciprocating pump should be tested occasionally by removing the pistons and rods and running a line through the cylinders. This test should be made routinely within the first year after the ship is commissioned, and at any time thereafter that you have reason to suspect misalignment.

To make a rough check on the alignment of a reciprocating pump, pull the rod packing from each end of the
pump and measure the clearances between the rods and the cylinder head throat bushings. Clearances should i be measured with the piston in three different positions: ', (1) at the top of the stroke, (2) at the center of the ' stroke, and (3) at the bottom of the stroke. If your measurements show that the clearance is not uniform, the \ pump is out of alignment. Throat bushings which have been worn out of round indicate either past or present misalignment of the pump.

To align a reciprocating pump, remove the pistons and the rods and run a wire line through the cylinders, as shown in figure 2-6. Fasten one end of the line to a finger piece secured to the bottom of the liquid cylinder, and the other end to a beam rigged above the steam cylinder. Center the line at the bottom and the top of the liquid cylinder, so that the line represents the long axis of the liquid cylinder. Adjust the position of the steam cylinder until it is in line with the liquid cylinder. The steam cylinder should be brought into line by means of shims placed at the bottom or at the top of the stay rods.

If misalignment is caused by distortion of the bulkheads or of the pump foundations, the trouble may be corrected by fitting shims in such a way that the foundation bolts and other securing bolts will not force the pump out of alignment.

Safety Precautions

The following safety precautions concerning reciprocating pumps have been specified by the Bureau of Ships:

1. Never attempt to jack over a pump while the steam valve to the pump is open.

2. Do not use the emergency or auxiliary feed pump for purposes other than those connected with the service of boilers or the use of feed water, except in emergency.

3. Before opening a steam cylinder or a steam chest, be sure that all drains are open; and be sure that the steam and exhaust root valves are wired closed.

rods i: es sk wsitk >r of i:

J/ye. orm. t: ch ha' preser.
•ns aE'; 2rs, a; e toi inder n cyi-.f the
team The
s of ?ds.
liJf-be la-he
TEMPORARY BEAM
STEAM CYLINDER
STAY RODS
WATER CYLINDER
CALIPERS

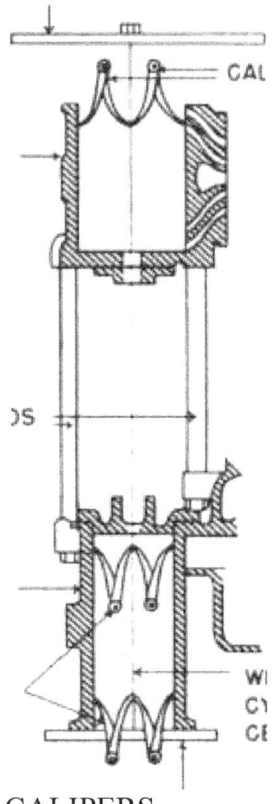

CALIPERS
WIRE REPRESENTING CYLINDER AXIAL 1 CENTER LINE
WCX)DEN BOARD
Figwr* 2-4.—Checking th« alignmtnt of a rMlprecoting pump.

4. Before opening the water cylinder or the valve chest of a pump handling water at a temperature in excess of 120° F, be sure that the suction and discharge valves are wired closed and that the cylinder and the valve chest are drained.

5. Always open the steam cylinder drain valves and the steam chest drain valves when the pump is shut down; leave them open until the pump is again in operation and has been cleared of condensate.

CENTRIFUGAL PUMPS

Centrifugal pumps are widely used on board ship for pumping nonviscous liquids. Centrifugal pumps are not positive displacement pumps. When a centrifugal pump is operating at a constant speed, the amount of discharge (capacity) varies inversely with the discharge pressure.

Capacity and discharge pressure can be varied by changing the speed. However, centrifugal pumps should be operated at or near their rated capacity and discharge pressure whenever possible. Impeller vane angles and the sizes of the pump waterways can be designed for maximum efficiency at only one combination of speed and discharge pressure; under other conditions of operation, the impeller vane angles and the sizes of the waterways will be too large or too small for efficient operation. A centrifugal pump, therefore, cannot operate properly at excess capacity and low discharge pressure, or at reduced capacity and high discharge pressure.

It is important to remember that centrifugal pumps are not self-priming. The casing must be flooded before a pump of this type will function. For this reason, most centrifugal pumps are located below the level from which suction is to be taken. Priming can also be effected by using another pump to supply liquid to the pump suction —for example, the feed booster pump supplies suction pressure for the main feed pump. Some centrifugal pumps have special priming pumps, air ejectors, or other devices for priming.

Where two or more centrifugal pumps are installed to operate in parallel, you should be particularly careful to avoid operating the pumps at very low capacity. Under these conditions, it is possible that a unit having a slightly lower discharge pressure might be pushed off the line and so be forced into a shut-off position. Although centrifugal pumps can operate at zero capacity, there is danger of overheating the unit if it is operated for any length of time with no discharge.

Most centrifugal pumps—and, in particular, boiler feed pumps, fire pumps, and others which may be required to operate at low capacity or in a shut-off condition for any length of time—are fitted with recirculating lines from the discharge side back to the source of suction supply. The main feed pump, for example, has a recirculating line going back to the deaerating feed tank. An orifice allows the recirculation of the minimum amount of water required to prevent overheating of the pump. On boiler feed pumps, the recirculating line must be kept open whenever the pumps are in operation.

When operating a centrifugal pump, remember that there must always be a slight leak-off through the packing in the stuffing boxes, in order to keep the packing lubricated and cooled. Stuffing boxes are used either to prevent the leakage of liquid from the pump, or to prevent the entrance of air into the pump; the purpose served depends, of course, upon whether the pump is operating with a positive suction head or is taking suction from a vacuum. If a pump is operating with a positive suction head, the pressure inside the pump is sufficient to force a small amount of liquid through the packing, when the gland is properly set up. On multistage centrifugal pumps, it is sometimes necessary to reduce the pressure on one or both of the stuffing boxes. This is accomplished by using a bleed-off line which is tapped into the stuffing box, between the throat bushing and the packing.

If a pump is taking suction at or below atmospheric pressure, a supply of sealing water must be furnished to the packing glands to ensure the exclusion of air; and some of this water must be allowed to leak off through the packing. Most centrifugal pumps use the pumped liquid as the lubricating, cooling, and sealing medium; however, an independent, external sealing liquid is used on some pumps.

In some of the newer boiler installations (1200 psi and above), some water from the discharge side of the feed
booster pump is passed through a cooler and is then used to cool the packing cavities on both the main feed pump and the feed booster pump. Since this water goes back to the deaerating feed tank after it has been used to cool the packing, this arrangement also provides recirculation for the feed booster pump.

When operating a centrifugal pump, be sure to open the vents as often as necessary to release entrained air. If the pump requires frequent venting, leaving the vents cracked so as to allow a continuous leakage of air and water may be desirable. The discharge from the vents should be piped to the proper drain system or to the bilges, as appropriate.

Operating Troubles

Some of the operating difficulties which you may have to deal with in centrifugal pumps are given below, together with their probable cause.

If a centrifugal pump DOES not deliver any liquid, the trouble may be caused by (1) insufficient priming; (2) insufficient speed of the pump; (3) excessive discharge pressure, such as might be caused by a partially-closed valve or some other obstruction in the discharge line; (4) excessive suction lift; (5) clogged impeller passages; or (6) the wrong direction of rotation.

If a centrifugal pump delivers some liquid but operates at INSUFFICIENT CAPACITY, the trouble may be caused by (1) air leakage into the suction line; (2) air leakage into the stuffing boxes, in the case of pumps operating at less than atmospheric pressure; (3) insufficient speed of the pump; (4) excessive suction lift; (5) insufficient liquid on the suction side; (6) clogged impeller passages; (7) excessive discharge pressure; or (8) mechanical defects such as worn wearing rings, impellers, stuffing box packing, or sleeves.

If a pump does not develop enough discharge pressure, the trouble may be caused by (1) insufficient speed of the pump; (2) air or gas in the liquid being pumped;

or (3) mechanical defects such as worn wearing rings, impellers, stuffing box packing, or sleeves.

If a pump WORKS for a while and then fails to DELIVER LIQUID, the trouble may be caused by (1) air leakage in the suction line; (2) air leakage in the stuffing boxes; (3) clogged water seal passages; (4) insufficient liquid on the suction side; or (5) excessive heat in the liquid being pumped.

If a motor-driven centrifugal pump takes too much POWER, the trouble will probably be indicated by overheating of the motor. The basic cause of the difficulty may be (1) operation of the pump at excess capacity and insufficient discharge pressure; (2) excessively high viscosity or specific gravity of the liquid being pumped; or (3) misalignment, a bent shaft, excessively tight stuffing box packing, worn wearing rings, or other mechanical defects.

Vibration of a centrifugal pump is often caused by (1) misalignment; (2) a bent shaft; (3) a clogged, eroded, or otherwise unbalanced impeller; or (4) lack of rigidity in the foundation. Insufficient suction pressure may also cause vibration, as well as noisy operation and fluctuating discharge pressure, particularly in pumps handling hot or volatile liquids.

Maintenance and Repair

The following discussion of maintenance and repair applies, in general, to most of the centrifugal pumps on board ship. There are, however, a number of different designs of centrifugal pumps in use; for specific information on any particular pump, therefore, you should study the manufacturer's instruction book furnished with the unit.

Inspections of operating pumps. —As a routine maintenance procedure, you should make regular and frequent inspections of each pump in operation. At least once an hour, check to be sure that all temperature and pressure

gages are indicating proper conditions; and that there is a proper leakage from the stuffing boxes.

Lubrication. —It is extremely important that the bearings of a centrifugal pump are properly lubricated. Most turbine-driven pumps use a pressure-lubrication system of the type described in chapter 3. Motor-driven pumps and some turbine-driven pumps having ball bearings are fitted for grease lubrication. Sleeve bearings are often lubricated by means of oil rings. Be sure that you understand the type of lubrication system used on each pump. Be sure that each pump is being properly lubricated at all times during operation.

Stuffing box packing. —The packing in centrifugal pump stuffing boxes should be renewed about once every 2 months. When replacing packing, be sure to use packing of the specified material and the correct size. Stagger the joints in the packing rings so that they will fall at different points around the shaft. Pack the stuffing box loosely and set up lightly on the gland, allowing a liberal leakage. With the pump in operation, tighten the gland and gradually compress the packing. It is important to do this gradually and evenly, in order to avoid excessive friction which would cause overheating and possible scoring of the shaft or the shaft sleeve. It is usually necessary to put in more packing rings, after the first ones have been compressed.

On some centrifugal pumps, a lantern ring is inserted between the rings of the packing. When repacking stuffing boxes on such pumps, be sure to replace the packing beyond the lantern ring. Also, be sure to pack the stuffing box in such a way that the packing will not block off the liquid seal connection to the lantern ring, after the gland has been tightened.

Figure 2-7 shows a stuffing box on a centrifugal pump. Notice that the packing is so arranged that the lantern ring will line up with the liquid seal connection when the gland is tightened.

Shaft sleeves. —On most centrifugal pumps, the shaft
is protected at the stuffing box by sleeves which are either screwed or keyed to the shaft. Sleeves should be examined whenever the pump is opened for inspection or repair The sleeves must be properly fitted to the shaft, to keep water from leaking between the shaft and the sleeves. If leakage does occur, fit a fiber or asbestos washer between the end of the sleeve and the shaft shoulder and fill all clearances with white lead or red lead.

The outer surface of the sleeve must be kept smooth and free from scores, in order to prevent excessive wear of the packing. A sleeve which is only slightly worn or scored should be sent to the machine shop for a finishing cut to smooth the surface. A badly worn sleeve should be replaced.

Water flingers. —Water flingers are fitted on shafts outboard of the stuffing box glands, to prevent water

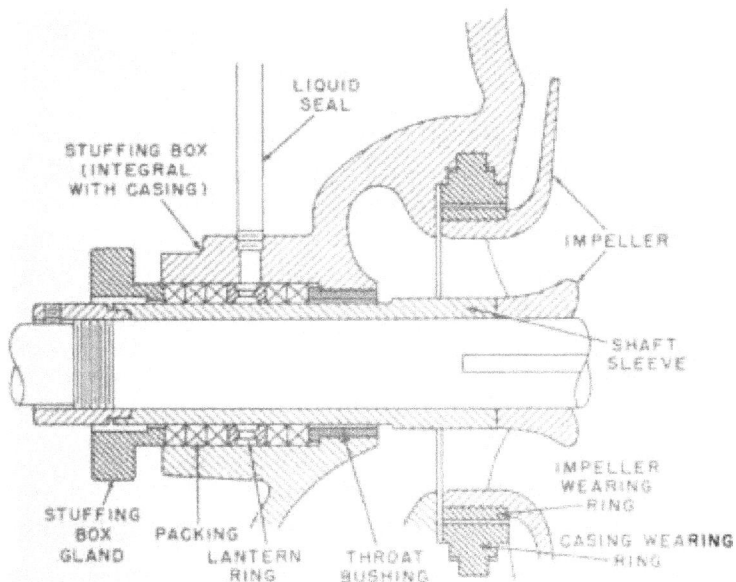

Figure 2-7.—Stuffing box on esntrifugal pump.

from following along the shafts and entering the bearing housing. The water flingers must be tightly fitted to the shaft.

Wearing rings. —Most centrifugal pumps have both impeller wearing rings and casing wearing rings, similar to those shown in figure 2-8. The smaller ring fits over the hub of the impeller and turns with it; the larger

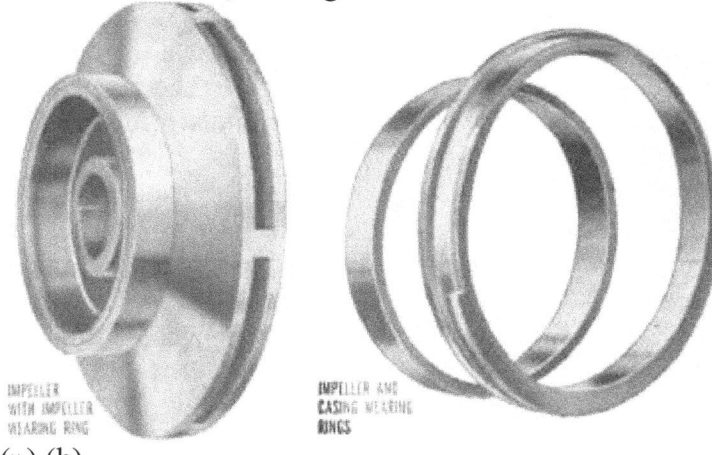

(») (b)

Figure 2-8.—Centrifugal pump impeller and wearing rings.

ring is stationary, being attached to the casing and prevented from turning by the semicircular fiange or shoulder on one side. A close clearance must be maintained between the two wearing rings, in order to prevent excessive leakage from the high-pressure side to the low-pressure side of the impeller. Some pumps have wearing rings only on the casing; in this event, the close clearance must be maintained between the impeller itself and the casing wearing ring.

Wearing ring clearances must be checked frequently, and must be maintained within specified tolerances. The allowable wear in these rings varies according to the type of pump and the service for which the pump is used. For boiler feed pumps, you should renew the wearing rings when the clearance shown in the manufacturer's plans is exceeded by 100 percent. As a general rule, you should not renew the wearing rings unless the amount of wear is at least 0.015 inch. If it is necessary for you to replace wearing rings, be sure to follow the manufacturer's instructions carefully. Improper fitting of the rings or incorrect assembly of the pump can result in serious damage.

Bearings. —Pump bearing clearances should be checked at least once every 3 months. Clearances must be maintained as shown on the manufacturer's plans, or according to the tables given in BuShips Maniud, chapter 40. Worn bearings cause excessive wear of the wearing rings, and may cause misalignment of the pump. Bearings on centrifugal pumps should be rebabbitted when bridge gage readings or leads show that the maximum allowable wear has occurred.

Bushings. —Clearances of all bushings along the shaft should be checked whenever the pump is opened for inspection or repair, and the bushings should be renewed when necessary. Bearing wear is very likely to cause wear in the bushings.

Driving units and accessories. —The maintenance and repair required for auxiliary turbines and accessories such as governors, flexible couplings, bearings, etc., is described in chapter 3 of this training course. For information on the care and repair of motors for motor-driven units, consult the appropriate manufacturer's instruction book and BuShips Manual.

Tests and Inspections

The following tests and inspections should be made on centrifugal pumps, and the results should be entered in the appropriate check-off list or log:

Daily : Turn idle pumps by hand.

Weekly: Run the pump under power. Lift all relief

valves by hand. Check the operation of the discharge check valves. Determine the condition of the lubricating oil.

Quarterly: Test all relief valves by steam, water, or oil, as appropriate. Measure the thrust bearing clearance. Check the axial position of the pump impellers. Check bearing clearances by leads or bridge gage readings. Examine and set up on all foundation bolts; secure all foundation dov^^el pins. Check all water-lubricated bearings and shafts for wear and scoring. Clean the lubricating system and renew oil or grease.

Annually : Open the pump, turbine, and reduction gear casings for inspection and cleaning. Check clearances of throat bushings and of impeller and casing wearing rings; renew bushings and rings if necessary. Examine all impellers, diffusers, turbine rotors, turbine blading, carbon packing, shafts, and shaft sleeves.

The tests and inspections required for the turbine end are listed in chapter 3 of this training course.

Safety Precautions

The following safety precautions must be observed in connection with the operation of centrifugal pumps:

1. See that all relief valves are tested at appropriate intervals. Be sure that relief valves function at the designated pressures.

2. Never attempt to jack over a pump by hand while the steam valve to the pump is open or while the electric power is on.

3. Do not tie down the overspeed-trip, the speed-limiting governor, or the speed-regulating governor. Do not in any way attempt to render these devices inoperable. Be sure that speed-limiting and speed-regulating governors are properly set,

4. Do not use any boiler feed system pump for any service other than boiler or feed water service, except in emergency.

5. Observe all safety precautions appropriate to the operation of the driving unit.

ROTARY PUMPS

Rotary pumps, like reciprocating pumps, are positive-displacement pumps. The theoretical displacement of a rotary pump is the volume of liquid displaced by the rotating elements on each revolution of the shaft. The CAPACITY of a rotary pump is defined as the quantity of liquid (gpm) actually delivered under specified conditions. Thus, the capacity is equal to the displacement times the speed (rpm), minus whatever losses may be caused by slippage, suction lift, viscosity of the pumped liquid, amount of entrained or dissolved gases present in the liquid, etc.

The rotating elements in a rotary pump may consist of gears, lobes, vanes, screws, cam-and-plunger arrangements, or other devices for trapping the liquid at the suction side and forcing it through the discharge outlet. Rotary pumps having three high-pitch screws are widely used on naval vessels for pumping liquids of relatively high viscosity, such as fuel oil and lubricating oil; they are well suited to this purpose, provided the speed of the pump is adjusted to the viscosity of the liquid being pumped. When pumping high-viscosity liquid, the pump must be run at reduced speed in order to allow time for the pump to fill.

Operating Troubles

Sometimes a rotary pump will fail to deliver liquid, or will deliver some liquid but be unable to reach the required capacity. A pump may deliver liquid for a while, and then lose suction or fail to deliver. Under some conditions, a rotary pump will require excessive power in order to operate at the proper capacity. If any of these troubles occur, or if the pump operates at adequate capacity but is noisy in its operation, you should stop the pump as soon as practicable and locate the source of the trouble. Some common causes of faulty operation in rotary pumps are discussed below.

Insufficient priming. —Rotary pumps are generally self-priming. However, a pump operating under a high suction lift may become air-bound or vapor-bound, and may not be able to prime itself.

Clogged suction or discharge lines. —Check all valves in the suction or discharge lines, to be sure that they are open. See if a broken valve spring or a detached or damaged valve disk is causing the trouble. Check the suction strainer (if fitted) to be sure that it is not clogged.

Air leakage. —Air leakage into the suction side of the pump or into any part of the suction piping or the suction manifold will cause improper operation of the pump. Check the pump stuffing box and the packing on all valve stems in the suction line and suction manifold, to be sure that air is not leaking into the suction side of the system. Be sure that the end of the suction pipe is properly submerged in the liquid supply.

Vapor lock. —Excessive suction lift or excessive temperature of the pumped liquid may cause partial vaporization of the liquid, and thereby cause vapor lock in the suction piping. Vapor lock is most likely to occur in long suction lines which have many loops and bends (fig. 2-9),

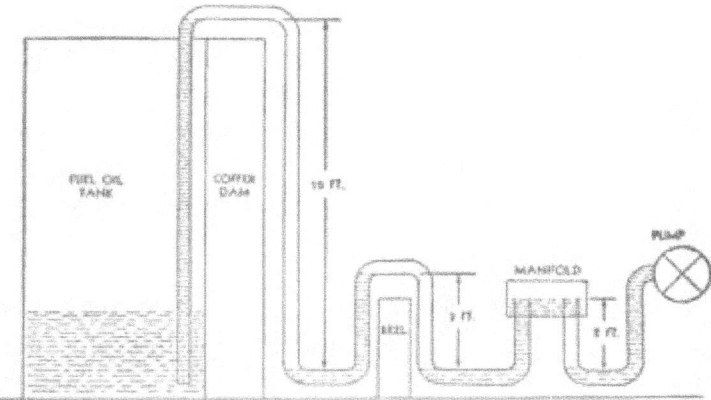

Figure 2-9.—Type of piping arrangement likely to cause vapor lock.

particularly if the pump is handling oil which has recently been taken on or oil which has been heated. Vapor lock is indicated by noisy operation and by pounding in the pump. To remedy this condition, line up another tank, using another manifold if possible, and shift suction. When the pump operates properly from this tank, shift back to the original tank and run the pump slowly. Reducing the speed of a rotary pump may help prevent the development of vapor lock.

Entrained gas. —Air or other gas which is dissolved in the liquid being pumped does not affect the pump's capacity at zero suction lift. As the pressure in the suction line decreases, however, the dissolved gas is freed from the liquid and thereby becomes entrained.

Air or other gas which is entrained in the liquid being pumped affects the capacity of the pump even at zero suction lift; and, as the pressure in the suction line decreases, the effect of entrained gas becomes even more pronounced. Entrained gas causes the same kind of noisy operation and pounding in the pump as is caused by vapor lock. Reducing the speed of the pump or heating the oil being pumped may help to remedy the situation.

Excessive viscosity. —If the liquid being pumped is too heavy or too viscous for the design of the pump, an excessive amount of power will be required to operate the unit. If practicable, the viscosity of the pumped liquid should be reduced by heating; and the speed of the pump should be reduced.

Improper direction of rotation. —If a motor-driven pump fails to develop discharge pressure when it is first operated after major repairs have been made to the motor, check to be sure that the pump is rotating in the proper direction.

Mechanical defects. —There are a number of mechanical defects which may cause serious operating difficulties in rotary pumps. If a pump does not operate

properly, check it for misalignment, tight packing, binding of the rotors, a bent shaft, an improperly adjusted relief valve, and other defects.

Maintenance and Repair

This discussion of maintenance and repair applies specifically to triple-screw, high-pitch rotary pumps. Fuel oil service, fuel oil booster and transfer, and lubricating

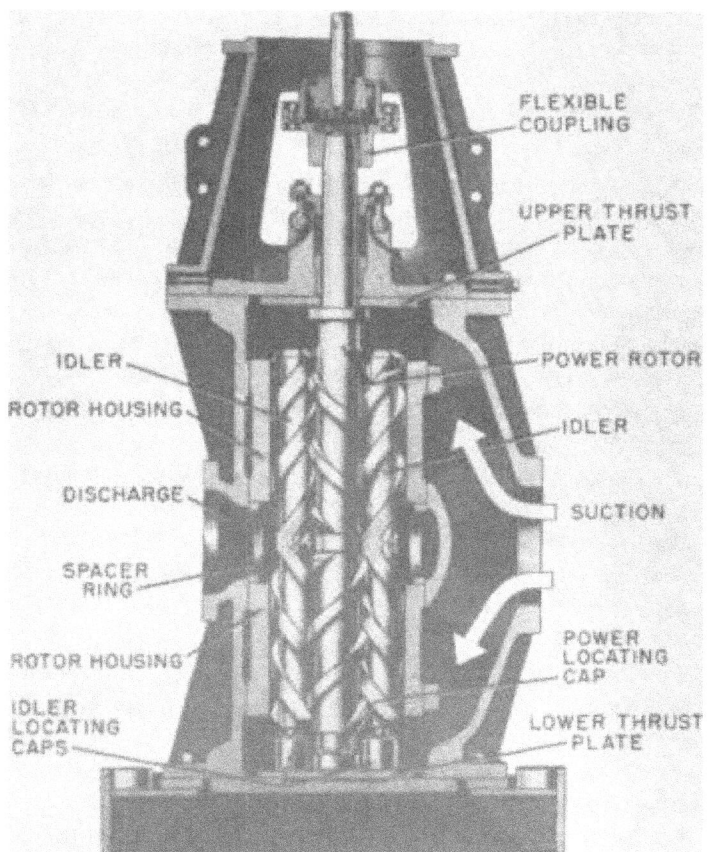

Figure 2-10.—Triple-tcrew high-pitch pump (cutaway vi«w).

oil service pumps are commonly of this type For more detailed information concerning any particular unit, you should consult the appropriate manufacturer's instruction book. Information on the maintenance and repair of driving turbines and accessories is given in chapter 3 of this training course.

Figure 2-10 is a cutaway view of a typical fuel oil service or lubricating oil service pump. The internal parts of this pump are shown in figure 2-11.

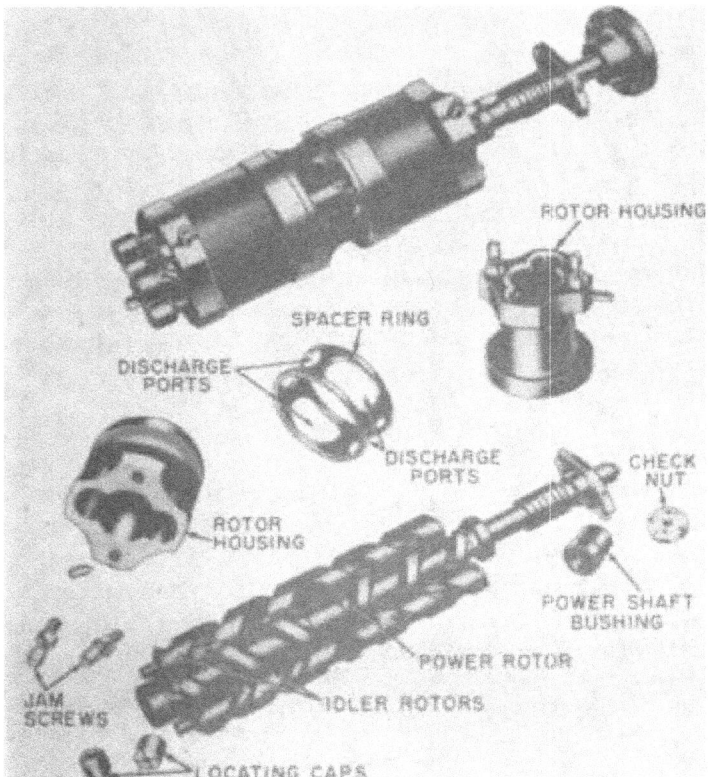

Figur* 2-11.—Internal parts of a triple-screw high-pitch pump.

Disassembling the pump. —To remove the driving unit, break the steam, water, and drain connections at the turbine end. Break the gland seal connections at the pump end. Disconnect the coupling. Remove the bolts which secure the spacing frame to the pump casing, and lift the driving unit off intact.

After the driving unit has been removed, the pump itself can be completely disassembled without any disturbance to the mounting or to the suction or discharge lines. Next remove the remaining half of the coupling, the packing gland, and the upper casing head. Then withdraw the rotors. The idlers must be supported as the rotors are being withdrawn from the housing, since the rotors are held in mesh only by the housing.

When the rotors have been withdrawn, the rotor housings are accessible. These housings fit snugly in the bore of the casing; the upper housing is separated from the lower housing by a spacer ring which has openings, or ports, for discharge. Each housing is positioned axially by jam screws which bear against the thrust plate on the top or bottom casing head; and each is positioned circumferentially by a guide pin. You must remove the guide pins before you can remove the rotor housings. Both housings and both guide pins should be marked so that the pump can be reassembled properly; the guide pins are individually fitted, and are not interchangeable. Special tools are provided for the removal of the guide pins and the housings. You can take the spacer ring out by hand, after the upper housing has been removed.

Assembling the pump. —Before you begin to assemble the pump, clean and inspect all parts. Be sure that the settings of the lower jam screws are correct, as indicated in the manufacturer's instruction book or on the blueprint of the pump.

To assemble the pump, lower the bottom housing into the casing. Be sure that it is seated firmly, and that the guide pin slot in the housing registers with the pin hole in the casing. Install the guide pin and its securing pipe plug. Put in the spacer ring, and then the upper housing. Line up the guide pin slot with the pin hole in the casing, and put in the pin and the securing plug.

Set up the jam screws in the upper housing, being sure to position them so that when the upper casing head is installed there will be the correct clearance between the head of each jam screw and the thrust plate. Insert the rotors, and see that they turn freely. If the rotors bind, you can be pretty sure that either the housing guide pins or the housings themselves are not properly installed.

Install the upper casing head, making sure that the thrust plate and the bushings are correctly located; and that the bushing is properly secured by its stop pin. Install the packing and the half coupling, and you are ready to connect the driving unit to the pump.

Installing new rotors. —If it is necessary to install new rotors in a triple-screw pump, you will have to establish the proper settings for the idler locating caps. The cap settings must be established before the rotors are installed. Since the rotors must be in proper mesh while the locating caps are being adjusted, the rotors must be inserted in one of the housings while the adjustments are being made.

The locating cap on the power rotor is keyed to the end of the shaft, and is therefore fixed in position. The idler locating caps must be driven on the ends of the idler shafts and then adjusted to the proper position. The base of each cap has a tapped hole into which a threaded pulling tool can be inserted for this adjustment.

The idler locating caps must be set so that they lie in a common plane with the power rotor locating cap, when the rotors are properly meshed; and they must be set so that each idler rotor is centered axially in relation to the clearance between its threads and the threads of the power rotor. In order to prevent direct metallic contact of the meshing threads, the amount of end play allowed the idler must be less than the clearance between the meshing threads.

After the idler locating cap settings are properly established, the caps must be riveted to the ends of the idler shafts.

Shaft packing. —This type of pump has only one stuffing box; it is located in the upper casing head and is subject only to suction pressure. An external pipe connects the pump discharge and the stuffing box, allowing the pumped oil to seal the stuffing box and lubricate the bushing which forms the seat for the packing and acts as an upper bearing. A valve in the liquid seal line is factory-adjusted to provide the correct amount of oil flow. The adjustment of this valve should NOT be changed. If the setting is disturbed, the valve should be reset so as to allow a slight leak-off from the gland when the pump is operating under the conditions for which it was designed, with the packing in good shape and the gland stud nuts tightened just enough to seat the packing firmly. Do not, under any circumstances, allow the pump to be operated with this valve closed.

The procedure for repacking the stuffing box is the same as that described for centrifugal pump stuffing boxes.

Tests and Inspections

The following tests and inspections should be made on all rotary pumps, and the results should be entered in the appropriate check-off list or log:

Daily : Turn idle pumps by hand.

Weekly : Run the pump under power. Lift all relief valves by hand. Check the operation of the discharge check valves (if installed). Check the condition of the lubricating oil; in particular, check for the presence of water in the lube oil.

Quarterly: Test all relief valves by steam, water, or oil, as appropriate. Measure thrust bearing clearances, and check the position of the pump rotors. Check all foundation bolts and dowel pins. Close the suction valve at the pump and note the amount of vacuum pulled; if

the pump fails to produce the required vacuum, it should be opened and repaired as necessary. Check the lube oil system and renew oil and grease.

Annually: Open the pump, turbine, and reduction gear casings for inspection and cleaning. Measure clearances of all internal wearing parts; renew parts if necessary.

Perform all daily, weekly, quarterly, semi-annual, and annual tests and inspections required on the turbine end, as described in chapter 3.

Safety Precautions

The following precautions must be observed in the operation of rotary pumps:

1. See that all relief valves are tested at appropriate intervals. Be sure that relief valves function at the designated pressures.

2. Never attempt to jack over a pump by hand while the steam or power is on.

3. Do not tie down the overspeed trip, the speed-limiting governor, or the speed-regulating governor. Do not in any way render these devices inoperable.

4. Never operate a rotary pump with the discharge valve closed, unless provision is made for adequate discharge.

5. Observe all safety precautions appropriate to the operation of the driving unit.

FORCED DRAFT BLOWERS

Although centrifugal-type forced draft blowers may still be found on some older ships, propeller-type blowers are much more commonly used and are being installed on almost all new construction. This discussion will be concerned with propeller-type blowers only; for information on centrifugal blowers, you should consult BuShips Marnial, chapter 53, and the appropriate manufacturers* instruction books.

Figure 2-12.—Throt-ffog* proptlUr-type fgrctd draft blow«r.
PROPELLER, ist STAGE
PROPELLER. 2nd STAGE-
PROPELLER. 3rd STAGE*
OIL BAFFLE
SEAL RING —
STATIONARY "guide VANES
UPPER 3EARING

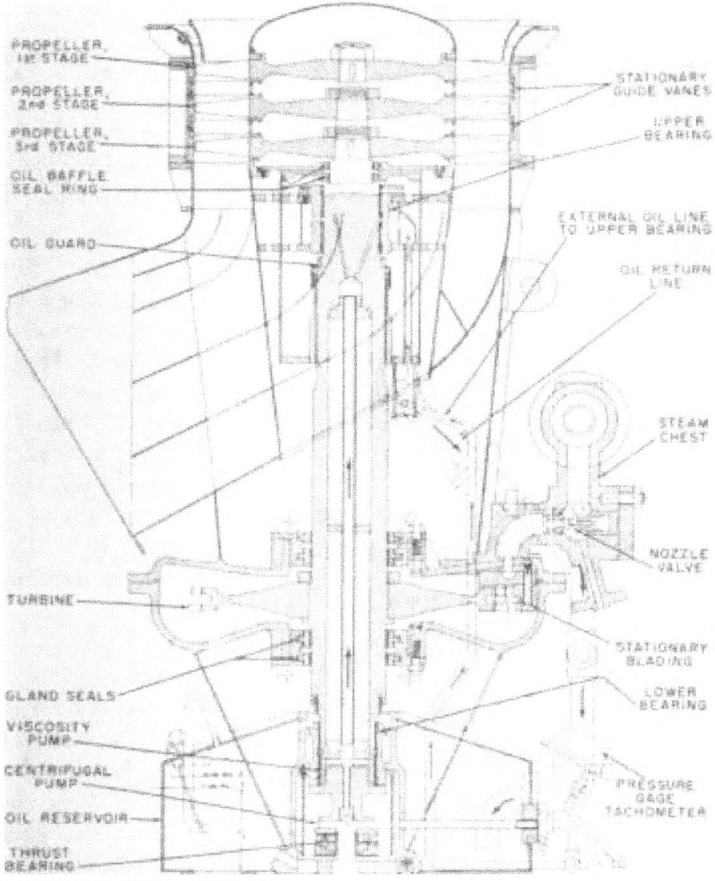

LOWER ^BEARING
PRESSURE GAGE TACHOMETER

Figure 2-13.^Three-ttage propeller-type forced draff blower (sectional view).

A modern, vertical, three-stage, propeller-type forced draft blower is shown in figures 2-12 and 2-13.

Forced Draft Blower Lubrication

Since blowers operate at very high speeds, correct lubrication of the bearings is absolutely essential. Pressure lubrication systems are used on all modern turbine-driven blowers, with appropriate modifications, in each case, to suit the design of the blower.

Many horizontal forced draft blowers use a simple gear pump in the lubrication system. The gear pump is turned by the main shaft, but is geared down to about one-fourth the speed of the turbine. The lube oil is pumped from the oil reservoir through the oil filter and the oil cooler, to the bearings. Oil then drains back to the reservoir by gravity.

A simple gear pump does not furnish oil to the bearings when the blower is being turned in the wrong direction. If an idle blower is forced to turn backwards, therefore, the bearings are likely to be severely damaged. In order to avoid this danger, a recent alteration requires that the simple gear pump in some horizontal blowers be replaced with a special type of gear pump which continues to pump oil, without change in the direction of oil flow, when the direction of rotation is reversed. This alteration applies to many vessels (particularly destroyers) having horizontal forced draft blowers.

Some vertical blowers are fitted with a gear pump and a lubricating system similar to that used for horizontal blowers. However, many vertical blowers have a somewhat different type of pressure lubrication system, in which the gear pump is replaced by a centrifugal pump impeller and a helical-groove viscosity pump. The centrifugal pump impeller is on the lower end of the main shaft, just below the lower radial bearing. The viscosity pump (sometimes called a friction pump) is on the shaft, inside the lower part of the lower bearing, just above the impeller.

The general arrangement of this type of lubrication system is shown in figure 2-14. In this system, unlike the system previously described, the lubricating oil does not go through the oil strainer or the oil cooler on its way to the bearings. Instead, oil from the reservoir is constantly being circulated through an external filter and a cooler and then back to the reservoir.

In this type of lubricating system, a viscosity pump is used to supply oil to the bearings when the blower is operating at low speed. The viscosity pump is essentially a shallow, helical groove or thread on the lower part of the shaft, inside the lower bearing shell. As the shaft rotates, the lubricating oil is picked up in the shallow groove and is carried to the tube or passageway in the shaft which leads to the upper bearing. The viscosity

UPPER BEARING
OIL GUARD
OIL TO
UPPER BEARING

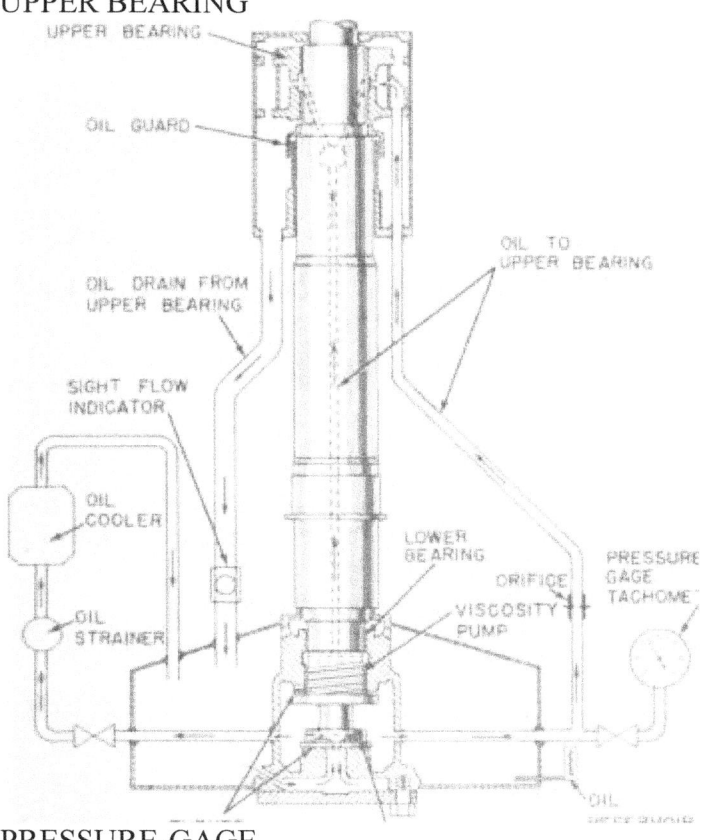

PRESSURE-GAGE
TACHOMETER
\'7b7 STRAINER
THRUST BEARING SURFACES
CENTRIFUGAL PUMP IMPELLER
RESERVOIR THERMOMETER
Figure 2-14.—Lubrication system for vertical forced draft blower.

pump is necessary because the pumping action of the centrifugal pump impeller is dependent upon the rpm of the shaft; and at low speeds the centrifugal pump cannot develop any appreciable oil pressure.

When the blower is operating at high speed, the centrifugal pump alone would provide more pressure than is needed for lubrication. Under these conditions, the viscosity pump serves to limit the pressure at which oil is delivered to the bearings. This function is an important one, since excessive oil pressure would cause flooding of the bearings, with consequent loss of oil from the lubricating system,

A system of air balancing is generally used to equalize air pressures between the oil side of a bearing and its outer housing, so that oil will not be drawn into the air stream.

In figure 2-14, you will notice that a tachometer is connected to the lubrication system and is referred to as a PRESSURE-GAGE TACHOMETER. This instrument is actually a pressure gage which is calibrated in both psi and rpm. It depends for its operation upon the fact that the oil pressure built up by the centrifugal pump impeller is directly proportional to the speed of the impeller; and the speed of the impeller is, of course determined by the speed of the main shaft. Thus, the instrument can be calibrated in both psi and rpm; and the rpm readings will indicate the speed of rotation of the main shaft. This type of tachometer can be tested on an ordinary deadweight pressure-gage tester.

Maintenance and Repair

Forced draft blowers require relatively little maintenance or repair, under normal operating conditions, provided they are adequately lubricated at all times. Proper maintenance of the lubrication system is, of course, essential.

The lubricating oil in the reservoir must be kept clean, and the reservoir must always be filled to the proper level
with oil of the specified weight and grade. An oil sample should be taken routinely once a week, and oftener if you suspect that the oil is contaminated. The oil should be changed whenever it shows an undue amount of sediment or water. After the old oil has been drained, the inside of the reservoir should be wiped clean.

The edge-filtration type of filter should be cleaned at least once each watch. At regular intervals, the filter should be dismantled and cleaned with an approved cleaning fluid; this should be done whenever the oil is renewed in the reservoir, and oftener if necessary.

The oil cooler also requires occasional cleaning and repair. Zincs in the oil cooler should be checked at least once a month, and renewed when necessary.

The oil grooves in the bearings must not be altered without approval of the Bureau of Ships. The passageways to the oil grooves, and all bearing drain holes, should be cleared of sediment whenever the bearings are taken down for overhaul.

Certain routine tests and inspections must be made in order to keep a check on the general condition of all forced draft blowers. Idle blowers should be turned by hand once a day, and an entry should be made in the appropriate log or check-off list. Automatic shutters should also be moved by hand once a day. After each period of steaming (and, in any case, at least once each quarter) all blower units and foundations should be carefully inspected ; and loose or broken rivets, nuts, and bolts should be tightened or renewed.

Forced draft blowers should be carefully observed during operation, and should not be run if there is excessive vibration or unusual noise. Vibration may be caused by worn or loose bearings; a bent shaft; loose or broken foundation bolts or rivets; or other defects. All such defects should be remedied as soon as possible, in order to prevent a complete breakdown of the blower unit. If a blower continues to vibrate after all defects have been corrected, the fan, the turbine rotor, or the shaft is prob-

ably out of balance. A Davey vibrometer (carried on tenders and repair ships) may be used to balance the unit without removing it from the ship.

It is essential that all clearances be maintained within the specified limits. Figure 2-15 shows the important clearances which must be checked on the vertical forced draft blower previously illustrated (figs. 2-12 and 2-13). The clearance indicated at each location applies only to this particular blower, and is likely to vary from one unit to another; be sure to consult the manufacturer's instruction book for the blowers on your own ship.

Minor repairs to fan blades may be made on board ship, in case of emergency, but major repairs cannot be made without special instructions from the Bureau of Ships. As a matter of routine maintenance, fan blades should be wiped down with an oily rag from time to time, to remove dirt and dust. You should never allow anyone to apply paint to a blower fan, or to any other rotating part of the unit.

The care and maintenance required for the driving turbine is discussed in the chapter on auxiliary turbines in this training course

Safety Precautions

The following safety precautions must be observed in connection with the operation of forced draft blowers:

1. Before starting a blower, always make sure that the fan is free of dirt, tools, rags, and other foreign material.

2. Do not try to move automatic flaps or shutters by hand if another blower serving the same boiler is already in operation.

3. When only one blower on a boiler is to be operated, make sure that the automatic shutters to the idle blower are closed.

4. Never try to turn a blower by hand when steam is being admitted to the unit.

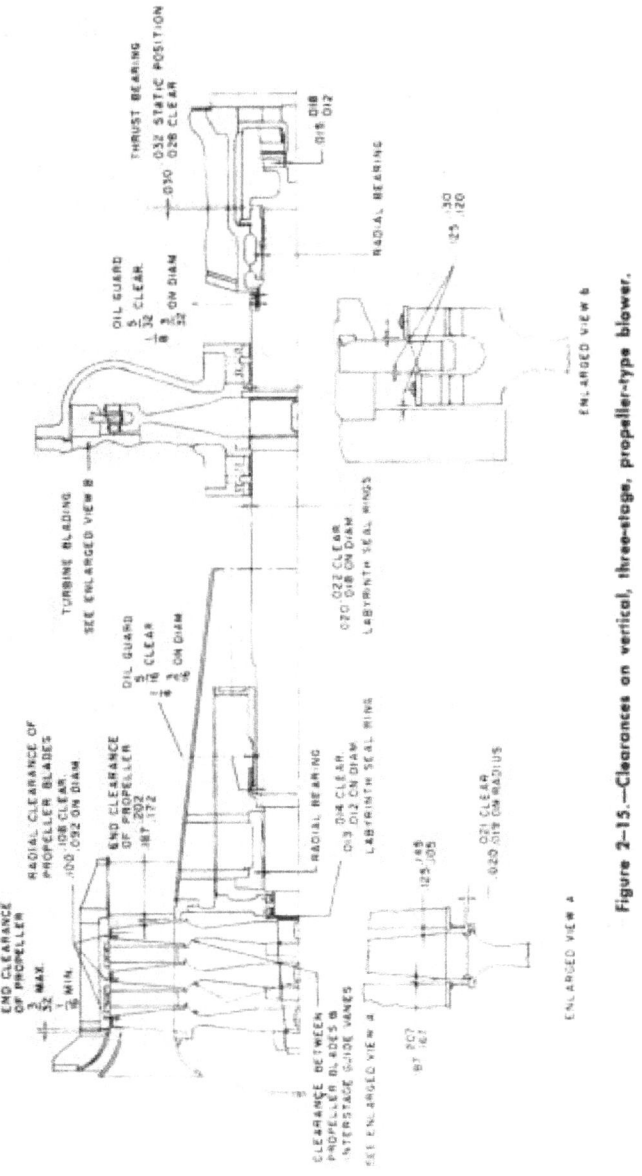

Figure 2-15.—Clearances on vertical, three-stage, propeller-type blower.

49

5. Never tie down the speed-limiting governor. Be sure that it is in good operating condition at all times; and that it is properly set.

6. Observe all appropriate safety precautions concerning turbine operation.

QUIZ

1. What reciprocating pump is particularly likely to become vapor-bound due to loss of suction head?

2. What condition may make it necessary to prime a reciprocating pump?

3. If a reciprocating pump has been operating properly and suddenly loses discharge pressure on one stroke, what is the most likely cause of the trouble?

4. Why should snifter valves or air chambers never be installed on emergency feed pumps?

5. What is the most common cause of groaning in the water end of a reciprocating pump?

6. How frequently should idle reciprocating pumps be moved by steam?

7. Where is the logical place to start an overhaul of a reciprocating pump?

8. How can you calculate the end-gap clearance for split steam piston rings, if this clearance is not given in the manufacturer's instruction book?

9. If excessive pressure builds up in the spaces between the rings and the steam piston, what can be done to remedy the trouble?

10. When a reciprocating pump is being aligned, how is the steam cylinder brought into line with the w^ater cylinder?

11. Why should a centrifugal pump be operated at or near its rated capacity and discharge pressure?

12. What danger is involved in operating a centrifugal pump at zero capacity?

13. How frequently should stuffing box packing be renewed?

14. What can be done to prevent leakage between the shaft and the sleeves, on a centrifugal pump?

15. When should the wearing rings on centrifugal boiler feed pumps be renewed?

16. How frequently should idle pumps be turned by hand?

17. What term is used to define the volume of liquid displaced by the rotating elements of a rotary pump on each revolution of the shaft?

18. In a rotary pump, what is the relationship between capacity
and THEORETICAL DISPLACEMENT?

19. Why must a rotary pump be run at a relatively low speed when it is handling a high-viscosity liquid?

20. How are the rotor housings positioned, in a triple-screw high-pitch rotary pump?

21. What two purposes are served by the viscosity pump in the lubrication system of a vertical forced draft blower?

22. How frequently should the edge-filtration type of filter in a forced draft blower lubrication system be dismantled and cleaned?

23. How frequently should idle blowers be turned by hand?

CHAPTER

AUXILIARY TURBINES AND ACCESSORIES

On steam-driven vessels, most fireroom pumps and forced draft blowers are driven by auxiliary steam turbines. A thorough knowledge of these turbines and their accessories is required for a complete understanding of the pumps and blowers. In the last chapter, we were concerned primarily with the pump end or the blower end of the unit; in this chapter, we will take up auxiliary turbines, their lubrication systems, and accessories such as flexible couplings, reduction gears, and governors.

Actual operating instructions for auxiliary turbines are not included in this chapter, since you are more likely to be concerned with trouble shooting, maintenance, and repair than with routine operation. However, as a BT1 or BTC, you must be entirely familiar with the correct procedures for starting, stopping, and securing all fire-room machinery; and you must be able to supervise the men who are operating the machinery. Operating information will be found in the manufacturers' instruction books and in BuShips Manual, chapter 50. Operating procedures for auxiliary turbines are also discussed in Boil-erman 3 and 2, NavPers 10535-C.

TURBINE CLASSIFICATION

Turbines may be classified as impulse turbines or REACTION TURBINES, depending upon the manner in which the steam causes the turbine rotor to move. When the

rotor is moved by a direct push or "impulse" from the steam impinging upon the rotor blades, the turbine is said to be an impulse turbine. When the rotor is moved by the force of reaction, the turbine is said to be a reaction turbine.

NOZZLE

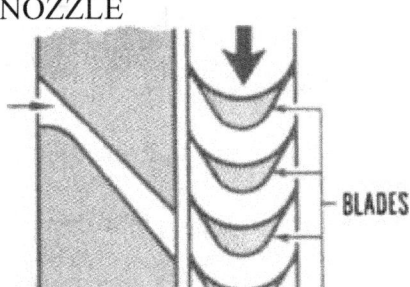

Rgwr* 3-1 .—Impulse turbine nozzle and blades.

The angle at which the steam hits the moving blades and the shape of the moving blades are the two main factors which determine whether the rotor is moved by a direct impulse or by reaction to an impulse. Figure 3-1 shows the nozzle and blade arrangement in an impulse turbine. Figure 3-2 shows the fixed blades and the moving blades in a reaction turbine.

ENTERING STEAM

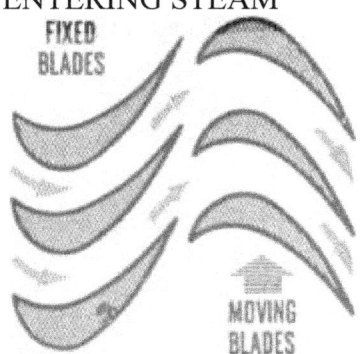

Figure 3-2.—Rxtd and moving bladtt in a reocfion turbine.

In an impulse turbine, the steam expands in the nozzle and so loses pressure but gains velocity. In the moving blades, the steam loses velocity but the pressure remains constant. Actually, an impulse turbine utilizes both the impulse of the steam jet and, to a lesser extent, the reactive force which results from the fact that the curving blades cause the steam to change its direction.

In a reaction turbine, the steam enters through a row of fixed blades which direct the flow of steam to the moving blades. As you can see in figure 3-2, the fixed blades and the moving blades are very similar in shape. Steam expansion takes place in both sets of blades.

A reaction turbine is moved primarily by (1) the reactive force produced on the moving blades when the steam increases in velocity, and (2) the reactive force produced on the moving blades when the steam changes direction. However, some of the motion of the rotor is actually caused by the impact of the steam on the blades; to a certain extent, therefore, the reaction turbine operates on the impulse principle as well as on the reaction principle.

In the remainder of this chapter, we will be concerned only with impulse turbines, since reaction turbines are seldom if ever used to drive auxiliary units. Impulse turbines may be classified according to the manner of staging and compounding, and according to the manner of steam flow.

Staging and Compounding

In an impulse turbine, a stage is defined as one set of nozzles and the succeeding row or rows of moving and fixed blades. Another way of defining a stage is to say that it includes the nozzles and blading in which only one pressure drop takes place. (In an impulse turbine, remember, the only pressure drop takes place in the nozzles. Therefore, the number of sets of nozzles in an impulse turbine indicates the number of stages.)

An impulse turbine which has one set of nozzles and only one row of moving blades is called a simple impulse TURBINE. Simple impulse turbines do not utilize the available energy of the steam as efficiently as multistage turbines; however, they have the advantage of being simple in design and construction, and are often used for small auxiliary units. The simple impulse stage is usually called a Rateau stage.

One way to increase the efficiency of the turbine is to add another row (or even two more rows) of moving blades to the rotor. Figure 3-3 shows a turbine which has two rows of moving blades. This type of turbine is called a velocity-compounded turbine because the residual velocity of the steam leaving the first row of moving blades is utilized in the second row of moving blades; and, if a third row is added, the remaining velocity of the steam is utilized in the third row. The fixed blades, which are fastened to the casing rather than to the rotor, serve to direct the steam from one row of moving blades to another.

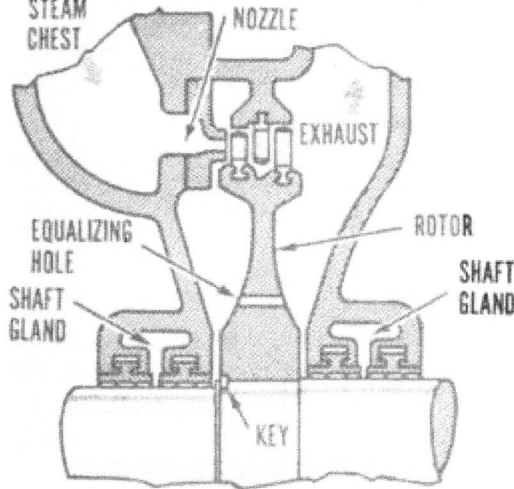

Figure 3-3.—Velocity-compounded impulse turbine.

The velocity-compounded impulse turbine shown in figure 3-3 has only one pressure drop and therefore, by definition, only one stage. This type of velocity-compounded impulse stage is usually called a Curtis stage. Many of the auxiliary turbines used for pumps and forced draft blowers consist of one Curtis stage.

It should be noted that velocity-compounding can also be achieved when only one row of moving blades is used, provided the steam is directed in such a way that it passes through the blades more than once. We will take up this point in greater detail in the discussion of types of steam flow.

Another way to increase the efltciency of an impulse turbine is to arrange two or more simple impulse stages in one casing. The casing is internally divided by diaphragms so that the residual steam pressure from one stage is utilized in the following stage. This type of turbine is known as a pressure-compounded turbine because a pressure drop occurs in each stage, as the steam expands through each set of nozzles. A pressure-compounded turbine is often called a Rateau turbine, since it is essentially a series of simple impulse (Rateau) stages arranged in sequence in one casing. Pressure-compounded turbines are not commonly used for small auxiliary units.

Various combinations of velocity-compounding and pressure-compounding are used in propulsion and generator turbines. These combinations will be discussed in the chapter in this course which deals with propulsion machinery.

Classification by Steam Flow

Auxiliary turbines may be classified according to the direction of steam flow and according to the repetition of steam flow

The direction of steam flow may be axial, radial, or helical. In general, the direction of flow is determined by the relative positions of nozzles, diaphragms, rotating blades, and fixed blades.

Most auxiliary turbines are of the axial-flow type — that is, the steam flows in a direction which is roughly parallel to the turbine shaft. All of the turbines illustrated thus far in this chapter are axial-flow turbines.

In a RADIAL-FLOW TURBINE, the steam enters in such a way that it flows radially either toward or away from the axis of the rotor. Radial-flow turbines are not generally used for propulsion turbines, but are sometimes used for driving smaller auxiliary units such as pumps.

In the HELICAL-FLOW TURBINE, shown in figure 3-4, the steam enters at a tangent to the periphery of the rotor and impinges upon the moving blades. These blades, which are usually called buckets, are milled into the rim of the rotor. The buckets are shaped in such a way that the direction of steam flow is reversed in each bucket, and the steam is directed into a redirecting bucket or reversing chamber mounted on the inner cylindrical surface of the casing. The direction of the steam is again reversed in the reversing chamber; and the continuous reversal of the direction of flow keeps the steam moving helically.

Several nozzles are usually installed in this type of turbine, and for each nozzle there is an accompanying set of redirecting buckets or reversing chambers. Thus the reversal of steam flow is repeated several times for each nozzle and set of reversing chambers.

Now let's consider the classification of a helical-flow turbine, with respect to staging and compounding. It is a single-stage turbine because it has only one set of nozzles and therefore only one pressure drop. It is a velocity-compounded turbine because the steam passes through the moving blades (buckets) more than once, and the velocity of the steam is therefore utilized more than once. The helical-flow turbine shown in figure 3-4 might be said to correspond roughly to a turbine in which velocity-compounding was achieved by the use of four rows of moving blades.

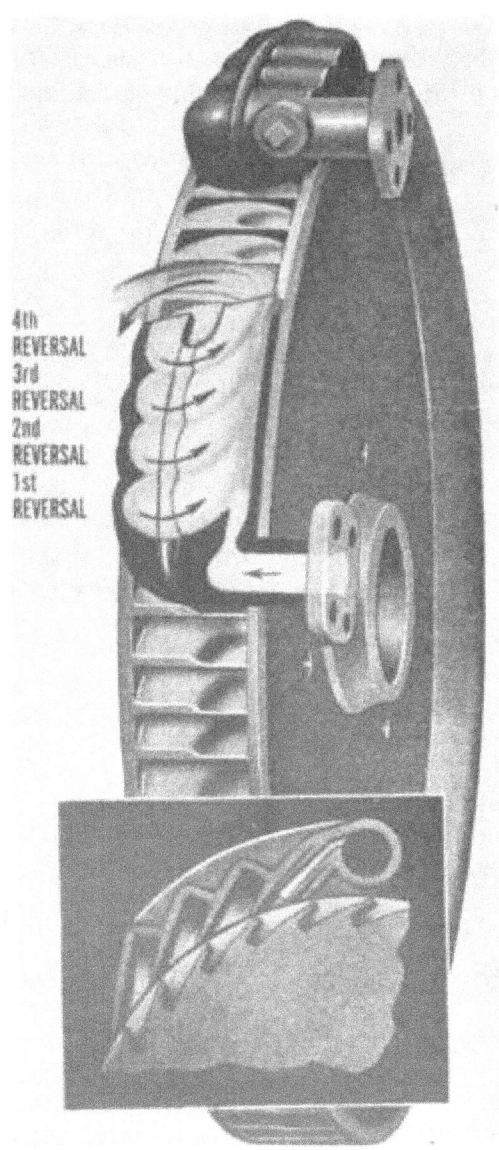

4th
REVERSAL
3rd
REVERSAL
2nd
REVERSAL
1st
REVERSAL

Figure S-A.—Helicol-flow turbine.

Helical-flow turbines are used for driving pumps and forced draft blowers.

Turbines are classified as being single entry or re-entry, according to the repetition of steam flow. If the steam passes through the blades only once, the turbine is called single entry. All multistage turbines are single entry.

Re-entry turbines are those in which the steam passes more than once through the blades. The helical-flow turbine shown in figure 3-4 is a re-entry turbine. Another type of re-entry turbine has one large reversing chamber instead of a number of smaller buckets or redirecting chambers. Some re-entry turbines are made with two reversing chambers.

PRESSURE LUBRICATION SYSTEMS

On very small auxiliary turbines, the bearings may be of the self-oiling type. These bearings have one or two oil rings which hang on the turbine shaft and revolve with it (although at a slower rate). The rings dip into the oil reservoir and carry oil up to the journal. From there, the oil spreads to lubricate the bearings.

On larger auxiliary turbines, a pressure lubrication system is used to lubricate the radial bearings, thrust bearings, and—in some instances—the governor bearings. Pressure lubrication systems do NOT provide lubrication for flexible couplings, governor linkages, and some governor bearings; these parts of the unit must be lubricated separately.

A pressure lubrication system requires a lube oil pump. As a rule, the lube oil pumps used for auxiliary units are positive displacement rotary pumps of the simple gear type. The lube oil pump is generally installed on the turbine end of a forced draft blower unit, but may be on either the driving end or the driven end of pump units. The lube oil pump is driven by the turbine shaft, through reduction gears.

Some forced draft blowers use a centrifugal pump, sup-

plemented by a viscosity pump, for lubrication of the unit. This type of lubrication system is explained in the discussion of forced draft blowers in chapter 2.

A pressure lubrication system designed for fuel oil service pumps, fuel oil booster pumps, and lubricating oil service pumps is shown in figure 3-5. It is similar in principle to the lubricating systems used on many other units.

In this system, the bottom section of the reduction gear casing forms the oil reservoir. The shaft which carries the gear-type oil pump on one end and the governor on the other is geared to the pump shaft, which is in turn geared to the turbine shaft. The lubricating oil passes

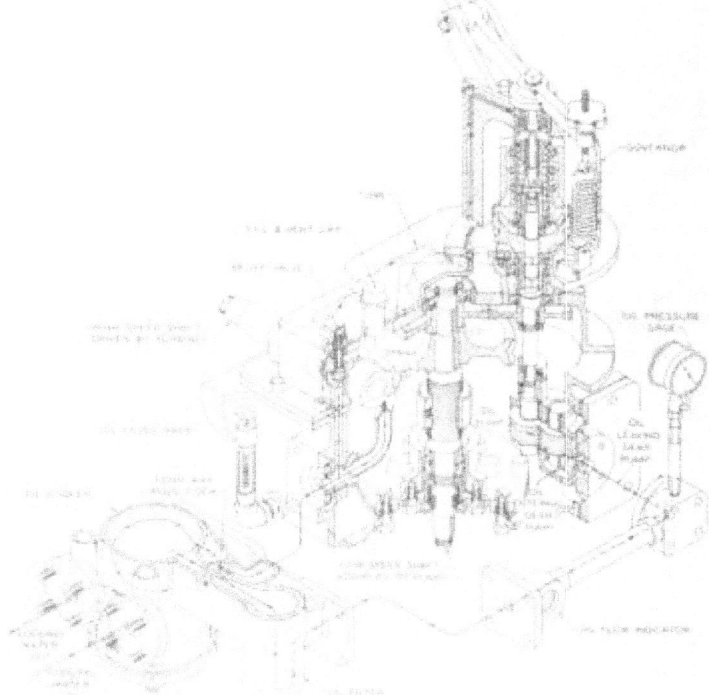

Figur* 3-5.—Prcssur* lubrication >yst*m for turbine-drivtn unit.

through an oil flow sight, an edge-filtration type filter, and an oil cooler. Oil is then piped to the bearings on the turbine shaft, to the governor, and to the worm gear on the pump shaft. The bearings and gear on the oil pump and governor shaft are lubricated by oil which drains from the governor and passes back into the oil reservoir.

Selection of Lubricating Oil

It is extremely important that the correct grade and weight of oil be used for turbine lubrication. Units having worm-type reduction gears should be lubricated with Navy Symbol number 3080 oil; for all other auxiliary turbines and reduction gears. Navy Symbol number 2190T should be used unless another type of oil is definitely specified in the manufacturer's instructions.

All new lubricating oil should be centrifuged or strained through a fine screen (60 mesh or finer) before being introduced into a lubrication system.

Care of the Lubrication System

Every precaution must be taken to keep the lubricating oil clean and free of water, dirt, and other foreign matter. Weekly samples should be taken from the lubrication system of each machine in the fireroom; these samples should be inspected when they are drawn, and should then be placed in a rack, allowed to settle, and inspected again for signs of sediment and water. A regular schedule of sampling and observing the lubricating oil should be adhered to.

The lubricating oil in the reservoir should be kept at the proper level. The oil should be changed whenever it shows signs of contamination. Large batches of used oil, such as would be drained from forced draft blowers and ship's service generator turbines, should be transferred to a lube oil settling tank and then put through the centrifugal purifier, in accordance with instructions given in chapter 45, BuShips Manual. Lube oil that is contaminated with salt water must be discarded.

The reservoir should be wiped clean before the new oil is added, and all piping, pumps, coolers, etc., should be cleaned to the maximum extent possible. Caution: Never use waste to clean any part of a lubricating system.

In refilling the unit, you should first fill the reservoir to the HIGH mark and run the unit slowly until the cooler, the piping, and all other parts of the lubricating system are entirely filled. Then shut down the unit, and add as much oil as necessary to reestablish the level at the high mark on the gage.

The edge-filtration type of filter should be dismantled and cleaned with an approved cleaning fluid at periodic intervals. The lube oil cooler may also require occasional cleaning. The zincs in the cooler should be inspected monthly, and replaced whenever necessary.

Defects which may cause oil to be thrown from the ends of the bearings include excessive bearing clearances, worn or damaged oil seals and deflectors, clogged orifice plates, and clogged oil grooves and drain holes in the bearings. If oil is being thrown, you must determine the cause and remedy the condition as soon as possible. As a temporary measure to prevent further loss of oil, if oil is being thrown from all bearings and if you cannot find any other cause of the trouble, the lubricating system relief valve may be set slightly lower than its usual 10 psi gage. The lower setting will result in less oil being sent to the bearings.

If the supply of oil to the bearings is interrupted, stop the turbine immediately and take whatever steps are necessary to reestablish the circulation of oil. Check the oil level in the reservoir. Be sure that the strainers are clean. Check the lines to be sure that there are no leaks or obstructions. If a four-way valve is installed, be sure that it is in position to allow the proper flow of oil. If necessary, increase the oil pressure by increasing the setting of the lubricating system relief valve.

Caution: If it is necessary to stop a turbine because of overheated bearings, slow the turbine but keep it
turning over slowly until the bearings and the journal have had time to cool. If the turbine is stopped suddenly, without this cooling-off period, the bearing metal may freeze to the shaft.

MAINTENANCE AND REPAIR

Since auxiliary turbines vary widely in design and construction, the following information on their maintenance and repair must be considered as general, rather than specific, in nature. More detailed instruction will be found in chapter 50 of BuShips Manual and in the instruction books furnished by the manufacturers. You should not attempt to operate, service, or repair any auxiliary turbine without having a thorough understanding of the manufacturer's instructions for that particular unit.

Radial Bearings

The bearings which support the rotor and maintain the correct radial clearance between the rotor and the casing are usually split-sleeve, babbitt-lined bearings. When the maximum allowable wear has occurred in these bearings, spare bearings should be fitted and the worn bearings should be rebabbitted by a Navy yard or other repair facility. Rebabbitting of bearings should not be undertaken by ship's force except in case of emergency.

It should be noted that some radial bearings on modern, high-speed auxiliary turbines are of the "thin-shell" type. Thin-shell bearings cannot be rebabbitted, and so must be replaced when the allowable wear has been exceeded.

There are three possible ways of measuring the amount of wear of radial bearings on auxiliary turbines: (1) by means of bridge gages, if these are furnished with the unit; (2) by the direct-measurement or crown-thickness method; and (3) by taking leads. Figure 3-6 shows how to measure bearing wear with a bridge gage (A) and by the crown-thickness method (B).

To use a bridge gage, you must first remove the upper half of the bearing housing and the upper half of the bearing shell. Place the bridge gage on the lower half of the bearing housing, in the manner shown in figure 3-6 (A), and measure the clearance at the top and at one side

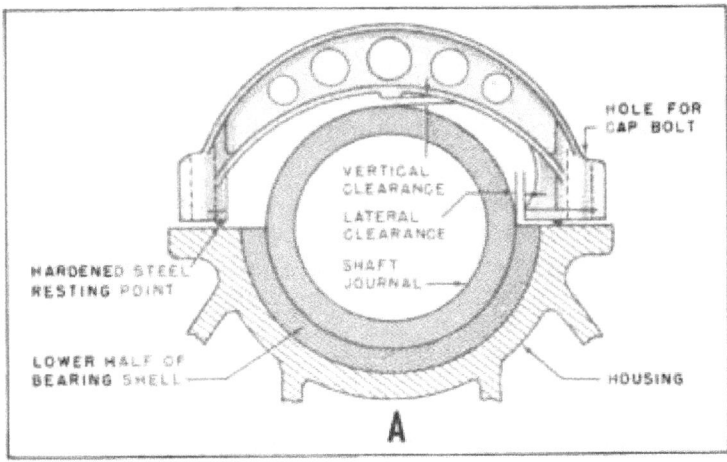

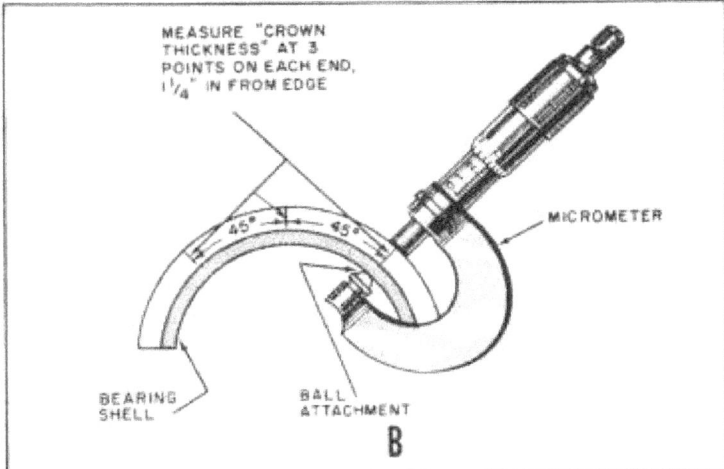

Figure 3—6.—How to measure bearing wear by (A) bridge gage, and (B) crown« thicknett method.

with feeler gages. Compare the obtained readings with the readings taken when the bearing was installed. When practicable, a 24-hour period should be allowed after the unit is shut down before bridge gage readings are taken, in order to allow lubricating oil to drain from the bearings.

The crown-thickness or direct-measurement method of measuring bearing wear gives the most accurate indication of actual wear; but it can be used only on bearing shells on which the outer diameter is a true machined surface.

The crown-thickness measurements are taken on the working half of the bearing, at the point of wear. In order to measure bearing wear by this method, original crown-thickness readings must be established when the unit is known to be in proper alignment, with all bearings at their designed clearances. If the bearing is for a horizontal shaft, the lower half of the bearing shell should have three radial lines scribed at each end—one at the center, and one on each side, 45° away from the center scribe line; and the upper half of the bearing shell should have one scribe line at each end, at the center. If the bearing is for a vertical shaft, the ends of the bearing shell should be scribed at four equidistant points.

To take crown-thickness readings, use a micrometer to measure the thickness of the bearing shell at each scribe mark, about V/^ inches in from the end of the shell. Compare your obtained readings with the original readings, which should be stenciled adjacent to the scribe marks. Figure 3-6 shows a micrometer being used to measure the crown thickness of a bearing shell. Notice that a ball-type anvil is used in the micrometer in place of the usual flat anvil. After converting to the rounded anvil, either adjust the micrometer to zero or subtract the diameter of the steel ball from the micrometer readings, in order to obtain the actual crown thickness.

The method of measuring radial bearing wear by taking leads is shown in figure 3-7. You remove the upper half of the bearing and lay several lengths of soft lead wire across the top of the journal. The leads should extend most of the way around the exposed half of the journal. Be sure to allow some space for the elongation of the leads which will occur when pressure is applied. Replace the upper half of the bearing shell and the upper half of the bearing housing, and set up tightly on

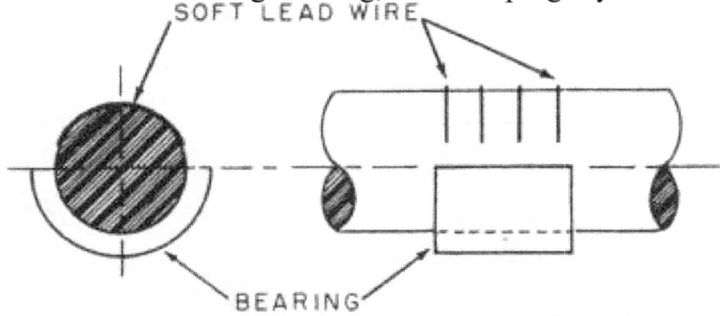

Figure 3-7.—Measuring bearing wear by taking leads.

the nuts. Remove the upper half of the bearing again, and remove the squeezed leads. Measure the thickness of each lead at several points. If the measurements are uniform, compare the obtained clearance with the designed clearance. If the leads vary in thickness, this indicates an uneven bearing surface or some type of misalignment.

Radial bearings and their journals should be spotted in with Prussian blue whenever you have reason to think that the bearing surface is uneven, and whenever new or rebabbitted bearings are being fitted. High spots on the bearing surface or on the journal should be removed.

The designed oil clearances for all bearings are generally shown in the manufacturer's instruction book. If you do not have an instruction book for the unit, or if you have one but it does not contain this information, consult the oil clearance tables in chapter 40 of BuShips Manual.

Thrust Bearings

Thrust bearings are used to hold the shaft in its correct axial position. Thrust bearing clearances (except in the case of pivot-shoe type thrust bearings) should be measured with feeler gages, if the bearing surfaces are accessible. Another method is to move the shaft back and forth as far as it will go and use a dial indicator to measure the amount of axial movement (thrust clearance). The obtained clearance should be compared with the designed clearance shown on the manufacturer's plan. In general, about 0.008 to 0.012 inch of end play is allowed in auxiliary turbine thrust bearings.

The end play in some types of thrust bearings may be adjusted within reasonable limits by changing the thickness of shims. On Kingsbury pivoted segmental type thrust bearings, these shims are placed between the base ring and the bearing housing. Nonsegmental types of thrust bearings are either provided with shims or else have spaces for the insertion of shims. Thrust bearings of the ball type cannot be adjusted in this manner; when ball thrust bearings develop excessive clearances, the bearings must be replaced as a unit.

Turbine Blades

Proper clearances in the thrust and radial bearings of a turbine do not necessarily ensure that proper axial and radial clearances exist between stationary and moving blades in the turbine or between moving blades and nozzles. It is important, therefore, to check the clearances and to be sure that all blade clearances are maintained within the limits specified by the manufacturer. Some auxiliary turbines have an opening in the casing, through which clearances may be measured by means of tapered

gagres. Clearances on various types of auxiliary turbines should be measured as follows:

Single-entry axial-flow turbines. —Take the axial clearance between the nozzle face and the shroud band on the first row of blading.

Re-entry axial-flow turbines. —Take the axial clearance between the nozzle and the blading, and the axial clearance between the blading and the reversing chamber.

Helical-flow turbines. —Take the radial clearance between the nozzles and the blading, and the radial clearance between the blading and the redirecting or reversing buckets.

Single-entry radial-flow turbines. —Take the radial clearance between the nozzle and the blading.

Re-entry radial-flow turbines. —Take the radial clearance between the nozzle and the blading, and the radial clearance between the blading and the reversing chamber.

Shaft Glands

Either carbon packing or labyrinth packing—or, in some cases, both—may be used in shaft glands to prevent the leakage of steam from the turbine casing. Turbines which operate condensing usually have both carbon packing and labyrinth packing. Most turbines which operate noncondensing have only carbon packing; however, the turbines used to drive some of the newer forced draft blowers and main feed pumps have only labyrinth packing.

Carbon packing rings are mounted around the shaft and are held in place by one or more springs. As a rule, three or four carbon rings are used in each gland, with each ring being fitted into a separate compartment of the gland housing. Generally, each carbon ring consists of three segments; however, rings consisting of two segments or of four segments are also used. The ring segments are butt joined. Lugs or stop pins are used to keep the carbon rings from rotating with the shaft.

Carbon packing requires refitting or renewal when it is worn out of round or is otherwise in bad condition. When refitting worn carbon rings, be sure to install each ring in its proper groove. As a rule, the grooves and each segment of each ring are marked with numbers so that they can be properly matched for assembly. If they are not marked, be sure to mark them as you take the rings off.

Carbon packing must never be lubricated with oil or grease. The natural lubricant contained in the carbon, together with the condensate which collects between the shaft and the rings, provides all the lubrication required. Oil and grease are extremely damaging to the carbon. When you are working with carbon rings, be careful that you do not accidentally get oil or grease on the carbon.

Refitting or renewing carbon packing rings is a rather diflficult and tedious job. You can simplify the work considerably by making up a jig or a mandrel, similar to the one shown in figure 3-8, and a checking gage. The jig can be made from a square steel plate or from round bar stock; it should be approximately % inch thicker than the ring to be fitted, and about V/2 inches greater

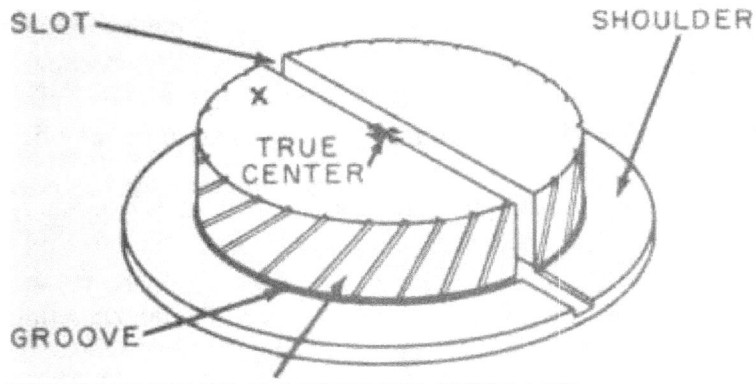

SLOT   SHOULDER

X
TRUE
CENTER

GROOVE

CUTTING EDGE ON TURNED SURFACE

Figure 3-8.—Jig for fitting carbon packing rings.

in diameter (or perimeter, if you are using a square plate) than the inside diameter of the ring. The upper part should be turned down to a diameter which is about 0.001 or 0.002 inch less than the final bore required for the carbon ring, and a cutting edge should be raised around the periphery of the turned surface. The cutting edge is made by making diagonal cuts about 0.003 inch deep, with a pitch of about \U inch, on the turned surface. The final diameter of the turned surface, including this raised cutting edge, should be the same as the required inside diameter of the carbon ring.

The width of the turned surface should be about Yg inch greater than the width of the carbon ring to be fitted. A small groove should be cut in the shoulder, as an extension of the turned surface; this groove will provide a space in which loose carbon particles can collect, so that they will not interfere with the fitting of the carbon ring.

Cut a yiB-inch slot through the turned surface, and continue it about inch into the shoulder or flange below the turned surface. One edge of the slot must pass EXACTLY through the center, forming a true diameter of the circular face of the jig; the other edge of the slot will therefore be slightly off-center. Be sure to mark that part of the face which forms a true half-circle after the slot has been cut. Remove all sharp edges from the turned surface of the jig and make sure that the edges of the slot are smooth.

The checking gage can be made by turning a piece of steel to the same diameter as the shaft to which the carbon ring is to be fitted. The checking gage should be Yq inch thicker than the width of the ring.

Assemble the segments of the carbon ring on the checking gage, and hold them in place with the garter spring. Check the fit of the joints and the clearance between the ring and the gage. If the joints do not fit properly, or if the bore is too large, remove the ring from the checking gage. Place each segment separately on the jig and

smooth the ends with fine sandpaper wrapped around a flat file or a block. Each segment must be placed on the true half of the jig, and the true centerline must be used as a guide when removing carbon from the end of the segment. Remove carbon from the ends of each segment until the assembled ring is TOO small to fit over the checking gage. Then assemble the entire ring, with its garter spring, on the fitting jig. Hold the ring against the bottom of the jig and turn it by hand until the cutting edge on the jig has removed enough carbon to allow the ends of the segments to make proper butt joints.

Remove the carbon ring from the jig and try it on the checking gage. If the ring is too small, put it back on the jig and remove some more carbon from the inner surface of the ring. If the ring is too large, replace each segment on the jig and remove carbon from each end until the ring is the correct size.

When carbon rings are fitted in the manner just described, the ends of the segments will make firm butt joints and there will be a small clearance between the inner surface of the ring and the shaft. On some auxiliary turbines, the carbon rings are designed to rub against the shaft and to have a clearance between the butting ends of the segments. When fitting these rings, make the jig with a diameter which is, including the raised cutting edge, the same as the diameter of the shaft. Rings cut on this jig should make contact with the shaft along the full length of each segment, and the ends of the segments should be filed down to make the proper butt clearance.

Oil Seal Rings

Various types of sealing arrangements are used on auxiliary turbines to prevent the escape of oil from bearings, gear casings, and other parts. Most commonly, a combination of oil guards and oil seal rings is used. The oil guards are similar in purpose and design to the water flingers installed on centrifugal pumps. The oil seal rings may be designed to make a friction contact with the shaft, or they may be of the labyrinth type.

Oil seal rings which make a friction contact with the shaft are used where the shaft passes out of an oil-flooded area. Rings of this type are subject to a certain amount of wear, and will require replacement from time to time. Note that these rings must be replaced, rather than repaired.

Labyrinth oil seal rings do not ordinarily become worn except when defective bearings cause the shaft to be displaced. If labyrinth rings do become worn, the lands may be reshaped with a hand chisel, in the manner shown in figure 3-9. The hand chisel must be backed up with a heavy bar; both the chisel and the bar must be ground to suit the angle of the lands.

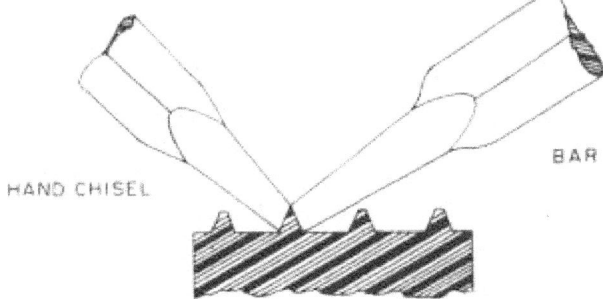

Figur* 3-9.—RMhaping lands en labyrinth packing rings.

With the tools in position, as shown, strike the hand chisel with a hammer. Move the tools to a new position, and strike the chisel again. As you work around the periphery of each land, be sure to place the tools so that each new position of the chisel overlaps the preceding position.

This reshaping procedure draws out the land, raising its height and restoring its edge. The lands should be drawn out sufficiently to give the clearance specified on the manufacturer's drawings. If the clearance is not obtainable from plans or instruction books, draw out the lands enough to give a clearance of 0.005 inch.

Flexible Couplings

On most turbine-driven pumps, the pump end is joined to the turbine end by means of a flexible coupling. Flexible couplings are designed to take care of slight misalignment between the driven and the driving shaft; however, it is important to remember that they cannot take care of any great amount of misalignment. Excessive misalignment will cause rapid wear of the coupling and damage to the pump and the turbine.

Most flexible couplings are lubricated by a self-contained oil supply. (Note that they are not lubricated by the pressure lubrication system which furnishes oil to the turbine bearings and the reduction gears.) The flexible coupling must be completely filled with the correct grade of heavy-weight oil when the unit is first started. At least once a month, the coupling should be inspected and more oil should be added if necessary. At least twice a year, flexible couplings should be opened, cleaned, and refilled with fresh oil.

Whenever a flexible coupling is opened up, be sure to inspect all gear teeth, pins, or other parts which may be subject to wear. Replace parts if necessary.

Reduction Gears

Reduction gears used with auxiliary turbines are either helical gears or worm gears. Helical gears are used where the shaft of the driven auxiliary is mounted parallel to the shaft of the turbine; worm gears are used where the shaft of the driven auxiliary is mounted at right angles to the shaft of the turbine.

Reduction gears for auxiliary units require relatively little maintenance or repair, provided they are properly aligned and properly lubricated. Excessive wear of the reduction gear bearings will cause incorrect tooth contact,

with possible serious damage to the gears. Bearings should be renewed or rebabbitted whenever necessary in order to maintain the correct tooth contact between the gears.

Overspeed Protection Devices

The three Kinds of overspeed protection devices which are used on auxiliary turbines are (1) the speed-regulating governor, (2) the overspeed trip, and (3) the speed-limiting governor.

The SPEED-REGULATING GOVERNOR, sometimes referred to as a CONSTANT-SPEED GOVERNOR, is used on constant speed machines to maintain a constant speed, regardless of load. Speed-regulating governors are usually set so that the turbine cannot even momentarily exceed 105 percent of normal operating speed. Speed-regulating governors are not so commonly used in the fireroom as are the speed-limiting governors.

Turbines equipped with speed-regulating governors may have, in addition, a safety device known as an over-speed TRIP. The overspeed trip shuts off the steam supply to the turbine after a predetermined speed has been reached, and thus stops the unit. Overspeed trips are usually set to trip out at about 110 percent of normal operating speed. Overspeed trips are not permitted on forced draft blowers built for the Navy; however, they are used on blowers in a few auxiliary vessels which were taken over by the Navy during wartime.

The overspeed protection device most commonly used on turbines for fireroom auxiliary machinery is the speed-limiting GOVERNOR. This is essentially a safety device for variable-speed units. It allows the turbine to operate under all conditions from no-load to overload, up to the speed for which the governor is set; but it does not allow operation in excess of 110 percent of normal operating speed. It is important to note that this type of governor is adjusted to the maximum operating speed of the turbine, and therefore has no control over the admission of steam until the upper limit of safe operating speed is reached.

In the naval service, speed-limiting governors are provided on turbines used for driving almost all types of pumps and forced draft blowers. Most speed-limiting governors are of the centrifugal-weight type; however, some oil-pressure type governors are also used.

It is essential that speed-limiting governors be properly set. At least once a quarter, and more often if necessary, speed-limiting governors must be tested under the supervision of the Engineer Officer, and the results must be entered in the appropriate log. If you are instructed to make a test of a speed-limiting governor, you should do it in the following manner:

Start the turbine, and run it slowly until it is properly warmed up. Gradually increase the speed, and take readings with a tachometer. As the turbine approaches full speed, observe the action of the governor. The governor should come into action smoothly, and should operate the governor valve to control the speed of the turbine. When the governor appears to have control, carefully open the throttle valve somewhat wider. Check to be sure that the governor does not allow the turbine to increase its speed to more than 5 percent above the maximum operating speed. If the governor fails to function at the proper speed, the governor setting must be adjusted. As a rule, adjustment is made by changing the tension on one or more springs. Be sure to consult the manufacturer's instruction book for the correct method of adjusting the setting on each speed-limiting governor.

Speed-limiting governors must be maintained in good operating condition at all times, and must always be properly lubricated. All governor parts must be kept clean. The governor valve seats must be checked whenever the turbine is dismantled for inspection or repair. Loose valve seats should be welded or silver soldered in place; cut or eroded valve seats should be replaced.

After any work has been done on a speed-limiting governor, the unit must be tested before it is put back into service.

Constant-Pressure Pump Governors

Constant-pressure pump governors operate on much the same principle as spring-loaded reducing valves, except that in the constant-pressure pump governor it is the discharge pressure of the pumped liquid, rather than the discharge pressure of the steam, which exerts an upward pressure on a diaphragm and so maintains a constant pressure in the discharge line. It is important to note that the pressure of the steam on the discharge side of the governor is NOT constant, but varies as necessary to keep the pump discharge pressure constant.

The two constant-pressure pump governors most commonly used in the Navy are the Leslie and the Atlas governors. A third type, formerly in use on many ships, is now being replaced by Leslie or Atlas governors. Although the following discussion pertains specifically to Leslie governors, it applies in general to the Atlas type as well, since the two are very similar in design and operation.

A constant-pressure pump governor for a lubricating oil service pump is shown in figure 3-10. The governors used on fuel oil service pumps and on main feed pumps are of the same type, except that the size of the upper diaphragm and the amount of spring tension are different on governors used for different services.

The adjusting spring exerts pressure downward on the upper diaphragm, through a crosshead and mushroom arrangement. When there is no pump discharge pressure, the spring forces the upper diaphragm and the upper crosshead down. A pair of connecting rods connect the upper crosshead with the lower crosshead; thus, the lower crosshead is moved down with the upper one. When this happens, the lower diaphragm is displaced downward by the lower mushroom, and the auxiliary valve disk is forced open. The auxiliary valve is supplied with steam from

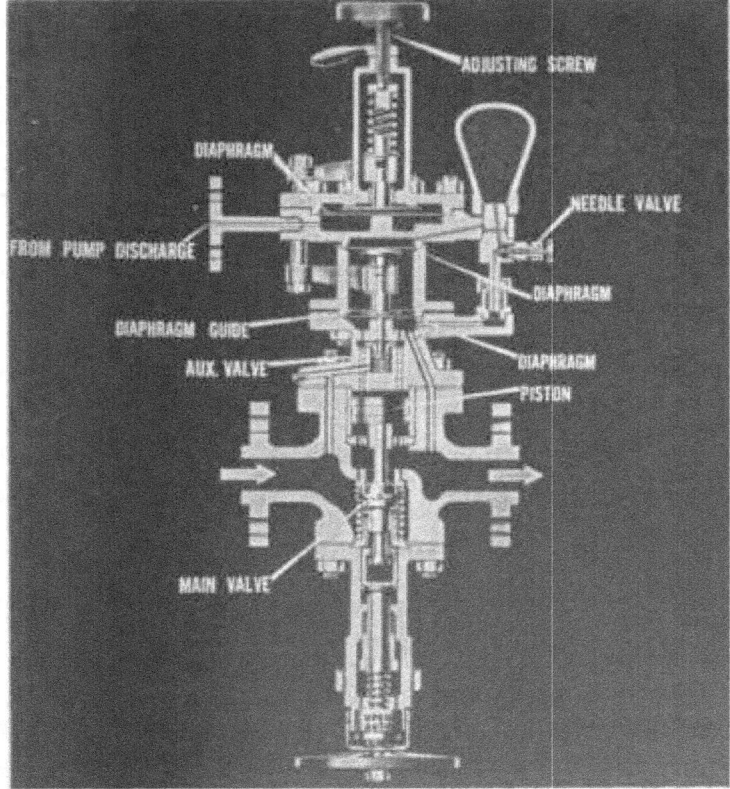

Figur* 3-10.—L*sli« constant-prMsur* pump governor.

the inlet side of the governor; and when the auxiliary valve is open, steam passes through ports to the top of the operating piston. The steam pressure on the top of the operating piston forces it downward and opens the main valve, thereby admitting more steam to the pump turbine and thus increasing the speed of the turbine.

The increased speed of the turbine is reflected in an increased discharge pressure from the pump. This pressure is exerted against the under side of the upper diaphragm, through an actuating line. When the force below

the diaphragm is greater than the force represented by the tension of the adjusting spring, the diaphragm is displaced upward. This causes a reduction in the size of the auxiliary valve opening and allows a spring to start closing the main valve against the now reduced pressure on the operating piston. When the main valve starts to close, the steam supply to the turbine is reduced, the pump is slowed, and the discharge pressure is decreased.

Steam from the main valve outlet is led through ports to chambers above and below the sealing diaphragms. A needle valve and a surge chamber (or steam chamber) are used to delay and minimize the effects of main valve outlet pressure changes upon the pressure above the top sealing diaphragm. They thus serve to dampen excessive reactions of the auxiliary valve to variations in pump discharge pressure.

With the unit under constant-load conditions, the operating piston takes a position which holds the main valve open by the required amount. A change in load conditions results in momentary hunting by the governor, until it finds the new position required to maintain pressure at the new load.

Two adjustments are required on the constant-pressure pump governor: (1) the adjusting screw must be set to regulate the discharge pressure of the pump; and (2) the needle valve must be set to minimize hunting. The adjustment of the needle valve should be made while the pump is operating under a light load at normal discharge pressure. Caution: The needle valve should NEVER be entirely closed.

The constant-pressure pump governor may be set for either automatic control or manual control. Under normal operating conditions, the automatic setting is used. When you need to control the pump manually, you must turn the bottom handw^heel to the open position, so that the main valve wiW be held fully open, allowing a full flow of steam to the turbine. With the valve in this position, you control the speed of the turbine by hand throttling.

Dirt and other foreign matter are sometimes carried with the steam into the governor, causing the governor to operate in a sluggish or unreliable manner. To correct this condition, you will have to take the governor apart and clean it.

Unscrew the connector union, which is located directly below the needle valve. Remove the bottom nuts of the superstructure studs. Lift off the superstructure above the lower diaphragm, without disassembling it. The superstructure does not contain any moving parts except the diaphragms; as a rule, therefore, the only time that you will need to disassemble the superstructure is when you are replacing diaphragms. If you must disassemble the superstructure, be very careful to replace the diaphragms and the mushroom and crosshead arrangement in their proper positions. Make sure that the connecting rods are free in their guides.

Unscrew the auxiliary valve seat, using the special wrench supplied for that purpose, and take out the auxiliary valve and the auxiliary valve spring. Remove the top cap and lift out the piston and cylinder liner. Remove the bottom cap, and take out the main valve and the main valve spring.

Clean all parts with kerosene. See that the main valve and the auxiliary valve seat properly, and that the piston rings are perfectly free in the groove. Clean the seat for the cylinder liner, so that the cylinder liner will not project above the face of the top flange of the main body. Clean the bore of the main valve guide. If necessary, regrind the main valve while the piston is in the cylinder.

When reinstalling the auxiliary valve seat, be sure that the joint faces of the top cap as well as of the auxiliary valve seat are perfectly clean. Do not use graphite or any other compounds. Screw in the auxiliary valve seat firmly, using the wrench provided. Tap the wrench with a hammer to ensure a very tight joint. Allow a clearance of 0.001 to 0.002 inch between the stem of the auxiliary valve and the seat of the diaphragm; the stem must never

project above the diaphragm seat. When tightening up the superstructure studs, pull down evenly without using excessive strain.

When renewing diaphragms and gaskets, use only standard replacement parts. Correct thickness of the diaphragm must be maintained. Never install a gasket beneath a diaphragm.

Turbine Casing Joints

Turbine casing joints must be made up metal-to-metal, without any type of sheet packing. Before making up casing joints, scrape and clean the joining surfaces, polish them with crocus cloth, and inspect them for burrs and bruises. Coat the surfaces with a thin layer of boiled linseed oil or a mixture of boiled linseed oil and graphite, manganesite, or Copaltite. The joints may also be made up with Usudurian, or some other approved type of unvul-canized rubber packing. When tightening the bolts, be sure to follow the correct bolt-tightening sequence.

Full instructions for making up turbine casing joints are given in chapter 41 of BuShips Manual, and in the manufacturer's instruction book for each unit.

Alignment

The alignment between the shaft of the driving turbine and the shaft of the driven unit should be checked before the unit is first put into operation, and at any time that you have cause to suspect misalignment. The shafts are checked for alignment at the flexible coupling.

Figure 3-11 shows the two ways in which the shafts may be misaligned; part (A) shows misalignment caused by the two shafts not being in a common plane; part (B) shows misalignment caused by lack of parallelism of the faces of the coupling halves.

To check the alignment, hold a straight edge across the side of the coupling, and check with feeler gages to be sure that the faces of the coupling halves are parallel. If the peripheries of the coupling halves are true circles of

Figur* 3-11.—^Twe typ«t of mitalignmant.

the same diameter, and if the faces are truly flat, exact alignment exists when (1) the distance between the faces is the same at all points, and (2) a straight-edge will lie squarely across the rims at any point. If the faces are not parallel, the feeler gage readings will be different at different points. If one coupling half is higher than the other, the straight-edge will not lie squarely across the rims.

Occasionally, you may find units in which the coupling halves are not true circles, or are not identical in diameter. When aligning the shafts on such units, be sure to make proper allowances for any differences between the coupling halves. The trueness of the coupling halves may be checked by holding one half stationary and revolving the other half, checking alignment at each quarter turn of the half being rotated. The half previously held stationary should than be rotated and checked in the same manner.

The clearance between the coupling halves is generally given in the manufacturer's instruction book. It is usually about inch, but may vary according to the size of the unit, the length of the shafts, the operating temperatures, and other factors.

The manufacturer's instruction book may specify certain allowances to be made when the unit is cold at the time of alignment. If no such specifications are given, you can assume that temperature changes will not affect the alignment of the unit.

TURBINE VIBRATION

Vibration in a turbine is a positive indication that the unit is not in proper operating condition. If the trouble is not remedied immediately, bearings and packing clearances will become excessive; this will result in loss of oil and steam, and in further damage to the bearings and packing. If the turbine is continued in operation for any length of time after it begins to vibrate, the entire unit may be completely disabled.

Turbine vibration may be caused by (1) loose, worn, or poorly lubricated bearings; (2) rubbing or binding of parts; (3) improper alignment; (4) loose or broken foundation bolts; (5) binding of the carbon packing; (6) a bent or bowed shaft; or (7) unbalance in either the driven unit or the driving turbine.

If a turbine vibrates, you should investigate all possible causes of the trouble as soon as possible. If the turbine still vibrates after all causes except unbalance have been investigated, you can assume that the turbine rotor requires balancing. A running balance may be made with a Davey vibrometer, while the unit is in place; or the unit may be removed from the vessel and balanced at a naval shipyard. The maximum readings on the Davey vibrometer scale should not exceed 0.003 inch.

TESTS AND INSPECTIONS

At frequent intervals during each watch, check the bearings, oil flow sights, and all pressure and temperature gages on all operating auxiliary turbines. Listen for sounds which would indicate damaged or defective bearings. At least once during each watch, operate the speed-limiting or speed-regulating governor valve stem by hand. Caution: When operating the governor valve stem, be careful not to overspeed the unit.

The following tests and inspections must be performed DAILY on idle turbines, and appropriate entries made in the log or check-off list:

1. Turn idle turbines by hand.

2. If a hand pump is provided, use it to circulate oil through the lubrication system of each idle turbine.

The following tests and inspections must be made WEEKLY, and the results entered in the appropriate checkoff list or log:

1. Inspect the valves, cocks, and joints of steam, exhaust, and drain lines.

2. Operate all valves not in use; oil the valve stems if necessary.

3. If an overspeed trip is installed, lubricate the linkage.

4. Run the turbine with steam, if practicable.

5. Test the overspeed trip by overspeeding the unit.

6. Operate the turbine casing relief valve by hand.

The following tests and inspections must be made QUARTERLY, and appropriate entries made in the log or check-off list:

1. Test the speed-limiting governor. (This must also be done whenever a turbine is put back into service after prolonged idleness, and whenever the speed-limiting governor has been dismantled for any reason.)

2. Examine the interior of the casing, if possible. (Peepholes are provided on many auxiliary turbines.) Inspect the interior of the casing for evidence of corrosion.

3. Sound the casing with a hammer, to detect cracks.

4. Examine the casing for loose or broken bolts.

5. Check the setting of the turbine casing relief valve by steam.

6. Clean the steam strainers.

7. Check the shoes of the thrust bearing for clearance and for the condition of the bearing surface.

8. After examining the thrust bearing, blow through it with clean air.

9. Check all sleeve-type bearings for clearances and for the condition of the journal and bearing surfaces.

10. Check the calibration of all gages.

The following tests and inspections must be made every SIX MONTHS, and the results entered in the appropriate check-off list or log:

1. Check the condition of the carbon and labyrinth packing.

2. Open, examine, and clean all ball bearings.

The following tests and inspections must be made annually, and the results entered in the appropriate checkoff list or log:

1. Open the turbine and reduction gear casings for inspection and cleaning.

2. Examine the turbine rotor, the blading, the shaft, all bearings, and the reduction gears.

SAFETY PRECAUTIONS

As a First Class or Chief Boilerman, it will be your responsibility to see that all appropriate safety precautions are observed in the operation of fireroom auxiliary turbines. The following precautions must be observed by all personnel:

1. Always turn the turbine rotor by hand before admitting steam to the casing.

2. Never lash down an overspeed trip, a speed-limiting governor, or a speed-regulating governor. Do not in any way attempt to render these safety devices inoperable.

3. Be sure that the turbine casing relief valve is set to lift at the proper pressure, and that it is operable at all times.

4. Before starting a turbine, be sure that it is free of foreign objects.

5. Always test the overspeed trip (if installed) when putting a turbine into service.

6. Never start up an auxiliary turbine before the turbine and the steam lines are properly drained. Turbine casualties and serious injury to personnel have resulted from inadequate drainage which allowed large slugs of water to be carried over with the steam.

QUIZ

1. In an impulse turbine, what constitutes a Stage?

2. In what part or parts of an impulse turbine does a pressure drop occur?

3. What is a Rateau Stage?

4. What is a Curtis Stage?

5. What type of oil should be used in the pressure lubrication system of a unit having worm-type reduction gears?

6. What defects are likely to result in oil being thrown from the ends of the bearings?

7. If it is necessary to stop a turbine because of over-heated bearings, why should you slow the turbine and then keep it turning over slowly for a while, before actually stopping it?

8. When you take bridge gage readings on a turbine bearing, with what do you compare the obtained readings?

9. When you are taking bearing clearances by leads, how far around the journal should each lead extend?

10. If it is not possible to take a thrust bearing clearance with feeler gages, how can you measure the clearance?

11. How much end play is generally allowed in auxiliary turbine thrust bearings?

12. What are the critical clearances on a helical-flow turbine?

13. How often should flexible couplings be opened, cleaned, and refilled with fresh oil?

14. How frequently must speed-limiting governors be tested?

15. What two adjustments are required on a constant-pressure pump governor?

16. When you are aligning the shaft of a driving turbine with the shaft of a driven auxiliary, why is it necessary to find out whether the coupling halves are true circles, and whether they are identical in diameter?

17. How often should idle turbines be run by steam?

•5

CHAPTER

FUEL OIL SERVICE EQUIPMENT

In this chapter, we will take up three items of fuel oil service equipment: fuel oil heaters, fuel oil meters, and fuel oil strainers.

FUEL OIL HEATERS

There are four types of fuel oil heaters approved for naval use. Type A is a shell and tube type of heater in which the oil circulates inside the tubes. Type B, generally referred to as the G-FIN heater, is widely used aboard combatant vessels. Type C is an extended-surface tube type of heater, in which the oil circulates outside of the tubes. The Type D heater is a shell and tube type of heater in which the oil circulates outside of the tubes.

Fuel oil heaters are normally cleaned when boilers are being overhauled. If performance indicates that a heater is fouled, however, it should be cleaned at the earliest opportunity. Mechanical methods of cleaning are approved for Type B and Type D heaters only. Chemical cleaning by either one of the two Fitzgerald methods is approved for all types of heaters.

Vaporization Cleaning Method

The following procedure is used to clean heaters by the vaporization method. Use this method only when the heaters undergoing cleaning can be set ashore.

1. Drain the oil heater thoroughly. In the multipa. i type, remove the plug from each pass in the bottom head to drain as much oil from the heater as possible.

2. Make all connections as shown in figure 4-1 and see that all couplings are tight and all valves closed. Notice that there are connecting lines between the generators and the fuel oil heater inlet and between the fuel oil heater outlet and the generators. In the latter, this connection should be the lowest point of the head in the horizontal type, to prevent sludge

^F.O. HEATER

FILUNfi PLUQ

REUEF VALVES SET AT 50 POUNDS

PRESSURE GAGE 30 POUNDS TO 40 POUNDS AT START. 10 POUNDS TO 20 POUNDS IN OPERATION

/ PLUG,*,

FLEXIBLE METALUC HOSE
VENT WEU CL EAR OF PERSONNEL -g
250° WHEN EVAPORATING A q, tQjf)
HOT OVER 150* WHEN Ul _ |- if HI
RECEIVING jffl ,,,«p ly
THERMOMETER NOT OVER 250° WHEN EVAPORATING
VALVE LEGEND 00VAPOR TO F.O. HEATER
®(E) CONDENSATE FROM
F.O. HEATER 0(D) VENT VALVE
dXT) STEAM TO COIL$
(h)(j) COIL DRAIN
GEN. NO. 1

GAGE GLASS

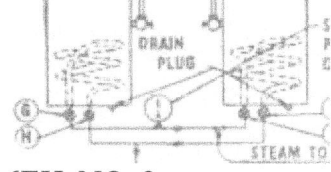

6EH. NO. 2
FILUN6 PLUG
-STEAM COIL PRESSURE GAGE NOT OVER SO POUNDS
FLEXIBLE
METALLIC
I) HOSE STEAM TO COIL S x

COIL DRAIN TO F.O. HEATER DRAIN TRAP

Figure 4-1.—Vaporization method of cleaning fuel oil heaters.

accumulation from blanking off the lower tubes. The vent line must lead clear of the operating personnel.

3. On generator No. 2, open valve E from the bottom of the heater, and valve D to the vent line. Keep valve F to the top of the heater closed.

4. Fill generator No. 1 nearly to the top of the gage glass with trichloroethylene. This filling is done through the filling plug, using a funnel.

5. On generator No. 1, crack steam valve C and drain valve H ; then admit steam to the heating coils. The chemical has a very low latent heat, and the amount of steam required is correspondingly small. Valve G should be so regulated that the pressure on the heating coil gage does not exceed 50 psi.

6. Build up pressure on generator No. 1 to a gage of from 30 to 40 psi.

7. Open valve A on generator No. 1 and allow the vapor to flow to the oil heater. Maintain a vapor pressure of from 10 to 20 psi by regulating valves A, G, and H. The vapor condenses in the heater, breaking down and carrying with it any deposits in the heater. The condensed and contaminated chemical flows to generator No. 2 and any uncon-densed vapors are led through the vent line.

8. When the chemical in generator No. 1 is evaporated to the bottom of the gage glass, secure the steam on this generator and allow the pressure to equalize in the system. The valving of the system should then be changed so that generator No. 2 is connected to the top of the oil heater, and generator No. 1 is connected to the drain of the oil heater and the vent. Steam may then be admitted to generator No. 2, and the process of cleaning continued.

9. After the last pass of the vapor has been made and the heater has been cleaned as outlined above, shut

down the steam and allow the pressure to equalize throughout the system. Close all valves and disconnect the heater from the generator. Connect a steam hose to the inlet side of the heater, and a vent line from the outlet side to a place clear of the operating personnel. Blow steam through the heater for about 5 minutes to remove all remaining contamination and vapor.

10. After the heater has been cleaned, you do not usually need to remove the heads for further cleaning or examination. When a very badly fouled heater is cleaned, you may find pieces of carbon in the heads or a slight amount of carbon in the tubes. Any carbon remaining on the tubes will ordinarily be dry, soft, and friable (easily pulverized) and will be carried off incident to service. To determine the degree of cleaning obtained with the amount of chemical circulated, remove the heads of the first few heaters of each type cleaned.

11. To remove contamination from the chemical, evaporate as much of the chemical as possible from the generator. The thermometer on the generator will indicate when the chemical is largely evaporated and the sludge is concentrated. The boiling point of trichloroethylene at atmospheric pressure is 187° F. When the temperature of the vapor reaches 250° F., the distillation of the chemical and the residue should be discontinued, and the remaining sludge drained from the generator. The hot sludge will contain considerable chemical, so care must be taken in the matter of exposing personnel to the fumes. To free the second generator of its contamination, evaporate the chemical from the second generator and manipulate the valving to direct the flow of vapor from the second generator to the first generator.

Safety Precautions

Condensation of the vapor is the essence of the cleaning process outlined above. As the cleaning proceeds, the heater will warm up and may get so hot that efficient condensation of the vapor will not take place. A large proportion of the vapor would then flow through the heater and out the vent. The temperature in the return generator is an indication of the degree of condensation taking place. This temperature should not exceed 150° F. In order to keep it to this limit, you have to reduce the rate of flow in the heater by throttling down on the vapor inlet valve or shutting down completely for a time to allow the heater to cool.

The amount of chemical that must be put through an oil heater to thoroughly clean it depends on the size and form of the heater and the degree of contamination. In the horizontal type about 10 gallons of chemical will prove sufficient to clean a badly fouled heater of about 100 square feet of heating surface. Take care, however, to see that the heater does not become overheated. In general, 10 gallons of vaporized chemical can be circulated 3 or 4 times through the heater in a period of not more than 2 hours.

Liquid Circulation Cleaning Method

Cleaning fuel oil heaters by the liquid circulation method may be undertaken by the forces afloat. Figure 4-2 illustrates a simple arrangement which may be used for cleaning by the liquid circulation method. The following instructions will serve as a guide:

1. Drain thoroughly all passes of the oil side of the heater, and connect the pump discharge to the lower oil-line branch of the heater, with a connection from the upper end leading to the drum. In this latter connection, locate a valve at or near the drum, with a pressure gage installed between the valve and the heater. The suction pipe arrangement is shown in figure 4-2.

PRESSURE GAUGE VALVE

THREADED PTE CONNECTION

30 OR SO^AL STEEL BARREL • OR DRUM

USE SMALL DIA. SUa' « TO PROVIDE VENT FOR DRUM

> ^

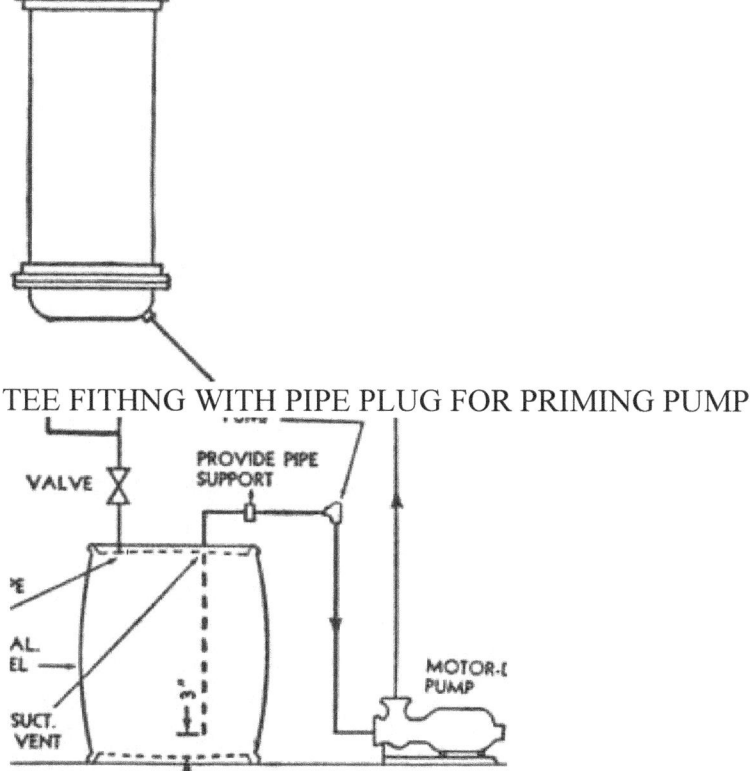

TEE FITHNG WITH PIPE PLUG FOR PRIMING PUMP

MOTOR.DRIVEN PUMP

Figure 4-2.—Oiagrain of th« liquid circulation cloaning mothed.

2. Since a given amount of the cleaning agent, equal to the capacity of the heater and the piping, will be pumped from the drum before any liquid is returned, you will have to estimate the capacity of such parts, in order to determine the minimum amount of cleaning agent required on starting. A drum of 30- or 50-gallon capacity is usually sufficient.

3. Start the supply and exhaust ventilation before cleaning, and maintain it throughout the cleaning process.

4. Remove the pipe plug in the T-fitting in the suction line and pour in enough cleaning agent to fill the suction line and prime the pump. Replace the plug, open the return flow valve to the drum, and start the pump.

5. Inspect the system for any leaks and correct these leaks. After the return flow to the drum has been established, slowly close down on the return flow valve in order to maintain a back pressure on the gage of 5 to 10 psi.

6. Admit low-pressure steam to the fuel oil heater to warm up the chemical to not more than 135° F., as the warmer the chemical, the more effective it will be in removing contamination. At this temperature, the amount of vapor formed is small, and only a small loss of chemical results.

7. Circulate the chemical through the heater for about 2 hours, depending on the condition of the heater. The heads of the first heater cleaned should be removed for examination to allow your men to determine the circulation time required for various degrees of contamination.

8. With low-pressure air, blow back to the drum the chemical remaining in the heater after circulation. The loss of the chemical by this cleaning method is very slight and only small replacements are required.

Two or three fuel oil heaters may be cleaned with the same chemical before a complete renewal or reclamation of the chemical is required. The pump packing should be inspected frequently, since the chemical attacks certain types of packing. 9. Disconnect the generator from the heater and connect the steam hose to the inlet side of the heater. Lead a hose from the outlet side into an empty oil drum and blow out with steam for about 5 minutes to remove any remaining chemicals or contamination. Slight carbon deposits remaining on the tubes will usually be dry, soft, and friable, and will be carried off incident to service.

A modification of the circulating method, wherein a mixture of kerosene, trisodium phosphate, and water is circulated prior to the use of the trichloroethylene, has been used successfully on some vessels. As a matter of fact, kerosene itself is capable of removing most types of sludges normally encountered.

Your inspections may reveal the failure of kerosene or trichloroethylene to remove harmful deposits. When this situation prevails, use a chemical solvent known as ortho-dichlorobenzene instead of trichloroethylene in the liquid circulation method only. Orthodichlorobenzene is an excellent solvent for the tarry binders cementing together the larger coke-like clusters, which, subsequently, are carried away by the circulating steam. Orthodichlorobenzene should be circulated cold; otherwise the instructions and precautions relating to the use of trichloroethylene should be followed when the orthodichlorobenzene chemical cleaner is being used. Take special precautions in cutting in the steam to the cleaned heater. You may expect a reduction in steam pressure requirements when appreciable amounts of deposits have been removed.

CI«on the Safe Way

Observe the following precautions when using liquid trichloroethylene:

1. Handle liquid trichloroethylene carefully. Do not bring it in contact with the skin in excessive amounts. It tends to take the oil out of the skin, giving it a white color and producing itching. If it is brought into contact with the skin and irritation results, the condition can be relieved by rubbing the affected parts with an animal or vegetable oil.

2. Avoid breathing trichloroethylene vapor. It is an anesthetic, although slower in action than ether or chloroform. Breathing the fumes in moderate

FUEL OIL HEATER

Type.

No..

(Date).

Done

Xot DoDe

1. Assemble drawings

2. Collect previous data...

3. Oil side:

(a) Examine, clean, repair, renew:

1. Tubes

2. Retarders

4. Steam side:

(a) Examine, clean, repair:

1. Tubes

2. Baffles

5. Flanges, gaskets, and gasket seatings:

(a) Examine, clean, repair, renew:

1. Bearing surfaces.

2. Gaskets

3. Studs, bolts, nuts, etc

6. Reassemble

7. Test, hydrostatic.

Figure 4-3.—Repair guide lit! for fuel oil heater.

amounts will cause headache, nausea, and lassitude. Breathing the vapor in excessive amounts will cause

DEATH.

3. If persons become affected in any way by the fumes, shut down the apparatus, ventilate the space thoroughly, and before proceeding with the work, locate the source from which the vapor is escaping.

4. In case any men are overcome by the fumes of tri-chloroethylene or other chemical solvent fumes, remove them to fresh air and summon medical aid.

When a fuel oil heater is to be overhauled, use the repair guide list illustrated in figure 4-3.

G-FIN FUEL OIL HEATER

The sectional G-fin fuel oil heater (fig. 4-4) is known as the Navy type-B unit. The heating element consists of a section of standard weight, iron pipe size, seamless steel tube, on the outer surface of which 18 longitudinal steel fins are attached. The fins extend along the straight length of pipe from shell connections to the return bend at the floating end of the unit. At this end, the heating element is supported within the shell by a steel ring around the fins. The shells or outer tubes, each containing a single heating tube, are arranged in banks and are connected at one end by a common tube sheet and a cover containing partitions for properly directing the flow of oil. Two G-fin tubes connected at one end by a U-tube return bend form one heating element. Four such elements are required for a heater with eight shells.

The internal surface of the heating element tubes serves as a steam space, all elements being connected in series by external return bends joined to the elements by unions. The male half union on the pipe side is an integral part of the shell plug; the other half is fillet-welded to the return bend.

At the end of the unit containing the inlet and outlet connections, each G-fin pipe is butt-welded to a forged

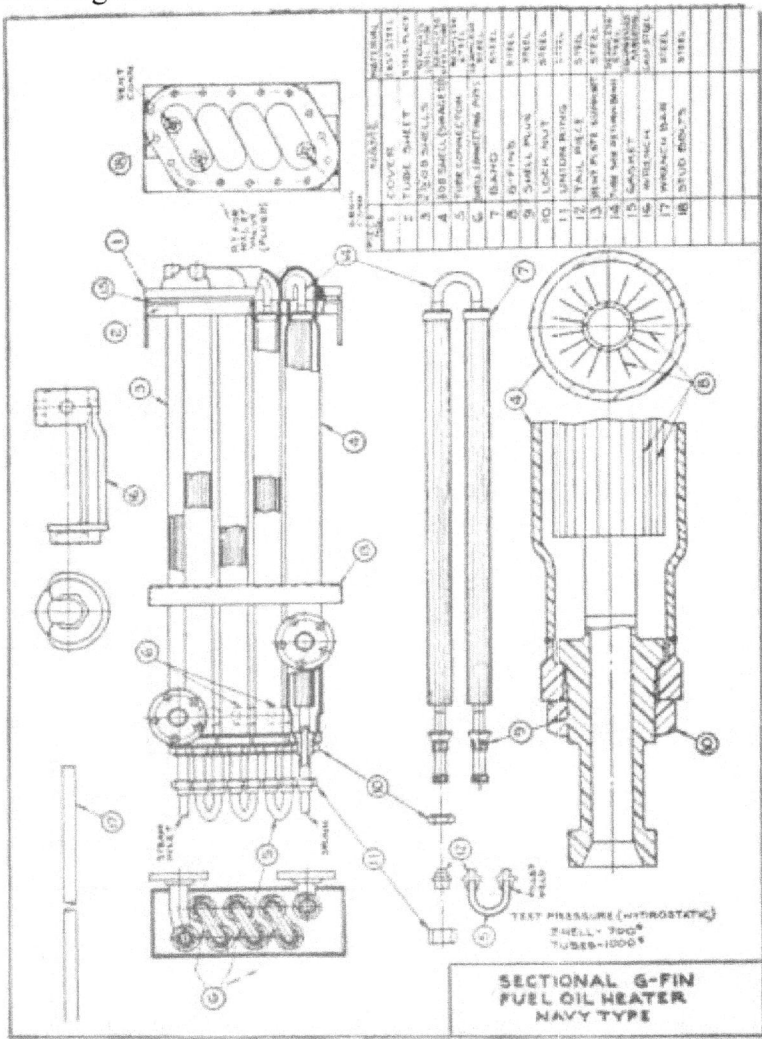

Figure 4-4.—Sectional G-fin fuel oil heater.

steel shell plug. This plug has a conical face which forms a metal-to-metal joint with a corresponding seat in the shell end. These joints are made tight by means of lock nuts which are easily accessible from the outside at all times. The threads on the lock nut match those on the male half union and the lock nut can therefore be screwed off over this half union.

The oil to be heated passes through the shell along the straight longitudinal paths formed by the G fins, in the direction opposite to that of the steam flow. To facilitate the flow of oil into and out of the shell, the pipes are left free of fins opposite the shell connections and around the U bend.

There are no internal joints for leakage of oil to steam, or steam to oil, since the steam flows through a continuous length of pipe with both steam inlet and drain connections external to the oil spaces of the heater.

The exterior surface of each heating element has more than 8 times the interior cylindrical surface. The transfer of heat from the steam to the oil is, therefore, much more effective than with a bare pipe of equal length.

The sectional heaters are usually arranged with two to four units in parallel for full and normal load operation. At fractional loads one or more of the units can be cut off the line to better maintain constant oil temperatures. When a four-section installation is clean, the capacity of the entire unit will be greater than required; in this case, three units are usually suflRcient to carry the full heater load. If only or % heating capacity is required, one or two of the sections may be used. Another advantage of this type of arrangement is that a damaged section may be cut out without shutting down the entire fuel oil heater.

The heating elements may be completely removed from the shells by disconnecting the union joints in the steam inlet and drain lines, releasing the lock nuts, and removing the rear cover. The half of the union adjacent to the lock nut will pass through the lock nut, since the threads

of these two parts are the same. The heating surface may be removed without disturbing the oil inlet or* oil outlet connections. The sludge or dirt which will accumulate can be easily scraped from the fin surface. A light scraping tool shaped like a hoe, and with a blade formed to fit between the two fins, will make a very suitable tool for cleaning.

FOULING OF FUEL OIL HEATERS

The fouling of fuel oil heaters is governed by three factors: (1) the rate at which the oil passes through the heater, (2) the temperature to which the oil is heated, and (3) the nature of the oil itself.

Overheating the oil is probably the greatest single cause of fouling. "Cracked" oils, particularly, break down under heat and deposit carbon residue. Such oils are reduced in viscosity from a rise in temperature more than are straight-run oils. Under these circumstances you must be especially careful not to heat cracked oils beyond the temperature that gives the correct viscosity for eflflcient atomization. When heaters are being operated at low outputs, take special care to prevent excess oil temperatures.

Since the usual rate of operation of fuel oil heaters is low, heavy fouling may not be detected until an attempt is made to operate the heater at its full-rated capacity. For that reason, your men on watch in the fireroom should know at all times the state of fouling of all heaters.

OIL LEAKS IN HEATERS

Always take special precautions to prevent the leakage of fuel from the oil side to the steam side of the heaters. This is necessary because oil that leaks into the steam side of the heaters may be carried through the feed system to the boilers. Drain inspection tanks are provided to determine the presence of oil in the heater drains. Inspect these drains once an hour. If the presence of

oil is noted, discharge the drains to the bilge and, if operation permits, isolate the heater and uncover and repair the source of the leak.

METHODS OF OPERATION

In the majority of installations in naval vessels the fuel oil heating installation is divided into several units providing sufficient fuel oil heating capacity to handle the heavier oils at full power; therefore, at low-power operation the full capacity of the heaters is greatly in excess of requirements. For this reason, as the load is reduced, the number of heaters or units in use should be decreased. As a general rule, one heater or heater section should be operated at maximum capacity before a second one is cut in; it is considered better practice to run one heater at full load rather than to operate two heaters under partial load. Vessels built to naval specifications provide a small steam bypass valve around the steam throttle valve of the heater to provide for adjusting oil temperatures under light load conditions.

In general, water should not be allowed to accumulate in the steam sides of fuel oil heaters. Drain pipes and fittings should be arranged in such a way that complete drainage can be had at all times, and the heater can be operated without steam blowing through. In naval vessels, this is provided for by a drain receiver fitted with a gage glass located below the level of the tubes, to provide a water seal. Care must be exercised to keep all traps in the drain lines in efficient working condition, and they should never be bypassed while the heater is in operation. When water is allowed to accumulate in an idle heater, corrosion is increased and the life of the heater is shortened. If such accumulation takes place while a heater is in operation, the tubes are partially submerged and the capacity of the heater is reduced. However, under especially steady light load conditions when using Navy Special Fuel Oil or the light grades of commercial bunker fuel (which require little heating to obtain the required burning viscosity), the following operating procedure is permissible:

1. Open the condensate drain valve a very small amount.

2. Apply only enough steam pressure to build up pressure slightly in excess of the drain system back pressure, which pressure in most vessels will be the deaerating feed tank pressure plus a small pressure drop in the drain system.

3. Regulate the steam supply to the heater by small changes in steam valve (or steam-valve bypass) to obtain desired oil temperature. Allowing condensate to accumulate in the steam side of the heater with the drain valve nearly closed would render ineffective a certain portion of the heating surface. Accordingly it will not be possible to keep the water level in the drain sight glass as in normal operation. As condensate accumulates within the heater, steam pressure needed to maintain a given oil temperature will increase.

4. With the heater partly filled with condensate, and when the operating steam pressure required is more than about 15 psi (gage) above the drain back pressure, the opening of the heater condensate drain valve should be increased and the heater steam pressure maintained at or below this point until full drainage and normal heater operation can be established.

TESTS OF HEATERS AND FUEL OIL SERVICE SYSTEM

At least once each quarter, or when a fireroom has not steamed for a period of more than a week, all parts of the fuel oil system on the discharge side of the fuel oil service pump— including the fuel oil heaters in that fireroom— should be subjected to an oil pressure equal to the authorized setting of the fuel oil service pump discharge relief valve. All joints should be carefully inspected for oil leakage. In vessels fitted with a recirculation system.

this system should be used during the foregoing test and the oil heated to a temperature corresponding to a viscosity of 200 Seconds Saybolt Universal.

When any part of the fuel oil service system has undergone repairs affecting either the strength or tightness, all parts of the system, including the heaters, should be subjected to an oil pressure equal to j50 percent of the design operating pressure; and the steam side of the heater, including the piping from the heater steam throttle valve to the drain discharge cut-out valve, should be subjected to a hydrostatic test equal to the design steam operating pressure of the heater. Care should be taken to prevent damaging the fuel oil heater relief valves. It may be advisable to remove the oil relief valves and install either blank flanges or plugs, depending on whether the connection is flanged or threaded.

All pressure tests of the fuel oil service system should be made with the hand fuel oil pump, where such a pump is installed. Idle pumps on a system undergoing test should have their discharge valves fully closed. The pump suction valve and other valves in the suction piping between the pumps and an adjacent fuel oil service tank should be kept open to insure that any leakage of the pump discharge valves will not impose excessive pressure in the service suction piping.

At least once each year, the steam side of all fuel oil heaters, including the piping from the heater steam throttle valve to the drain discharge cut-out valve, should be subjected to a hydrostatic test pressure equal to the design steam operating pressure of the heaters; and at 5-year periods this same test should be given, using a hydrostatic pressure of IY2 times the designed steam operating pressure.

Prior to conducting the foregoing tests, the necessary steps should be taken to ensure that all air is removed from the parts of the system under test. The results of all the foregoing tests should be logged so that entries can be made in the machinery history.

FUEL OIL METERS

The measurement of quantity of fuel used is highly important on naval vessels. The quantitative, hourly computation of fuel expenditure provides a daily check against the soundings of fuel oil tanks and promotes engineering analysis and economy. Meters for use in ships must be rugged, durable, and suitable for prolonged use with little adjustment.

Operation

The nutating piston or disk type oil meter illustrated in figure 4-5 operates by volumetric displacement of the liquid measured. Oil enters the meter from the left and passes upward through the coarse strainer into the gear space above the measuring chamber. It then enters the measuring chamber through the entrance (upper) port in the side of this chamber. Figure 4-6 shows the diaphragm (vertical dividing plate) in the measuring cham-

Figure 4-5.—Oil meter.

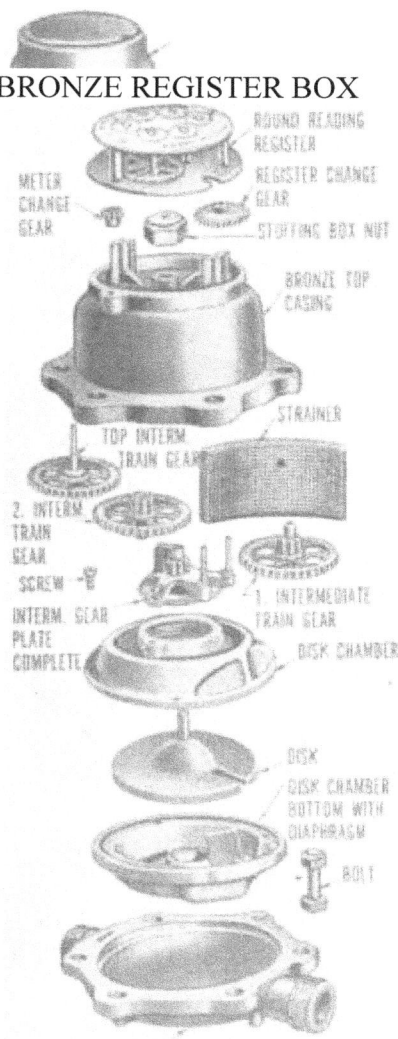

BRONZE REGISTER BOX

BRONZE BOTTOM CASING

Figure 4-4.—Exploded view of th« oil motor.

ber. Oil enters the disk chamber plate. The disk in the measuring chamber is free to nutate (rock around) about its lower spherical bearing surface, but is not free to rotate, being constrained by the fixed diaphragm which runs vertically through the slot in the disk. This nutational motion, which is illustrated in figure 4-7, is similar to that of a spun coin just before it settles flat on a table. It is to be noted that the action of the nutating disk is a process which has no definite cycle, in that there is no starting point nor finishing point.

The position (a) in figure 4-7 was taken as a starting position for the purpose of the explanation. If the meter

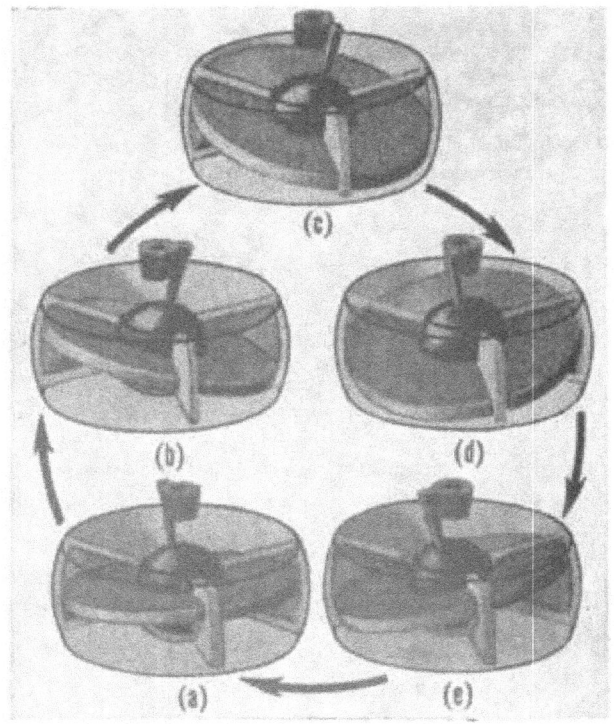

Figure 4-7.—Diagram of the principle of operation of a nutating piston.

104

starts with the disk tilted up on the. side of the chamber where the diaphragm is located, the fluid, as it flows around, will force the disk to nutate in a clockwise direction as viewed from above. As the highest point of the disk passes point (c) 180° from the dividing plate, the admission of the fluid to the bottom of the disk will be shut off, because of the position of the disk.

Meanwhile, fluid has commenced to enter above the disk. The force of the fluid above the disk will sweep the fluid below the disk around (clockwise as viewed from above) and out at the exhaust port in the bottom of the measuring chamber (just to the right of the diaphragm). As the disk nutates in the above described manner, the pin extending from the upper bearing sphere rotates the gear train driver block.

The nutating pin is restrained and guided by the roller (the cone-shaped bearing) which is concentric with the driver block and its shaft. This construction prevents the disk from ever attaining a neutral (horizontal) position. When the meter is being used to measure oil, the leakage or slippage around the edge of the disk will be negligible and can be accurately calibrated. The flow of fluid is even and continuous, with no pulsations.

Possible Troubles

Fuel oil meters should not show an error of more than 2 percent on a volumetric basis. Meters become unreliable from friction, leakage, erosion, and worn parts, or the moving parts in the measuring chamber may become encrusted from the liquid metered. A meter should be overhauled and tested if it fails to register properly. Unreliable operation is usually caused by one of the following:

1. The packing around the driving spindle of the top intermediate gearshaft exerts too much friction, because it has been tightened unnecessarily, or incorrectly packed.

2. Gears do not mesh properly, because of wear.

3. A bearing in the gear train becomes worn.

4. Foreign matter obstructs the submerged parts.

5. Foreign matter in the liquid may cause abrasive wear and change the dimensions of the chamber or the disk, especially the dimensions of the ball or bearing sphere of the disk, and the lower chamber bearing in which it rests.

6. Air passed through the meter will register like oil and cause high registration. Recalibration cannot correct for this type of inaccuracy—the only remedy is to keep air from passing through the meter.

The ordinary type meter can be opened at the bolted flange without disconnecting the meter from the line. The trouble is usually in the submerged parts and can be rectified by cleaning the parts and removing the foreign matter. More extensive repairs will depend on the conditions found and on the type of meter.

It is important that the machined inner surfaces of the measuring chamber be clean and smooth. If these surfaces become rough or coated with residue, they may be polished with crocus cloth, but must not be scraped.

The proper packing for the stuffing box for meters operating at temperatures of 150° F. or lower consists of cork washers and a little graphite. For temperatures above 150° F., packing should be asbestos graphite packing and care must be taken not to tighten the stuffing box nut more than is necessary to prevent leaks.

To test a meter, disconnect it from the line and pass through it a known volume or weight of the same liquid on which it is regularly used. Oil meters positively must not be tested on water, but on the same kind of oil regularly metered. Before starting the test, expel all air from the meter by running through it sufficient fluid to fill it completely. For correct test results, the control valves must be on the outlet of the meter. Measure the fluid for the test either by volume or weight, and compare the re-isult with that obtained from the meter register. Two

tests should be made, one with the full opening of the meter piping connection, and the other with a restricted outlet orifice to determine the accuracy at approximately 10 percent of maximum flow. Tests must be made at approximately the temperature prevailing in normal use. Meters measure by volume; therefore, if a test is made by weight, the weight per gallon at metering temperature must be accurately calculated, or the test will indicate an error which does not actually exist.

If a test shows that a meter is inaccurate, the details should be written up on a current ship's maintenance project Repair Record Card for the first Naval Shipyard availability period. The information should include the manufacturer's name, the type or model of the meter, the serial number, and the amounts of fluid registered and actually passed.

In using meters, remember to take the following precautions :

1. See that strainers in the line to the meter are kept clean at all times.

2. If oil is circulated through the system before the boilers are lighted off, care should be taken to bypass the meter.

3. Do not allow the meter to freeze, as the casing may crack.

FUEL OIL STRAINERS

Duplex high-pressure basket-type strainers are installed on the discharge side of the fuel oil service pumps, generally between the heater and the manifold. In this location the heated oil is very fluid and readily gives up the contained foreign substances. The flow of oil through the strainer should be from the center of the basket to the outside, so that the dirt and sediment will be deposited in the basket.

As a matter of routine, inspect and clean strainers once each day when in use, unless an abnormal pressure drop through the strainers, as indicated by the duplex gage.

shows that more frequent cleaning of strainers (once each watch under way) may be required if fuel has been received from foreign sources or has been in storage, or if there is a possibility that there has been a mixture of straight-run and cracked fuels. If dirty oil has been received on board, strainers will have to be cleaned more frequently. Have at least two spare baskets for each strainer, clean and ready to use, stowed on racks near the strainer.

QUIZ

1. Under what circumstances can fuel oil heaters be cleaned by the vaporization method?

2. In cleaning a heater by the vaporization method, what steam pressure is admitted to the heating coils?

3. What chemical is used in the vaporization method of cleaning fuel oil heaters?

4. What is the boiling point of trichloroethylene at atmospheric pressure?

5. When a heater is being cleaned by the vaporization method, what indicates the degree of condensation which is taking place?

6. In general, how many gallons of trichloroethylene will be required to clean a heater which has about 100 square feet of heating surface?

7. What method of cleaning fuel oil heaters may be undertaken on board ship?

8. What factors determine the amount of trichloroethylene which will be required for cleaning any particular heater?

9. When cleaning a heater by the vaporization process, why is it important to keep the heater from becoming too hot?

10. When cleaning a heater by the liquid circulation method, for how long a time should you circulate the chemical?

11. If kerosene and trichloroethylene fail to remove harmful deposits from a fuel oil heater, what chemical may be used for this purpose?

12. What first aid treatment should be given if trichloroethylene is brought into contact with the skin?

13. What will happen from breathing a large amount of trichloroethylene vapor?

14. What three factors govern the fouling of fuel oil heaters?

15. In what cleaning method may orthodichlorobenzene be used?

16. How frequently should the heater drains be inspected for the presence of oil?

17. Is it considered better to operate one fuel oil heater at full load, or two heaters at partial load?

18. At what temperature should the oil be for testing the fuel oil system?

19. What will happen if water is allowed to accumulate in idle heaters?

20. If a fireroom has not steamed for a period of more than one week, what parts of the fuel oil system must be tested before use?

21. What oil pressure must be used when the fuel oil system is tested for strength or tightness?

22. At the steam-side test of fuel oil heaters which is made every 5 years, what hydrostatic pressure is used?

23. What is the proper packing for the stuffing box of a fuel oil meter operating at a temperature of 150"* F. or lower?

24. How frequently should the strainers be inspected and cleaned?

BOILER FITTINGS AND INSTRUMENTS

As a First Class or Chief Boilerman, you will find it necessary to have a thorough understanding of the many fittings and instruments required for the operation and control of a boiler. Basic information on boiler fittings and control instruments is given in Boilerman 3 & 2, Nav-Pers 10535-C; in various chapters of BuShips Manual; and in the appropriate manufacturers' instruction books. In this chapter, we will take up some of the boiler fittings and instruments which, because of their importance, will require your particular attention: safety valves, soot blowers, feed water regulators, and superheater steam flow indicators.

SAFETY VALVES

Several types of safety valves are used on naval boilers, but all are designed to open completely (pop) at a specified pressure, and to remain open until a specified pressure drop (blowdown or blowback) has occurred. Safety valves must close tightly, without chattering, and must remain tightly closed after seating.

Operating Principles

The three types of steam drum safety valves most commonly used in the Navy are: (1) the huddling chamber type, (2) the nozzle reaction type, and (3) the jet flow type. Of these, the huddling chamber type is probably the most widely used.

A steam drum safety valve of the huddling chamber type is shown in figure 5-1. The initial lift or opening of the valve is caused by the static pressure of the steam in the drum acting upon the bottom of the feather. As

RELEASE NUT
COMPRESSION SCREW
LEVER PIN
COMPRESSION SCREW
LOCKING NUT
TOP SPRING WASHER
FEATHER GUIDE RETAINING RING
LOCKING NUT
TOP LEVER
L-EVER PIN
—DROP LEVER
SPRING.
BOTTOM SPRING WASHER

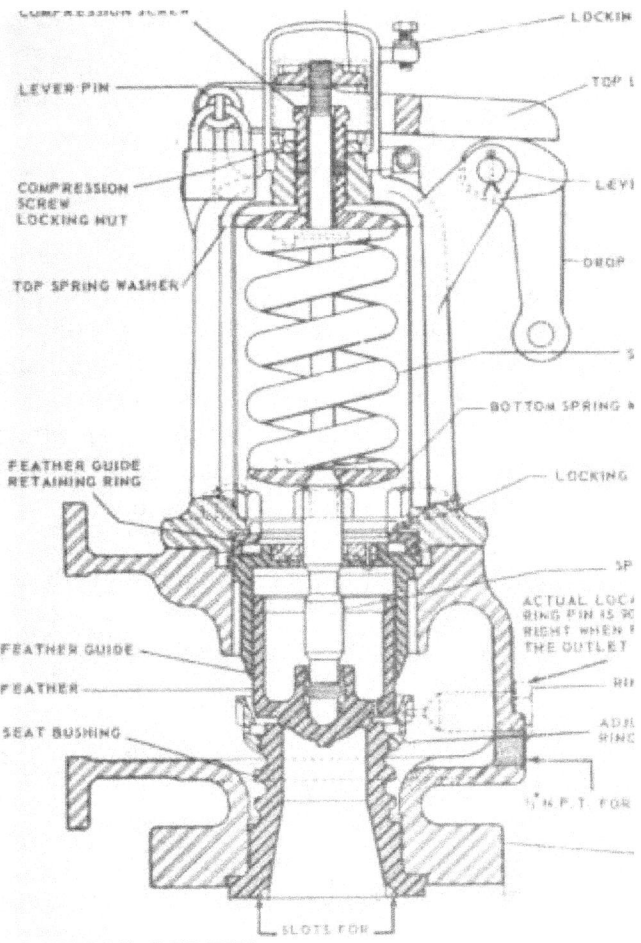

LOCKING SCREW
SPINDLE
ACTUAL LOCATION OF RING PIN IS 90* TO THE RIGHT WHEN FACING THE OUTLET
RING PIN
ADJUSTING RING
N.P.T. FOR DRAIN
BASE
.SLOTS FOR. WRENCHES

Figure 5-1.—Huddling chamber steam drum safety valve.

soon as the valve begins to open, a projecting lip or ring of larger area is exposed for the steam pressure to act upon. The huddling chamber, which is between the valve seat and a restricting orifice, fills with steam as the valve opens. The steam in the huddling chamber builds up a static pressure which acts upon the extra area provided by the projecting lip of the feather. The resulting increase in force overcomes the resistance of the spring, and the valve pops—that is, it opens quickly and completely. Because of the larger area now presented for the steam pressure to act upon, the valve reseats at a lower pressure than that which caused it to lift initially. After the specified blowdown has occurred, the valve closes with a slight snap.

The amount of tension on the spring determines the pressure at which the valve will pop. The position of the adjusting ring determines the shape of the huddling chamber, and thereby determines the amount of blow-down which must occur before the valve will reseat.

Figure 5-2 shows a steam drum safety valve of the NOZZLE REACTION type. The initial lift of this valve occurs when the static pressure of the steam in the drum acts upon the disk insert with force sufficient to overcome the tension of the spring. As the disk insert lifts, the escaping steam strikes the nozzle ring and changes direction. The resulting force of reaction causes the disk to lift higher, up to about 60 percent of rated capacity. Full capacity is reached as the result of a secondary, progressively increasing lift which occurs as an upper adjusting ring is exposed. The ring deflects the steam downward, and the resulting force of reaction causes the disk to lift still higher. Blowdown adjustment in this type of valve is made by raising or lowering the adjusting ring and by raising or lowering the nozzle ring.

The JET FLOW type of steam drum safety valve utilizes both the reaction and the velocity of the escaping steam. The static pressure on the disk overcomes the spring tension and causes the initial flow of steam. The escaping

Figure 5-2.—Neizle reaction steam drum tafety valve.

steam strikes against a nozzle ring and discharges into the body of the valve. The resulting reactive force lifts the disk higher, and thereby increases the area of flow and the velocity of the steam. As the velocity increases, some of the steam discharges through orifices which are located around the lower end of an adjustable guide ring; thus an upward force is created and the valve is lifted still higher. The position of the disk at the time of pop-

ping is such that the area of discharge is greater than the area of the nozzle; therefore, the accumulation of pressure which is required for full lift in the huddling chamber and nozzle reaction types of valves is unnecessary for full lift of the jet flow valves.

Blowdown adjustment in the jet flow valve is made by controlling the amount of steam which is allowed to flow through the orifices. The orifices are regulated by adjusting the positions of metering disks. Since the metering disks can be adjusted by externally accessible screws, blowdown adjustment is quite easily accomplished in this type of safety valve.

Safety valves are always installed on the superheater, as well as on the steam drum. Superheater safety valves are always set to lift and reseat at a lower pressure than the steam drum valves.

Spring-loaded superheater outlet safety valves are used on some uncontrolled superheat boilers. For temperatures below 700° F., these valves are very similar to the steam drum safety valves, except that the superheater valve springs and bonnets are made of special heat-resistant alloys. When used for temperatures above 700° F., the spring-loaded superheater outlet safety valves have bimetallic thermal compensators which automatically compensate for the changes in steam density which occur with variations in temperature.

Some boilers are fitted with pressure pilot-operated superheater safety valve assemblies. Each assembly usually consists of three connected valves: (1) a small spring-loaded steam drum safety valve, (2) an actuating valve (sometimes called a pilot valve) and (3) an actuated, or unloading, superheater valve. The spring-loaded drum safety valve and the actuating valve are installed on the steam drum. The stem of the drum valve is connected mechanically, by means of a lever, to the stem of the actuating valve. The actuating valve is connected by piping to the space above the disk in the superheater outlet unloading (or actuated) valve. In addition, a hand-operating valve is installed in the actuating line, so that the unloading valve on the superheater may be opened by hand if necessary.

A pressure pilot-operated superheater safety valve assembly is shown in figure 5-3. In this type of assembly, the unloading valve has no spring; it relies solely on pressure differential for its operation. Normally there is a static pressure above the disk of this valve, since some steam is allowed to pass through a small orifice in the disk.

When the drum safety valve is opened by steam drum pressure, the actuating valve is unseated. The opening of the actuating valve opens the actuating line to the atmosphere, and the sudden relief of pressure above the disk causes the superheater unloading valve to lift wide open.

When the pressure in the steam drum falls to the reseating pressure of the drum valve, both the drum valve and the actuating valve close. The closing of the actuating valve closes the actuating line to the superheater unloading valve. The pressure above the disk of the unloading valve builds up again very quickly, and the disk is reseated sharply and cleanly.

Another type of pilot-operated superheater outlet safety valve assembly is used on some high-pressure boilers. This assembly consists of two valves, rather than three: a pilot valve on the steam drum, and a spring-loaded superheater unloading valve. The pilot valve is similar to an ordinary steam drum safety valve, but is smaller. The pilot valve and the superheater outlet valve are interconnected by a pressure-transmitting line from the discharge side of the pilot valve to the underside of a piston attached to the spindle of the superheater valve. When the pilot valve pops at the steam drum pressure for which it has been set, the steam pressure w^hich is transmitted from the pilot valve to the superheater valve is suflScient to lift the superheater valve. If for any reason the pilot

ujcr

3t;

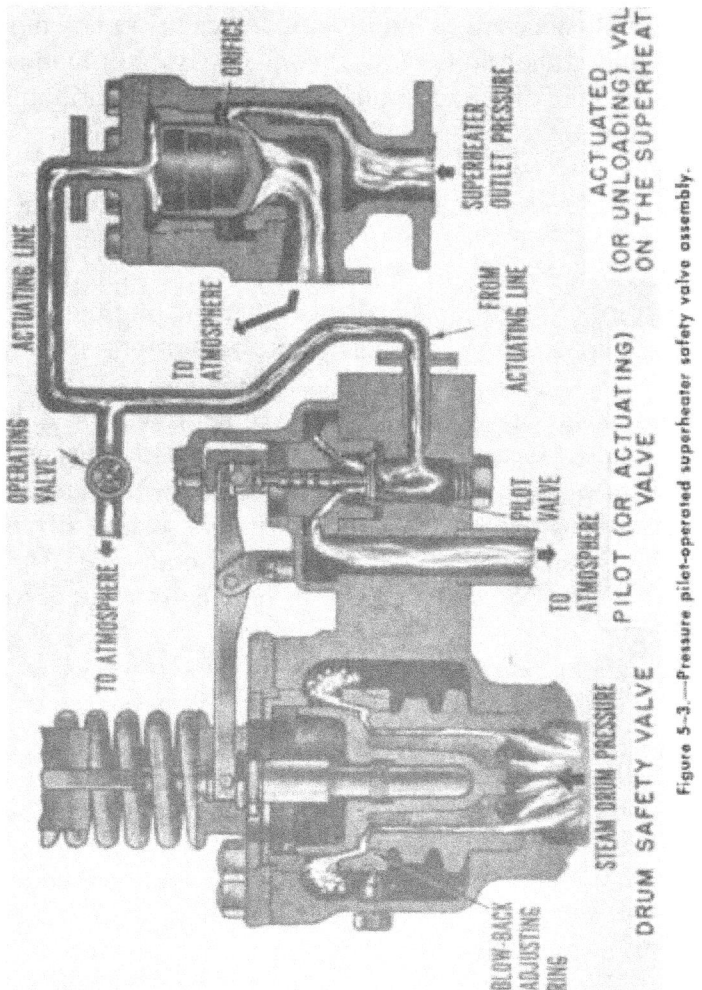

Figure 5-3.—Pressure pilot-operated superheater safety valve assembly.

valve fails to function, the superheater safety valve will open at a slightly higher pressure, like any regular spring-loaded safety valve.

Safety Valve Maintenance and Repair

It is essential that all safety valves be maintained in the best possible operating condition at all times, and that all tests, inspections, and adjustments be made as required. However, you should NOT make unnecessary repairs or adjustments to safety valves. As a general rule, safety valves should be left strictly alone as long as their operation is satisfactory; disturbance or alteration of the internal parts often causes unsatisfactory operation in a valve which was previously working all right.

After a period of service, it is likely that safety valves will simmer when they are within V/y to 2 percent of their popping pressure. Such simmering does not require corrective action, provided the valve does not simmer at the boiler operating pressure.

The popping and reseating pressures which are stamped on the valve name plate or given in the manufacturer's instruction book are specifically authorized by BuShips. If the exact authorized popping pressure cannot be obtained, a popping pressure which is within 3 psi of the authorized pressure should be considered satisfactory, provided all valves on the boiler pop in the proper sequence.

For continued satisfactory operation of safety valves, it is important that the essential original dimensions be maintained. Worn parts should be replaced when necessary. Grinding in of valve parts should be kept to a minimum, since any grinding in involves the removal of metal, and, therefore, a departure from the original dimensions. A safety valve may be ground in lightly once or twice without appreciably affecting the clearances; but heavy or repeated grinding will change the dimensions and cause unsatisfactory operation of the valve. In many cases, neither the equipment nor the technical skill re-

quired for a complete overhaul of safety valves is available on board ship. Whenever possible, safety valves in need of repair should be overhauled by a naval shipyard or some other repair activity. Since emergency repairs must sometimes be made, however, it is important that a full allowance of boiler safety valve repair parts and tools be kept on hand at all times; and its equally important that you have some understanding of the correct procedure for repairing safety valves.

If it is necessary for you to grind in a safety valve, or to make other repairs to it, be sure to consult the manufacturer's instruction book, repair manual, and drawings. This discussion is intended to show some of the more important aspects of safety valve repair, with special emphasis on valves of the huddling chamber type. However, the Navy uses several different types of safety valves, and no single discussion can be equally applicable to all types. Before attempting any repair to safety valves, therefore, be sure that you have a thorough understanding of the particular valves you are working on.

Straightening the spindle. —The spindle on a safety valve must be kept very straight and true with respect to its essential working surfaces, so that the spring force may be transmitted to the feather without any lateral binding.

A dial indicator is generally used to check the trueness of the spindle. The total indicator reading (or double eccentricity) at the point where the spring washer seats on the spindle should not exceed 0.006 inch. At the point where the spindle passes through the spring compression screw, the total deviation from trueness should not exceed 0.003 inch. If a spindle is found to be not true at either of these points, it should be straightened or replaced.

Grinding in the seat bushing. —If a safety valve seat bushing is badly worn or damaged, the valve should be removed and the bushing remachined. Remachining may be accomplished either with the seat bushing in place in

the valve body, or with the bushing removed from the body.

A seat bushing which is not damaged enough to require machine work may be ground in with a cast iron lap. (Note that this process is referred to as GRINDING IN, rather than LAPPING, even though a lapping tool is used. As a rule, the removal of small amounts of metal is referred to as grinding; lapping involves the removal of larger amounts of metal, such as would be necessary if the surface contained relatively large irregularities.)

Figure 5-4 shows how a lap may be used to grind in the seating surface of the seat bushing in a huddling chamber safety valve. The working surface (A) of the lap should be accurately and smoothly machined to form a 45° angle with the guiding surface (B). Surface (A)

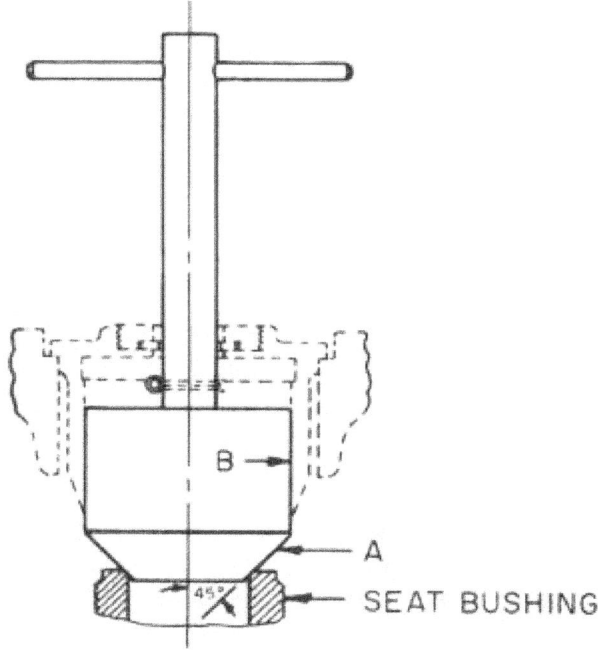

Figure 5-4.—Lgp for grinding in seat bushing.

must be truly centered with respect to surface (B), so that the final seat of the bushing will be truly round and accurately formed to the 45*" angle.

Rotate the lap back and forth about a quarter of a turn; shift the position of the lap from time to time, so that the lap will be moved gradually, in increments, through several rotations. Be careful not to bear sidewise on the handle as this might make the seat out-of-round. Use Navy "fine" lapping and grinding compound; this is a 220-grain compound, Standard Stock No. G51-C-1645. Replenish the compound frequently while working on the seat.

From time to time, spot in the seat to find out how the work is progressing. First wipe off all compound from the seat and from the lap. Then put a very small amount of light lubricating oil on the surface of the lap, and rub the lap against the seat. A slight amount of polishing occurs, which shows the location of the actual metal-to-metal contact. If the contact is uneven, correct it by further grinding in.

Grinding in the feather. —After the seat bushing has been ground in, spot in the feather. Remember that the correct location of the seat contact is different on different types of safety valves. In the nozzle reaction type of valve, the seating surfaces of the disk and the nozzle are flat. In the huddling chamber type of steam drum safety valve, the seat contact is narrow and quite low, as shown in figure 5-5. In the pressure pilot-operated superheater outlet safety valve assembly, the unloading (or actuated) valve has a feather which is similar in design to the feather in the huddling chamber type of steam drum valve. On the superheater unloading valve, however, the seat contact should be high on the feather, rather than low. The exact location of the contact must always be determined from the manufacturer's instruction book or repair manual.

Navy lapping and grinding compound with a grain of 600 (Standard Stock No. G51-C-1648-150) should be used

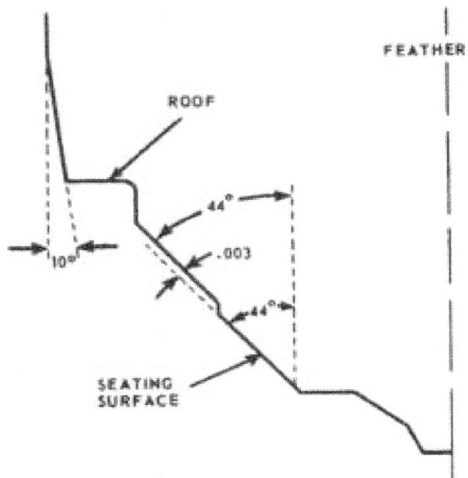

FigHr* 5-5.—Details of feather in huddling chamber safety valve.

for beginning to grind in the feather. This is the finest valve grinding compound listed in the Standard Stock Catalog. However, even this grade may be somewhat coarse for grinding in the feather of a safety valve; therefore, you should apply it very sparingly and work very carefully to avoid overgrinding.

In the huddling chamber type of safety valve, the seat angle on the seat bushing is generally 45° and the seat angle of the feather is usually 44^2°- The use of these angles gives the desired seating contact when the valve is ground in under factory conditions. However, it has been found that an angle of 45° on the seat bushing and 44° on the feather will generally give the best results when the valve is being ground in under service conditions.

When the seating surface of the feather has been ground in to the proper width, angle, and location, bring the seating surfaces of the feather and the bushing to a brilliant polish. To do this, you must first wipe the surfaces entirely clean; then apply a small amount of light

BORE DIAMETER STAMPED HERE

Figurs S-6. —S«at gag* applied to feather seat.

lubricating oil, and turn the feather to polish both surfaces.

Using the seat gage. —A seat gage is generally furnished for use in checking the dimensions and angles of the feather and the seat bushing. When using the seat gage, be sure that you apply the appropriate part of the gage to each part of the feather and the seat bushing. Figures 5-6, 5-7, 5-8, and 5-9 show the proper use of the seat gage.

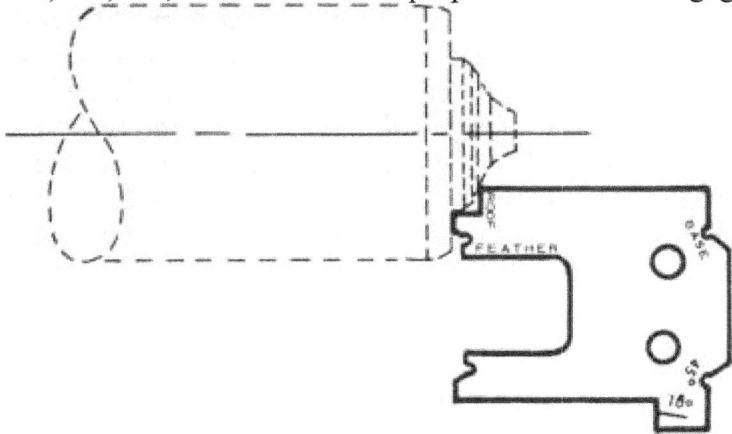

Figure 5-7.—Seat gage applied to roof of feather.

Figure 5-8.—S«at 909* oppiisd to 10* taper on ftathor.

Measuring and cutting the overlap. —Many huddling chamber safety valves operate on the back-pressure PRINCIPLE—that is, steam pressure from steam which is trapped on the upper side of the feather is utilized to assist the spring in forcing the feather back down on its seat. When the valve is blowing, a small amount

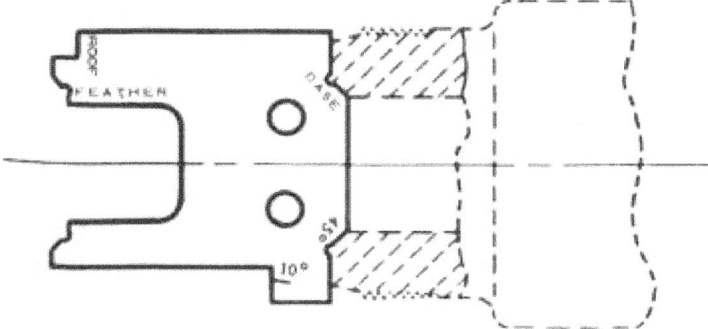

Figure 5-9.—Soat gag* opplied to soot of buthing.

of steam is bled from the huddling chamber into another chamber; this bleed steam escapes to the atmosphere between a floating washer and the overlap shoulder on the spindle. As the valve starts to close, the overlap shoulder reaches the top of the floating washer, thereby trapping the bleed steam in the chamber below, setting up a pressure on top of the feather, and closing the valve.

The distance from the top of the floating washer to the overlap shoulder is known as the overlap. This distance must be correct in order for the valve to function properly. After the seats have been reconditioned, or after new parts have been installed, the overlap must be checked and made standard. The size of the overlap varies according to the size of the bore and the pressure for which the valve is designed. Consult the manufacturer's instruction book or repair manual for the overlap size on any particular valve of this type.

An OVERLAP MEASURING COLLAR should be used to measure the overlap, as shown in figure 5-10. Place the collar over the spindle, and bring it down on top of the floating washer. Make sure that the feather is seated, and that

OVERLAP MEASURING COLLAR
LOCK SCREW

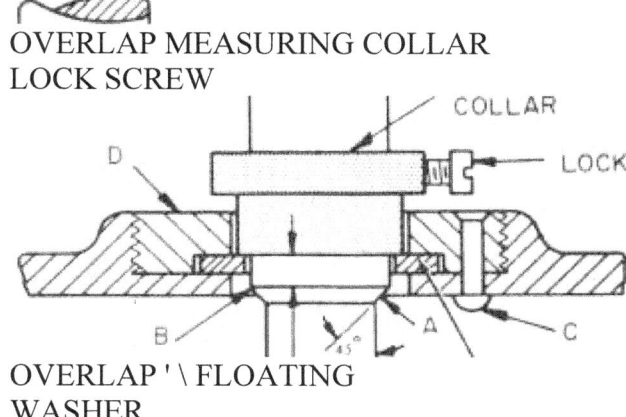

OVERLAP ' \ FLOATING
WASHER

Figure 5-10.—Method of measuring overlap.

the spindle is resting firmly in the pocket of the feather. Tighten the lock screw, so that the collar is held firmly onto the spindle, and remove the spindle and collar together. Then measure the distance from the bottom of the measuring collar to the edge of the overlap shoulder (indicated as B in figure 5-10). This distance is the overlap.

If the overlap is too large, machine the surface (A) at a 45° angle until the overlap is reduced to the proper size. Finish with an abrasive cloth, and break the edge (B) with an oil stone.

If the overlap is too small, machine the ball point of the spindle to permit the spindle to take a lower position. In machining the ball point, be careful to keep the same contour.

If the floating washer is not entirely free, drive out the rivet (C) and unscrew the retaining nut (D). Remove the floating washer, and clean all parts of the assembly. When reassembling, put in a new rivet.

Assembling the valve. —Before assembling the valve, clean the ball point of the spindle and the pocket in the feather, to remove any accumulation of rust or other foreign matter. Be sure that the spring seat makes a good contact with the lower spring washer. Check the blowdown adjusting ring to be sure that it turns freely. If the repair work on the valve has consisted only of a moderate amount of retouching the seats for tightness, the ring may be left in the same position as before the work was begun. If parts have been remachined or replaced, you should raise the blowdown adjusting ring a half a turn, in order to prevent multiple popping or chattering while the valve is being reset.

Wipe the seats of the feather and the bushing, and make sure that they are entirely free of dirt, lapping compound, etc. Assemble the feather, guide, and spindle. Be sure that the spindle threads are screwed completely through the threaded portion of the feather, so that the ball point of the spindle rests in the bottom of the feather pocket.

Assemble the top over the end of the spindle, with the spring and the washers in place, and tighten the flange nuts.

Screw the release nut back onto its approximate position on the spindle. Install the cap with the cam lever in place. Insert the top lever and its pin. Turn the release nut until a %-inch clearance is obtained between the top lever and the cam lever, when the cam lever is located over one of the side ribs of the top. Remove the cap, and install the cotter pin.

Testing and Setting Safety Valves

Boiler safety valves should be tested after each regular cleaning period or general overhaul of a boiler, before the boiler is put into service. Each individual valve of each group should be lifted by steam, and should be reset if it is at variance with the authorized popping and reseating pressures. Safety valves should be tested by steam at any other time when their operation indicates that testing is necessary.

If the safety valves of any boiler cannot be adjusted TO LIFT PROPERLY, THE BOILER MUST NOT BE USED FOR STEAMING UNTIL THE FAULT HAS BEEN CORRECTED.

The settings for the safety valves of all ships are issued by BuShips. The popping and reseating pressures of

ALL safety valves MUST BE AS SPECIFICALLY AUTHORIZED

BY BuShips. However, a popping pressure within 3 psi of the authorized pressure is considered satisfactory, provided all valves on the boiler pop in the proper sequence.

The pressure gage installed on each boiler should be used to set all safety valves installed on that boiler. The pressure gage should be carefully calibrated prior to testing, setting, or resetting any safety valve. Pressure gages on all boilers should be calibrated to agree as closely as possible, particularly within 50 psi above and below boiler operating pressure.

In order to test and set a safety valve, you must first gag all other safety valves on the boiler. The gag screw should be made handtight only. Do not use excessive force in applying the gag, as this might bend the spindle or damage the seating surfaces of the feather and the seat bushing.

On the valve being tested, be sure that the lugs on the upper spring washer do not touch the bonnet. Raise the steam pressure until the valve pops.

When setting safety valves, you should first set the popping point; then, if necessary, make the blowdown adjustment. On all types of safety valves commonly used in the Navy, the popping point is set by adjustment of the compression screw. Blowdown adjustment is made in various ways, depending upon the type of valve.

In the huddling chamber safety valve, blowdown adjustment is made by raising or lowering the adjusting ring. This ring is threaded right-hand on the seat bushing. It is important to remember that lowering the adjusting ring will decrease the amount of blowdown and RAISE the reseating pressure. Raising the adjusting ring increases the amount of blowdown and LOWERS the reseating pressure.

In the nozzle reaction safety valve, the UPPER adjusting ring is the principal means of blowdown adjustment. When adjusting this ring, remember that it is ABOVE the space through which the steam flows as the valve begins to lift. Therefore, raising this upper adjusting ring will DECREASE the amount of blowdown. The nozzle ring also affects the amount of blowdown. However, the nozzle ring is factory set, and seldom requires resetting in service. If it is necessary to adjust the nozzle ring, remember that RAISING the ring INCREASES the nozzle effect, and thus INCREASES the blowdown. Lowering the nozzle ring reduces the nozzle effect, and thus decreases the blowdown.

In the jet flow type of safety valve, the amount of blow-down is determined primarily by the adjustment of the metering disks. The position of each disk may be changed by means of a metering screw which is externally accessible. When making blowdown adjustment on this type of valve, be sure that all disks are the same distance from the main valve disk guide. To increase the blowdown, the metering screws should be turned to the right (screwed in); to decrease the blowdown, the metering screws should be turned to the left (screwed out). The nozzle ring in this valve also affects the amount of blowdown. However, this ring is factory set and seldom requires resetting in service.

When adjusting the blowdown on any type of safety valve, be sure to keep count of the number of notches and the direction that you move each ring. This will allow you to return the ring to its original setting, in case of error, and thereby prevent gross maladjustment of the ring.

It is important to follow the manufacturer's instructions when setting safety valves. In particular, be sure to note how much each adjustment will affect the popping pressure or the reseating pressure. For example, the manufacturer may specify a limit as to the number of notches that an adjusting ring may be moved for any one adjustment, and also the total number of notches that an adjusting ring may be moved altogether.

AUTOMATIC FEED WATER REGULATORS

Single-element, two-element, and three-element automatic feed water regulators are now being used in the Navy. Single-element regulators are probably more widely used than two- or three-element regulators; however, the multi-element regulators are being used increasingly on new construction.

Single-element regulators are generally classified as being thermal-mechanical, thermal-pneumatic, or thermal-hydraulic. Two-element regulators may be thermal-mechanical or thermal-pneumatic. Most three-element regulators are of the pneumatic type.

The essential difference between single-element and multi-element regulators lies in the

fact that single-element regulators are controlled ONLY by the existing water
level in the steam drum, whereas multi-element regulators are controlled by the existing water level PLUS one or two other factors. Because of these additional factors, multielement regulators are able to compensate for changes in boiler water volume which occur as a result of changes in the firing rate. These changes in boiler water volume, which are known as SWELL and shrink, cause the single-element regulator to lag behind the actual requirements of the boiler when the firing rate is changing rapidly. Since the single-element regulator depends solely upon the existing water level in the drum, it decreases the feed supply immediately after the firing rate is increased, and increases the feed supply immediately after the firing rate is decreased. Thus the single-element regulator tends to underfeed or overfeed the boiler, whenever rapid changes are taking place.

Two-element automatic feed water regulators are controlled by the existing water level in the steam drum and by the steam flow from the boiler. Thus, steam flow is used as a corrective factor to prevent underfeeding or overfeeding of the boiler under conditions of rapid change. The steam-flow correction may be made on the basis of rate or magnitude of steam flow, or on the basis of rate of change in steam flow.

The three operating conditions which influence most three-element automatic feed water regulators are: (1) steam drum water level, (2) steam flow from the boiler, and (3) feed water flow. Three-element regulators are even more effective than two-element regulators in compensating for swell and shrink. In addition, three-element regulators are able to maintain the correct water level in spite of rolling or pitching of the ship.

In some three-element regulators, the element which is sensitive to steam drum water level is of the thermal-mechanical type; in other systems, a pressure difference between two columns of water is used to measure the water level in the drum. Steam flow may be measured in terms of the pressure drop across the superheater, or in
terms of a pressure differential caused by the use of a special steam flow nozzle installed in the steam line. Feed water flow may be measured in terms of the pressure drop between the inlet and the outlet of the economizer, or in terms of a pressure differential caused by the use of a special feed water flow nozzle installed in the feed line. The measurements of the three variables (water level, steam flow, and feed water flow) are used to position various units in the pneumatic system, and this results in the automatic positioning of a feed-regulating valve.

Use of Feed Water Regulators

Where single-element automatic feed water regulators are installed, their use is mandatory when the ship is operating under battle conditions, when manual feeding of the boiler might become difficult or impossible. The single-element regulators must be cut in immediately when general quarters is sounded. They may also be used when the ship is cruising under normal conditions—and, indeed, it is essential that they be operated frequently enough to ensure their proper functioning under battle conditions.

Single-element regulators can control the water level within acceptable limits under relatively steady steaming conditions. Under severe maneuvering conditions, however it may be necessary to resort to manual feed control. Since complete reliance cannot be placed in a single-element regulator, a checkman must remain on station and be ready to take manual control, when the single-element regulator is in use under normal cruising conditions.

Multi-element automatic feed water regulators are much more successful in controlling the water level in the boiler. When two- or three-element regulators are installed, their use is required AT ALL TIMES. Primary reliance for the control of water level must be placed in the multi-element regulator, and NOT in manual operation of the feed check valve. When a multi-element regulator

is in service, it is not necessary to have a checkman stationed at the feed check valve. All boilers equipped with multi-element regulators are also equipped with remote water level indicators, which are installed at the lower level where they can be observed by the man in charge of the watch.

Emergency Operation

Each multi-element feed water regulator is equipped with an emergency device which takes control in the event of failure of the air supply or other casualty to the multielement system. In some systems, an emergency pilot valve becomes operative as soon as there is an air failure, causing the system's thermal-mechanical water level element to assume complete control. In other systems, the emergency device consists of a thermal-hydraulic single-element regulator which is entirely separate from the multi-element control system, except that it is activated by failure of the air supply in the multi-element system.

In the event of air failure or other damage to a multielement regulator, the boiler may be operated with the emergency device in service. In all cases, however, the emergency device is in essence a single-element regulator —that is, it operates only on the basis of the water level in the steam drum. Since this type of regulator is unsatisfactory under some operating conditions, it is important to restore the multi-element regulator to service as rapidly as possible.

If a multi-element feed water regulator cannot maintain the proper water level in the boiler, switch from AUTOMATIC operation to remote manual operation, and operate the regulator manually from the control panel. If the water level cannot be maintained by remote manual operation of the regulator, cut out the regulator, station a checkman at the feed stop and check valve, and control the water level manually. The feed water regulator must be repaired and restored to service as soon as possible.

Maintenance of Feed Water Regulators

It is essential that all automatic feed water regulators be maintained in proper operating condition at all times. Consult the manufacturer's instruction book for maintenance and repair procedures on the feed water regulators installed in your own ship. In general, the following points should be remembered in connection with feed water regulators:

1. Keep all parts of the system clean. Do not allow scale or dirt to clog the feed-regulating valve. Be sure the air filters in pneumatic systems are replaced as often as necessary.

2. Do not allow anyone to paint a generator, a thermostat, or any other part which is not designed to operate properly when painted.

3. Do not allow anyone to apply lagging, except where specifically authorized.

SOOT BLOWERS

Practically all soot-blower elements, except those of the retractable single-nozzle type, are fabricated from li^-inch or 2-inch iron pipe tubing having VENTURI-FORM nozzles disposed equidistantly along one side. Specifications require that the spacing of the nozzles should not be greater than every third tube if the axis of the element is at right angles to the axis of the tubes, or not greater than 3 inches if the axis of the element is parallel to the tubes.

The metal used in soot-blower elements depends upon the gas temperatures to which the elements will be exposed, so the elements are not interchangeable even if the nozzles are identical in size and spacing. If occasion for disassembling the soot blowers arises, take particular care to ensure that each element is reinstalled with its proper unit.

For best results in blowing tubes, the steam pressure in the soot-blower head must be

approximately 300 psi. This pressure can be checked by installing a pressure gage
at the connection provided on the head for this purpose. If necessary, the orifice in the head can be readjusted (on the adjustable type) or reamed (on the plug type) to give the correct pressure.

Sufficient time must be allowed for gradual and thorough warmup of the steam lines before steam is admitted to the soot blowers. If this precaution is not observed, the piping may be subjected to thermal shock which could result in failure of the piping.

Maintenance

Compliance with the following instructions will aid materially in keeping the soot blowers in proper operating condition:

1. Inspect the soot-blower supply piping and drains to ensure that installation is in accordance with plans. The supply piping to each soot-blower element should slope downward continuously from the element to the drain connection, which should be located at the lowest point in the line. There should be no pockets or horizontal sections in the soot-blower steam supply piping. The drains should be located in such a position that the discharge does not drain onto the drums, casings, piping, or machinery. Be sure, also, that the discharge does not constitute a hazard to personnel.

2. Soot-blower piping should be inspected at each major overhaul, and at any other time when there is reason to believe that unsatisfactory conditions exist. The inspection should be sufficiently thorough to determine definitely whether or not serious corrosion is taking place. Inaccessible portions of the piping usually may be hammer-tested with a light hammer. Affected sections of piping, valve seats, heads, or valves should be thoroughly cleaned, and when necessary, repaired or renewed.

3. All moving parts in the soot-blower heads must be inspected regularly to ensure proper operation. The gland packing should be inspected periodically and adjusted or renewed as necessary to prevent steam leakage.

4. Care should be taken that element steam valves seat properly, do not leak, and are not being held open by failure of personnel to return the operating mechanism to the closed position after blowing tubes. Existence of leaks can sometimes be detected by an odor of gas being emitted from the drain connection.

5. On soot blowers fitted with cams, the cam settings should be checked at least yearly with the boiler plans to ensure that steam is being admitted into the soot-blower elements only when the elements are rotating through their proper blowing arcs. The fore-and-aft position of nozzles should also be checked with plans and instruction books, to ensure that the nozzles are in proper position in relation to the tubes.

6. Frequent inspections should be made to ensure that the check valve installed in the scavenging air connection to the soot-blower elements remains clean and free from corrosion products or other foreign material which would prevent its proper operation. The function of this connection is to supply air to the soot-blower element above the steam valve, thus preventing combustion gases from backing up into the soot-blower heads or piping where the sulfur content of the gases combined with moisture may cause serious acid corrosion. Each check valve should be disassembled as often as necessary to ensure proper operation.

7. The drain valve in the soot-blower supply piping should be left open when soot blowers are secured. In preparing to blow tubes, sufficient steam should be blown through this valve to ensure removal of
all water from the supply lines; the valve is then closed while the tubes are actually

blown. After the soot blowers have been secured, the drain valve should again be opened.

8. On some installations a drain nipple has been provided at the low point in the supply piping. This nipple is normally provided with a '/s-inch diameter hole. Make sure that this hole remains open by inserting a probe therein and observing whether or not steam is emitted from it when tubes are being blown. On other installations the supply piping has been provided with a drain valve having either a hole drilled in the valve disk or a notched seat. These openings should be kept open in the manner described above.

9. If binding is experienced on an element which is rotated through 360°, a change in the idle position of the element may be made from time to time to effect closure of the valve at different element nozzle positions. This rotation of the element relative to the swivel tube should be accomplished as specified in the manufacturer's instruction book, and may be done only on units which have a blowing arc of 360°.

10. Elements which have become badly warped should be removed from the boiler and heated and straightened, as specified in the manufacturer's instruction book.

The SINGLE-NOZZLE RETRACTABLE-TYPE SOOT BLOWER,

shown in figure 5-11, is installed in the upper casing of the superheater furnace of M-type boilers. It directs a fan-shaped jet of steam toward the water-screen tube banks of the boiler. The element, when secured, rests in a shrouding between the inner and the outer casings of the boiler. In figure 5-11 you see the soot blower in the secured position. When the hand wheel of the soot blower is turned, the element is extended about 5 inches into the superheater furnace. As the element reaches the end of

its travel, the ports line up with the steam inlet connection and the steam is allowed to blow through the element nozzle and across the superheater water screen and the superheater tube nest. The angle of the nozzle discharge is about 58°. In some installations the nozzle is rotated while it is blowing steam.

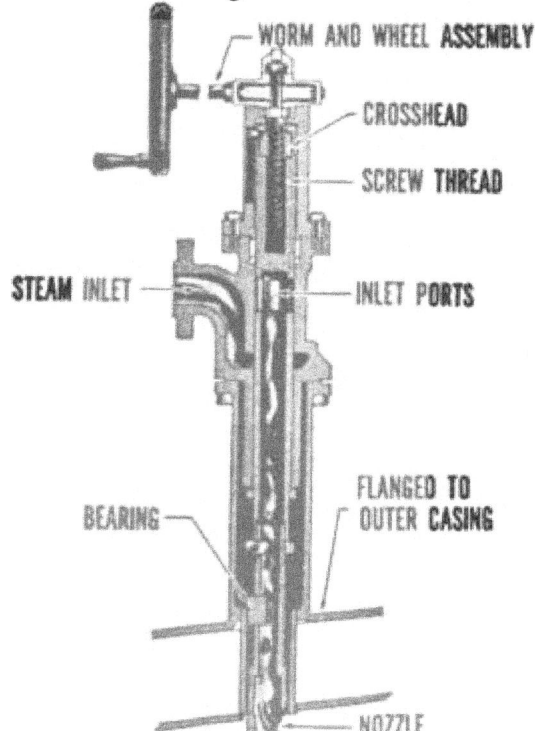

Figure 5-11.—Single-nozzle retractable toot blower.

Particular attention should be paid to the drainage, tightness of shut-off valves, and condition of internal sealing rings; and the blowing arc must be properly directed toward the superheater bank, and not toward the

water wall. Casualties to the pan brickwork may occur with this type of soot blower if it is improperly drained, if it leaks, or if it is improperly directed.

Frequency of Use

Soot blowers should be used at least once a watch while under way, twice a day during in-port steaming, and just prior to securing a boiler. On controlled superheat boilers, tubes must never be blown on the superheater side unless the superheater side is lighted off.

SUPERHEATER STEAM FLOW INDICATORS

Superheater steam flow indicators are used as warning devices to show when the rate of steam flow across the superheater has fallen below a predetermined safe limit. When superheater steam flow indicators are installed, they must be kept in service at all times.

Superheater steam flow indicators measure the difference in pressure between the superheater inlet and the superheater outlet. Since the pressure drop across the superheater is proportional to the rate of steam flow through the superheater, it is possible to use the pressure drop as an indication of the rate of steam flow. Superheater steam flow indicators are usually calibrated in inches of water, rather than in psi, because the pressure difference that they measure is relatively small.

Two types of superheater steam flow indicators are commonly used on naval vessels. Although both types respond to the pressure differential between the superheater inlet and the superheater outlet, they differ in the mechanism by which this difference is transmitted to an indicating dial.

The Bailey low flow superheater protection device, shown in figure 5-12, is an electronic telemetering system. It consists of three basic elements: (1) a transmitter, which is actuated by differential pressure; (2) an indicating receiver, which resembles a voltmeter; and (3) a

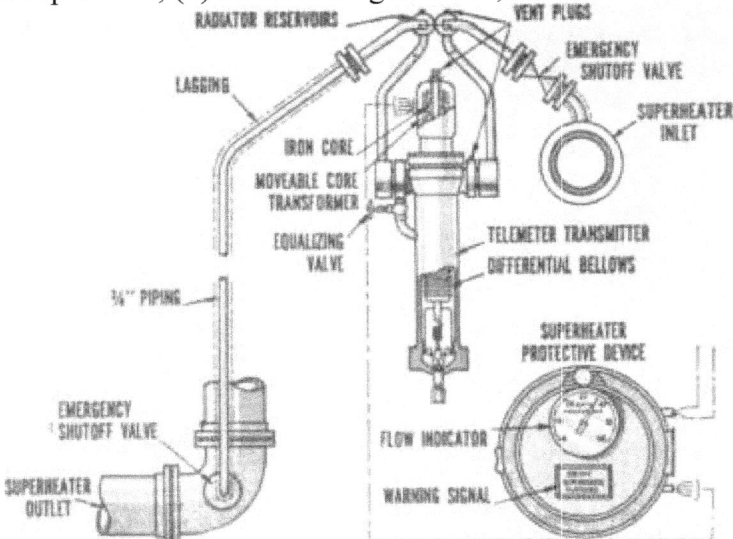

Figure 5-12.—Bailey low flow superheater protection device.

vacuum tube amplifier and motor control unit, which balances the circuit and prevents the supply voltage from affecting the reading on the gage.

Steam pressure from the superheater inlet acts upon one column of w^ater, and steam pressure from the superheater outlet acts upon a second column of water. In the transmitter, water from the inlet (high pressure) reservoir applies pressure against the outside of a bellows, while water from the outlet (low pressure) reservoir applies pressure against the inside of the bellows. The resulting movement of the bellows is transmitted by a rigid vertical rod to the movable iron core of a transformer which is located above the transmitter cylinder.

The pressure drop across the superheater, as represented by the position of the iron core, is transmitted electrically to the indicating receiver as a ratio of the output voltages from the two secondary windings of the transmitter. A reversing motor positions the indicating

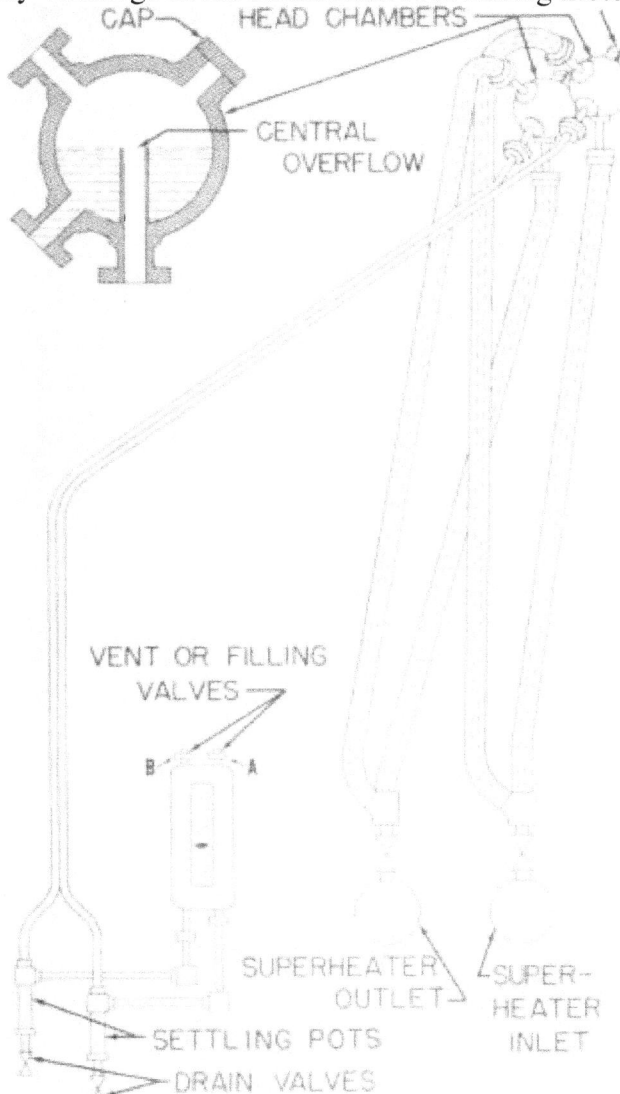

Figur* 5-13.—General arrangement of Yarway superheater steam flow indicator

pointer, and also operates a contact which, at a specified setting, closes an electrical circuit by which the warning signal "SECURE superheater burners" is illuminated.

The superheater protection device must be carefully primed before it is put into service, and all air must be removed from the system. The equalizing valve should be used to prevent damage to the bellows.

The general arrangement of the Yarway superheater STEAM FLOW INDICATOR is shown in figure 5-13. As you can see, steam pressure from the superheater inlet acts upon one column of water and steam pressure from the superheater outlet acts on the other column of water. Exposed piping connects each head chamber (reservoir) with the indicating unit.

The interior of the indicating unit is shown in cross section in figure 5-14. Notice that the water from the upper head chamber enters on one side of the diaphragm, and the water from the lower head chamber enters on the other side. The pressure from the upper head chamber is greater than that from the lower head chamber; therefore the diaphragm is moved. The diaphragm is connected by a pin linkage to a deflection plate, which moves in sensitive response to the movement of the diaphragm.

A permanent horseshoe magnet is rigidly mounted on that side of the deflection plate which is free to move. The poles of the magnet straddle a tubular well in which a spiral-shaped strip armature is mounted on jeweled bearings. A counterbalanced pointer is attached to the end of the armature mounting shaft.

When the deflection plate moves in response to variations in pressure, the magnet is made to move along the axis of the well. As the magnet moves, the spiral-shaped armature rolls in order to keep in alinement with the magnetic field between the poles. Thus a rotary motion is imparted to the armature mounting shaft; and the rotation of the shaft cause the pointer to move. The pointer hand moves over a brightly illuminated vertical dial which

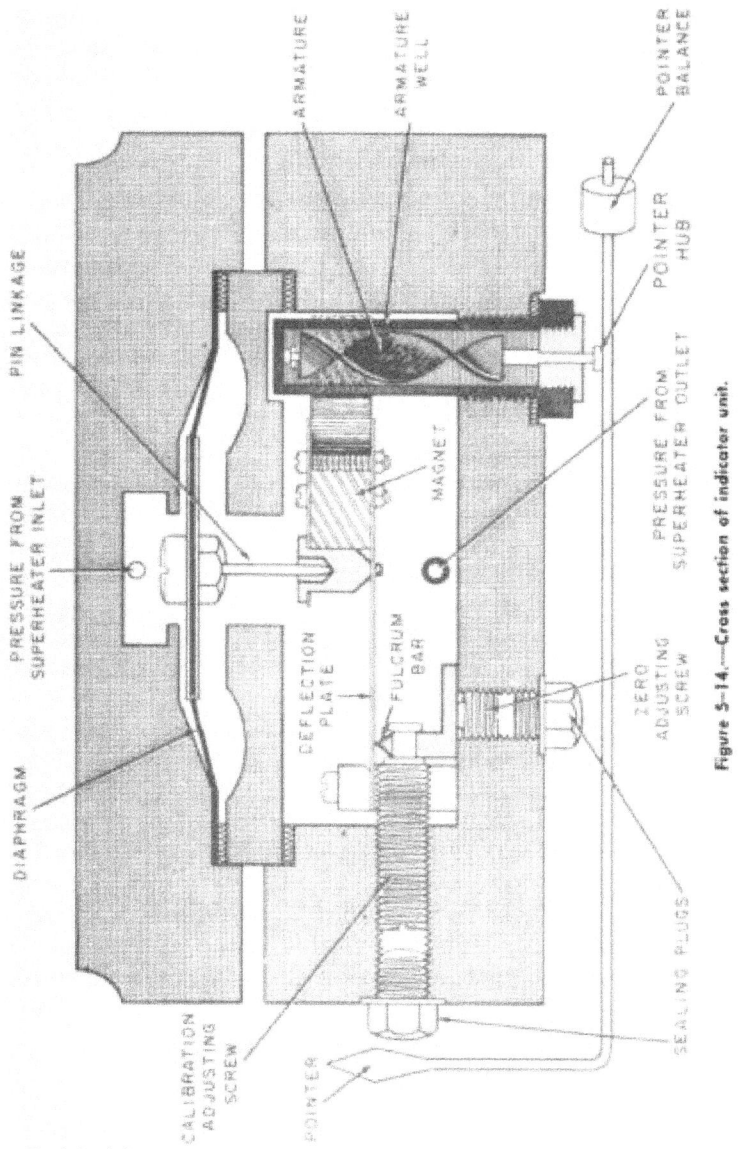

Figure 5-14.—Cross section of indicator unit.

is divided into green and red zones to represent safe and unsafe operating conditions.

The Yarway superheater steam flow indicator must be primed, its calibration checked, and its dial properly set before it is put into service. Instructions for performing these operations are given in the manufacturer's instruction books.

To cut in the indicator when the boiler is under steam, first crack the shut-off valve at the superheater outlet connection and then crack the valve at the superheater inlet connection. Allow a gradual build-up of pressures in the indicating system, then open both valves wide. If the boiler is not under steam, merely open the shut-off valves at the superheater outlet and the superheater inlet and allow pressure to build up in the indicating system as it builds up in the boiler.

If the superheater steam flow indicator is sluggish in operation, connect a water hose to each of the drain valves, in turn, and flush the piping thoroughly. Repeat the flushing operation as many times as necessary to remove all accumulated sludge from the piping. After flushing, be sure to close the drain valves before you remove the hose, so that prime will not be lost from the system.

If it is necessary to replace the light bulbs, the diaphragm, the deflection plate, or other parts of the indicating unit, be sure to follow the manufacturer's instructions carefully. Do NOT attempt to recalibrate the indicating unit unless you are sure that this is necessary. The indicator is calibrated at the factory, and seldom if ever requires adjustment in use, except after repairs have been made to the unit. If you are sure that recalibration is required, this may be accomplished by means of the calibration adjusting screw and the zero adjusting screw shown in figure 5-14.

QUIZ

1. Why does a safety valve reseat at a pressure which is lower than its popping pressure?

2. Which safety valves lift and reseat at the lower pressure— steam drum safety valves, or superheater safety valves?

3. If safety valves simmer when they are within 2 percent of their popping pressure, what corrective action is required?

4. What lapping and grinding compound should be used for grinding in the seat of a huddling chamber safety valve?

5. What lapping and grinding compound should be used for beginning to grind in the feather on a huddling chamber safety valve?

6. On a huddling chamber safety valve, what distance is referred to as the overlap?

7. What device should be used to check the dimensions and angles of the seat and of the feather or disk on a safety valve?

8. When should boiler safety valves be tested?

9. What pressure gage should be used for setting the safety valves on a boiler?

10. When gagging safety valves, how much should you tighten the gag screw?

11. How would you increase the amount of blowdown in a huddling chamber safety valve?

12. What is the principal means of blowdown adjustment on the nozzle reaction safety valve?

13. Would you raise or lower the upper adjusting ring on a nozzle reaction safety valve, in order to decrease the amount of blow-down?

14. How can you increase the amount of blowdown in a jet flow safety valve?

15. What two factors control the operation of a two-element automatic feed water regulator?

16. What three factors influence most three-element feed water regulators?

17. Under what conditions must single-element feed water regulators be used?

18. When must multi-element feed water regulators be used?

19. If a multi-element feed water regulator fails to control the water level in the boiler, what mode of operation should be tried first?

20. What steam pressure in the soot-blower head will give the best results in blowing tubes?

21. How often should soot blowers be used?

22. On soot blowers fitted with cams, how often should the cam settings be checked?

23. What measurement is used as an indication of steam flow in superheater steam flow indicators?

CHAPTER

BOILER WATER TREATMENT AND FEED SYSTEMS

The purpose of boiler water treatment is to prevent the formation of scale on the water side of the boiler, to reduce corrosion of boiler metal to a minimum, and to ensure, under all conditions of operation, against foaming and priming. To accomplish this, the following essential functions must be carried out: (1) preparation and maintenance of the purest possible feedwater, (2) chemical treatment of the boiler w^ater to maintain its composition within a desirable range, and (3) periodic inspection and cleaning of the watersides to maintain them in satisfactory condition. An indiscriminate or haphazard use of a boiler water compound or of boiler water treating chemicals will not give the desired results. Any treatment must be used intelligently and with a knowledge of the results to be obtained.

UNITS FOR REPORTING WATER ANALYSES

In the past, water analyses were reported in various units such as grains per gallon, percent normal, and milliliters per liter. However, the Navy has now adopted a standard system for reporting water analyses, and this system MUST be followed.

All boiler water tests and all feed water tests except the test for dissolved oxygen, are reported in terms of a unit called equivalents per million (EPM). The dissolved oxygen content of feed w^ater is reported in terms

of a unit called parts per million (PPM). It will be easier to understand equivalents per million if you first understand parts per million.

Parts per million is a weight-per-weight unit denoting the number of parts of a specified substance in a million parts of water. For example, 58.5 pounds of salt in 1,000,000 pounds of water represents a concentration of 58.5 parts per million (ppm). Note, also, that 58.5 ounces of salt dissolved in 1,000,000 ounces of water, or 58.5 tons of salt dissolved in 1,000,000 tons of water, represent the same concentration—that is 58.5 ppm. Similarly, 8 pounds of oxygen dissolved in 1,000,000 pounds of water represents 8 ppm.

Equivalents per million can be defined as the number of equivalent parts of a substance per million parts of water. (The word "equivalent" here refers to the chemical equivalent weight of a substance.) The chemical equivalent weight is different for each element and compound. The chemical equivalent weight of sodium chloride (common table salt) is 58.5. A solution containing 58.5 parts per million of this salt is said to contain 1 equivalent per million. If a substance has a chemical equivalent of 35.5, a solution containing 35.5 parts per million is described as having a concentration of 1 epm.

All water tests performed on board ship could be expressed in equivalents per million. However, the standard system requires that alkalinity, hardness, and chloride content be expressed in epm but that dissolved oxygen be expressed in ppm.

PRINCIPLES OF BOILER WATER TREATMENT

Salts in Boiler Water

The necessity for boiler water treatment arises from the fact that in all waters there are some dissolved salts. The salts dissolved in boiler water include those derived from sea water by evaporator carry-over and condenser leakage, those leached and dissolved from the coatings and interiors of piping and tanks, those deliberately added in the boiler compound, and combinations of these various salts. Some of these salts are corrosive, some are scale-forming, and some combat the objectionable properties of the corrosive and scale-forming salts.

Process of Scale Formation

The solubility of salts in water varies with the temperature of the water solvent. Some salts are much more soluble in hot water than in cold water, whereas the solubility of other salts decreases with increasing temperature. The latter salts are the ones which form boiler scale. In the cooler parts of the boiler the scale-forming salts may, to a large extent, be dissolved in the boiler water; but under the much higher temperature in the boiler tubes, the saturation point may be reached while the actual salt content is quite low.

Since the scale-forming salts crystallize most readily in the hottest water, and since the hottest water is that which is directly in contact with the metal surface, boiler scale forms directly on the metal of the boiler tubes. On the evaporative surface, bubbles of steam form as the water vaporizes. The salts contained in this vaporizing water cannot evaporate, and are forced back into the water envelope which closely surrounds the steam bubbles. Crystals of scale-forming salts will be deposited where the steam bubble, water, and tube surface make contact with one another. In this area (the hottest part of the system), the solubility of the scale-forming salts is at the lowest, and the amounts present at the maximum.

As the steam bubble leaves the metal surface, a ring of small scale crystals will be left to mark the location where the bubble was in contact with the boiler tube. With the separation of successive steam bubbles, the many rings of scale crystals become interlaced and the individual crystals increase in size, their growth being perpendicular to the tube surface. (Note : True scale is formed in this manner. It is possible, however, under conditions to

be discussed later, for suspended matter to settle from the boiler water and bake on the evaporative surfaces. Although such deposits may be as difficult to remove mechanically as true scales, they should be characterized as baked sludges, and not as scales.)

Scale-Forming Salts

Normally, calcium sulfate is the only scale-forming salt of serious importance in naval boilers. Calcium sulfate is soluble in boiler feed water but is virtually insoluble at the temperatures existing in boiler tubes, so that in the absence of proper treatment almost all of the calcium sulfate fed to the boilers will be deposited as scale. Calcium sulfate scale is so hard and tightly adherent that it is virtually impossible to remove it by mechanical cleaning and its chemical removal by normal boiler water treatment may require months.

Silicates form hard glass-like scales. Their occurrence usually is evidence of unusual feed water conditions such as may be caused by the use of cement wash for coating feed bottoms or the use of shore water for make-up feed water.

Calcium carbonate and magnesium hydroxide occasionally occur in boiler scale, but the solubility of these salts is so low at all temperatures that they are not of serious concern and are considered a normal constituent of the boiler sludge.

If naval boiler feed water is properly prepared the occurrence of scale should be very rare, and its discovery should be the signal for a prompt investigation of the conditions which permitted its formation.

Effect of Scale

The real hazard of scale lies in the fact that it may cause tube failures. Furnace temperatures are far higher than can be withstood by uncooled tubes. A layer of scale on the tube surface prevents normal heat transfer through the tube wall to the boiler water; with the resultant increase in the temperature of the tube metal, the tube may waste away from oxidation on the fire side or become heat blistered. A very thin layer of scale is sufficient to raise the tube-wall temperature to a point where tube failure will occur.

Baked Sludge

Heavy films of metal oxide or water-treatment sludges occasionally reduce heat transfer in water-wall and fire-row tubes to the extent that sludge baking and heat blistering occur. Baked sludge is difficult to remove by mechanical means and is not disintegrated by boiler compound. Excessive accumulation of sludge indicates that blow-down has been inadequate. The presence of sea water resulting from excessive condenser tube leakage is an occasional cause of heavy sludging. The products of corrosion of condensers and feed piping systems may also contribute to boiler sludging.

Process of Corrosion

Corrosion of boiler water-sides is a special case of electrolytic corrosion. Iron in contact with water tends to go into solution, and under the influence of an electric current the rate of solution is more rapid at the anodic (positive) areas. The inevitable chemical and physical variations in the surface of the boiler metal have corresponding slight electrical differences. The surface of the metal in contact with the water consists of a multitude of tiny electrolytic cells each with its anode and cathode. Iron tends to go into solution at the anodes, and atomic hydrogen goes from the anodes to the cathodes, where it forms a coating. The reactions at the anodes and the cathodes, respectively, are as follows:

1. Fe (tube metal)—2 electrons—Fe * (ferrous ion)
2. 2H* (hydrogen ion) +2 electrons=i2H (atomic hydrogen).

This effect cannot be prevented, but it can be reduced to a tolerable minimum by maintenance of the conditions

which keep a layer of atomic hydrogen, which is an electrical insulator, over the cathodic areas. This can be accomplished by simultaneously keeping boiler water alka-linities in the prescribed range and eliminating dissolved oxygen from the feed water and boiler water. The importance of these factors is elaborated below.

Effect of Low Alkalinity

Low alkalinity permits corrosion to proceed by removing the protective layer of atomic hydrogen:

1. 2H (atomic hydrogen) = Ho (molecular hydrogen gas).

For each molecule of hydrogen gas which escapes from a cathodic area, an atom of iron must go into solution at the anodic area to repair the protective film. Sea-water contamination is a potential cause of low alkalinity because the magnesium ion precipitates the alkaline constituent of the water:

2. Mg- + 20H=Mg(0H)o (precipitate).

Acid corrosion, that caused by low alkalinity, may be identified by general etching of the waterside surfaces and it may be accompanied by localized pitting.

Effect of Excessive Alkalinity

If the boiler water alkalinity is too high, strong caustic concentrations may develop in the film in contact with the evaporative surfaces. Such strong caustic solutions can dissolve the protective oxide layer from the surface of the metal and permit continuous reaction of the tube metal with water:

FeOf 2NaOH=Fe (ONa), + H..O Fe (bare metal) +H0O—FeO + Hz

This process can cause severe furrowing and grooving in fire-row and stud-wall tubes, where high evaporation rates establish a local high-concentration alkaline film.

Sources of Dissolved Oxygen

Aerated feed water is the principal source of oxygen contamination of boiler water. Atmospheric oxygen may be drawn into a boiler secured at steaming level and permitted to cool. Air will dissolve readily in feed water. If it is not removed by the condensers or deaerating feed tanks, it passes on to the economizer and generating-tube nest. It is imperative that the dissolved-oxygen content of the feed water be reduced to the lowest attainable level before it leaves the deaerating feed tank. The feed water must also be protected against air leakage during its travel through the suction lines.

Effect of Dissolved Oxygen

Dissolved oxygen accelerates electrolytic corrosion by combining with and removing the protective layer of atomic hydrogen from cathodic areas and thus renewing corrosion of iron in the anodic areas. Oxygen corrosion can be identified by the occurrence of scattered, localized pits and the absence of general corrosion in the intervening areas. Oxygen pitting is accelerated by increase in temperature, dissolved oxygen, and acidity (decrease in alkalinity). The normal introduction of feed water to the top drum of a steaming boiler usually reduces the dissolved oxygen to a tolerable residual, since the evolved steam scrubs the dissolved gases away; but there is considerable evidence that traces of oxygen can cause serious pitting in fire-row or stud-wall tubes, or in other relatively hot portions of the boilers.

Process of Deaerotion

Modern methods of deaeration take advantage of the solubility characteristics of gases. The solubility of oxygen increases with an increase in the pressure of the oxygen-bearing atmosphere above the water. It decreases with an increase in the temperature of the water. The solubility of oxygen in water at various temperatures and

pressures is illustrated in figure 6-1. At reduced pressures the solubility of oxygen decreases rapidly with increase in temperature and reaches zero at the temperature corresponding to the boiling point. The primary step in deaeration, therefore, consists in raising the temperature of the feed water to the boiling point corresponding

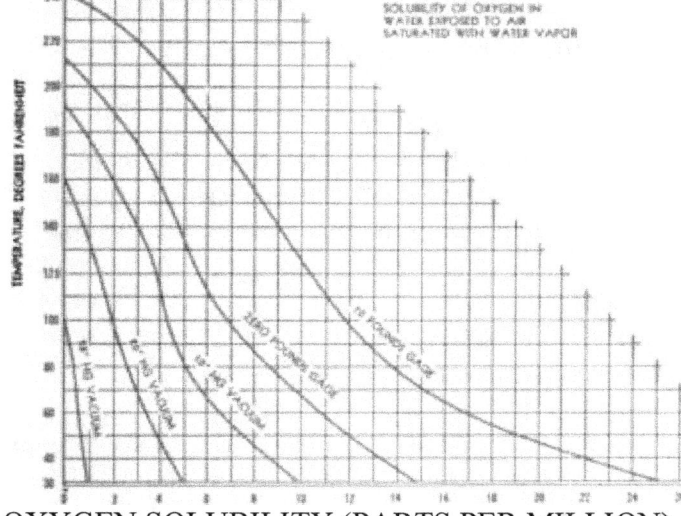

OXYGEN SOLUBILITY (PARTS PER MILLION)

Figure 6-1.—Oxygtn (olubility at various temperatures and pressures.

to the deaerating feed tank pressure. This renders the oxygen insoluble, but there is still the problem of purging the water of the last traces of oxygen and transporting the air-free water to the boiler without reaeration. The feed water systems used for solving these problems are discussed later.

Nature of Carry-Over

One of the principal objects of boiler water treatment is to prevent abnormal boiling conditions, such as foaming, carry-over, or priming. The greater the amount of dissolved solids in a water, the greater will be the tendency of the water to FOAM or froth when it is boiled violently. If the amount of suspended solid matter is considerable, any foam which forms is stabilized by the small solid particles and the film is increased in thickness. When this condition exists the steam bubbles may fail to break before entering the dry pipe. Small particles of water then will be carried along with the steam. This admixture of water and steam is called carry-over.

Under extremely fluctuating conditions caused by erratic firing, poor circulation, rough weather, etc., large quantities of water may be thrown into the steam space as though small intermittent explosions were occurring. Large amounts of water then will be carried into the dry pipe and give rise to the violent type of carry-over known as PRIMING. It is obvious that priming can be a source of great danger to the safe operation of any machinery.

CARE OF BOILER FEED WATER

Nature of Sea-Water Distillate

Practically all waters used for boiler-feed purposes contain some impurities which are undesirable. The distillate obtained from evaporation of sea water should not contain more than 0.065 epm chloride. Even then small amounts of all the other salts found in sea water will be found in the distilled water and will eventually enter the boiler with the make-up feed water. The soap hardness of water distilled from sea water normally will show less than 0.1 epm of hardness when the chloride content is 0.25 epm or less.

Undesirability of Shore Water

All shore waters contain some contaminating salts, the amounts of which depend on the character of the rocks,

sand, and earth that they have flowed over, and the extent and nature of municipal treatment. The impurity content of shore water will always be higher and, more important, will be chemically and proportionately different from that of the evaporated water which naval water-treatment is designed to control.

Water received from shore should not be used in boilers (except in emergency) without first being evaporated in the ship's distilling plant. If it becomes necessary to use shore water without distillation, the water should be tested and only a neutral water, low in hardness, should be accepted, even for emergency use.

A neutral water is one which is colorless with phenol-phthalein indicator and green with methyl-purple indicator. The evaporation of shore water may deposit on the evaporator tubes an extremely hard scale different from that resulting from the evaporation of sea water.

Sources of Feed Water Contamination

There are many ways in which impurities may find their way into the feed water system. Salt water is most likely to enter the feed system at the following places:

1. Main and auxiliary condensers
2. Distilling plant evaporators, condensers, and air ejectors
3. Drain collecting tanks, open funnel drain systems, and drain lines which run through bilges
4. Feed suction lines which run through bilges
5. Reserve feed tanks (leaky seams or rivets, and open sounding tubes)

6. Bottom blow valves (on idle boilers only)

Other types of contamination may occur when oil leaks into the steam side of fuel oil heaters or oil tank heating coils, when lube oil leaks from turbine bearings, and when air leaks into the parts of the system which are under vacuum. When there is evidence of any type of feed water contamination, every effort should be made to find the source and to correct the condition.

Before feed water suction is shifted from one reserve feed tank to another, the water in the new tank should be tested for hardness and chloride. Water having hardness greater than 0.5 epm should not be used as make-up feed; the hardness should normally not exceed 0.2 epm.

Water having a chloride content greater than 0.50 epm must not be used as make-up feed except in emergency. Normally the chloride content of make-up feed should not exceed 0.25 epm.

Prevention of Oil and Grease in Feed Water

In modern naval vessels, every precaution must be taken to prevent contamination of the feed water by oil or grease. If oil or grease contamination does occur, it will cause excessive foaming of the water in the steam drum; and the excessive foaming may, in turn, cause carry-over. In addition, oil may be deposited in the fire-row tubes, where it decomposes into a soft, carbon-like powder. Because this deposit interferes with heat transfer, the tubes are likely to overheat and to become corrugated or blistered. Oil in the boiler water may, under some conditions, lead to the formation of sludge balls. These balls, which are a mixture of oil and sludge, cannot be detected except when the boiler is opened up for inspection or repair. However, the sludge balls seem to collect the oil, and thereby minimize some of the troubles which would otherwise develop from the presence of oil in the boiler water.

The early detection of oil or grease in the feed system is of the utmost importance. Drains from fuel oil heaters and from heating coils in lubricating oil tanks and fuel oil tanks should always be passed through inspection tanks before they are discharged to the feed system. If oil or grease is detected in the drain inspection tanks, the drains must be diverted to the bilges until the source of contamination has been eliminated.

154

If oil or grease actually reaches the boiler water, it may be quite difficult to detect. After oil or grease has been exposed to the extremely high temperature in the boiler, any bubble or film of oil that appears in the gage glass is likely to be quite transparent and almost colorless. In addition, the oil may disappear temporarily when the gage glass is blown down, and may not reappear for some time. It is important, therefore, to take even the slightest indication of oil or grease as a serious matter. A very small drop of oil in the gage glass may indicate a large amount of oil in the boiler.

Avoiding Excessive Make-Up Feed

The amount of make-up feed used per hour in port and per mile under way should be checked daily, and every effort should be made to use as little make-up feed as possible. Even when the make-up feed water meets the established standards of purity, it is not as pure as the water that is already in the system; therefore, make-up feed water is to a certain extent a source of boiler water contamination. All boilers, piping, and valves should be kept tight and in good condition, in order to minimize leakage of water or steam from the system. The amount of make-up feed water used is one of the major factors considered in the evaluation of the ship's over-all engineering efficiency.

CARE OF BOILER WATER

Nature of Navy Boiler Compound

The system of water treatment used for naval vessels is based on the use of Navy boiler compound. This formula has been designed to meet the most severe conditions which are likely to be encountered in a well-managed ship. In addition, there has been provided an excess of scale-preventing chemicals. The factor of safety thus provided in the compound is not sufficient to warrant any laxity on the part of ship's personnel in the matter of keeping the purity of the boiler feed water at the highest attainable point.

The ingredients of Navy boiler compound are disodium phosphate (Na_2HPO_4), sodium carbonate or soda ash-($NaOCO_3$), and cornstarch. The first two chemicals work together in converting scale-forming salts into the relatively harmless sludges. The two in conjunction do a more thorough job than either could do by itself; in addition, they form a mixed sludge which, being made up of both phosphate and carbonate, has less tendency than either alone to agglomerate and form an adherent, cohesive mass. The sodium carbonate serves the additional purpose of providing the necessary alkalinity control. Actually, under boiler steaming conditions, most of the soda ash is converted to sodium hydroxide, while the remainder goes into the conditioning reaction described above. The cornstarch lends fluidity to the sludge resulting from the reaction of the other two ingredients, so that it does not pack in the mud drums, but may be sluiced out easily.

Navy boiler compound should be used as necessary to maintain zero hardness, and maintain alkalinity between 2.5 and 3.5 epm in the water of all boilers at all times. A ship unable to adhere to these instructions, or obtaining unsatisfactory conditions despite adherence to these instructions, should make a complete report to BuShips with a request for special authority to deviate from the indicated limitations.

It should be noted that the standard instructions for boiler water treatment do not apply to boilers operating at high pressures (1200 psi and above). Special instructions for water treatment are supplied to vessels equipped with these boilers.

Boiler Water Hardness

The occurrence of boiler water hardness usually is evidence that the make-up feed water contains excessive hardness (although false soap hardness may result from the presence of zinc supplied either by condenser corrosion or from galvanized storage tanks). The amount of compound which is necessary to maintain the proper alkalinity should simultaneously ensure zero hardness of the boiler water provided proper feed water is used. The source of any unusual hardness in the boiler or feed water should be searched for, found, and corrected. If the boiler water consistently shows hardness, and if the corrective measures that are taken fail to remedy this condition, a full report must be made to BuShips.

Chloride Content

The CHLORIDE concentration is used as an indication of the total amount of dissolved and suspended solids in the boiler water. The avoidance of the troubles caused by high concentrations of dissolved and suspended solids is accomplished by limiting the chloride concentration of the boiler water. The maximum chloride concentration permitted for boilers of various types is given below:

1. Water tube boilers, 2-inch or larger tubes—25 epm;
2. Water tube boilers, smaller than 2-inch tubes—15 epm.

When the chloride content of boiler water exceeds the proper limit, the excess chloride must be removed by blow-down, or the boiler must be secured, drained, and refilled.

Use of Surface and Bottom Blows

Proper and judicious blow-down is an aid in maintaining good boiler water conditions. In general, insufficient blow-down is given. The following points will serve as a guide in the use of surface and bottom blows:

1. A high chloride concentration indicates the presence in the boiler water of large amounts of dissolved salts and, frequently, of suspended solids such as sludges resulting from boiler water treatment. These can be removed partially by blow-down. When high chloride occurs, it is desirable to inspect the boiler interior to determine whether the high chloride is accompanied by sludge baking on the fire-row surfaces. When the chloride content approaches the maximum limit, it is most economical to remove the boiler from the line and empty it, refilling with feed water after the source of chloride contamination has been eliminated.

2. Boiler water also contains an appreciable amount of sludge products which result from corrosion of the metal surfaces in the feed water and boiler water systems. There is no shipboard method for measuring the quantity of corrosion products present in boiler water. However, a dirty or reddish brown color of the water indicates that such materials are present. Corrosion products should not be allowed to accumulate. Blow-down should be used to reduce the concentration of this material to a minimum.

3. The surface blow is more effective than the bottom blow in removing dissolved and suspended solids from steaming boilers, but only the bottom blow will remove solids which have actually settled from the circulating water. When foaming, carry-over, or priming occur, the surface blow always should be used. In order to obtain the most effective bottom blow the boiler should first be removed from the line and allowed to stand for about an hour. A short bottom-blow then should be given, followed by additional short blows at approximately half-hour intervals. Multiple short blows at intervals are much more effective in removing sludge than one long blow, but there is no advantage in using multiple blows if the boiler cannot be removed from the line.

CHLORIDE TEST

The procedure for making the chloride test is as follows:

1. Add 5 drops of chloride indicator to 25 ml of boiler water, or 100 ml of feed water, in a white porcelain casserole; the water will then turn blue-violet or red, depending on the degree of its alkalinity.

2. Add 0.05 normal nitric acid one drop at a time until the blue-violet or red color changes to a pale yellow; then add 1 ml excess of nitric acid.

3. Add reagent mercuric nitrate solution (0.025N) from the burette, meanwhile stirring the solution continuously, until the pale yellow of the sample disappears and a pale blue-violet color persists throughout. The rate of mercuric nitrate addition should be fairly rapid during the early part of the titration, then reduced to separate drops as the end point (blue-violet color) is approached.

4. Read the burette. The burette reading in milliliters equals the chloride content of the sample in epm. For example, a burette reading of 5.5 ml indicates a chloride concentration of 5.5 epm in the sample. If the chloride concentration exceeds 20 epm, it will be more convenient to use a smaller sample of boiler water rather than to refill the burette repeatedly. In such a case, dilute the smaller sample to 25 ml in the graduated cylinder with distilled water and mix well. The chloride concentration of the original sample can then be calculated as follows:

e m of chloride Burett^ reading (ml) X25

epm 0 c on e yQjyj^g solution diluted (ml)

For example, if a 5-ml sample is diluted to 25 ml, and the burette reading is 10 ml, then the chloride content is calculated as follows:

. , , ., 10 ml X 25 _^ epm of chloride= 5 ml

EQUIPMENT AND PROCEDURES FOR DISSOLVED-OXYGEN TEST

A dissolved-oxygen testing cabinet (fig. 6-2) is carried aboard vessels which perform feed water dissolved-oxygen tests. The principal items in the cabinet are three 400-ml reagent bottles, each with a sidearm aspirator and a 2-ml automatic-zero pipette for the following solutions: man-

Figure 6-2.—Dit«olved«oxygen testing cabinet.

ganous-sulfate (I), alkaline-iodide (II), and sulfuric-acio (III) ; a graduated 1-liter reagent bottle with an aspira tor, and a 10-ml automatic-zero burette for sodium thio-sulfate solution; two sample bottles with conical glass stoppers; a hot plate and a beaker for preparing starch solution; and a large casserole for making titrations. The cabinet also contains glass and rubber sampling tubing, a bottle of soluble starch, a starch-dropping bottle, a thermometer, a spatula, and a tube of lubricant.

Care of Oxygen-Testing Equipment

The manganous sulfate and alkaline iodide solutions are very concentrated, and even slight evaporation will clog the discharge tubes of their pipettes. The alkaline iodide solution will, in addition, attack the glass of its pipette and cause freezing of the stopcock if allowed to remain in the pipette for more than 8 hours. These difficulties can be avoided if both pipettes are cleaned promptly after each use. To clean the manganous sulfate pipette, fill the beaker with distilled water and raise it slowly onto the tip of the pipette with the stopcock opened, then lower the beaker and permit the stopcock tip to drain. Repeat this operation several times, using at least two changes of water. Perform the same rinsing operation on the alkaline iodide pipette, using at least two changes of distilled water. The stopcock plug should then be removed from the alkaline iodide pipette and allowed to remain covered with distilled water in the beaker until the next use of the cabinet.

Caution: The alkaline iodide and sulfuric acid solutions are strong, corrosive reagents. Care should be taken to avoid spilling either of these solutions on the person or the clothing of the operator, or in the cabinet. Mixing these concentrated solutions causes a violent reaction which results in spattering. Accidental mixing can be avoided by clearly marking the 400-ml reagent bottles from left to right with I (manganous sulfate), II (alkaline iodide), and III (sulfuric acid). An etched circle is provided on each bottle for such pencilled markings.

Collecting Dissolved-Oxygen Samples

All samples should be cooled to the lowest temperature possible. They must be cooled below the temperature of the surrounding air, and should preferably be brought to or below 70" F, during collection.

The sampling line should be as short as possible and free from loops or bends which might trap air. The line must terminate with an appropriate cooler. All connections and lines must be absolutely free from air leaks. The cut-off valve ahead of the cooler should be wide open while the sample is being collected. In this way, the cooler will be under full pressure and the rate of flow can be controlled by the valve at the outlet side of the cooler.

Attach a short length of rubber tubing (provided in the testing cabinet) to the outlet nipple of the sample cooler. Insert the glass tubing (also provided in the testing cabinet) inside the rubber tubing, and start the water flowing at a fast rate. Allow the hot sample to flow for at least 5 minutes to flush the line and the cooler. Then start a flow of cooling water through the cooler, and at the same time throttle back the flow of sample water to about 300 ml per minute, the rate at which the sample should be collected. You can determine this rate by noting the length of time it takes for your 300-ml sample bottle to fill; it should take about 1 minute.

When the sample water flowing from the cooler is as cool as possible, begin to collect the sample. Insert the glass tubing in a clean sample bottle, making sure that the end of the glass tubing is almost at the bottom of the bottle. Allow the sample water to flow continuously for at least 7 times the period required to fill the bottle the first time.

Keep the glass stopper wet in the water which overflows from the bottle. Slowly, without interrupting the flow, withdraw the glass tubing from the bottle and insert the stopper in the neck of the sample bottle immediately. Give the stopper a twist to make sure that it is properly fitted. Hold the stopper in place and invert the bottle. If a bubble (even a very small one) appears when the bottle is inverted, discard the sample and collect a new one. Follow the procedure described until you obtain a sample which is entirely free of bubbles.

Fixing Dissolved-Oxygen Samples

The method of fixing dissolved-oxygen samples which have been collected follows:

1. Replace the plug, which has been stored in water in the beaker, in the stopcock of the oxygen II pipette, and close the stopcocks on all three pipettes. Fill each pipette, using the aspirator bulb.

Caution: The capacities of the pipettes are so small that unless care is exercised in filling, excess reagent will be forced out of the vent hole. When pressure is released, excess reagent will siphon back into the bottle. Drain a small amount from each pipette into the 150 ml beaker, so that each pipette tip is full of

reagent without any air bubbles. Discard the drained reagent and rinse the beaker. Refill the three pipettes.

2. Remove the stopper from the sampling bottle and raise it onto the tip of the oxygen I pipette; open the stopcock on the pipette and allow the contents to drain into the sample.

3. As the level in the pipette drops into the tip, lower the sample bottle from the tip of the oxygen I pipette, closing that stopcock, and immediately raise the sample bottle onto the tip of the oxygen II pipette, opening that stockcock.

4. Allow the oxygen II pipette to drain completely into the sample bottle, lowering the sample bottle as the liquid level drops in the pipette tip. Replace the stopper in the sample bottle and close the stopcock on the oxygen II pipette.

5. Discard the excess solution around the neck of the stopper and then mix the contents thoroughly by holding the neck of the sample bottle between two fingers and with the thumb on top of the stopper, and swinging the sample bottle in a horizontal circle.

6. Allow the sample to stand until the precipitate settles and the sample is clear above the shoulder of the bottle.

7. Remove the stopper and raise the bottle onto the tip of the oxygen III pipette, opening the stopcock on the pipette.

8. When the contents of the pipette have drained into the sample, close the stopcock; replace the stopper; and again swirl the sample to mix it thoroughly.

9. When all of the precipitate has dissolved and the sample is clear, it is fixed and may be exposed to the air.

Dissolved-Oxygen Titration

The sample should be titrated immediately after it has been fixed. Titration should be completed within 15 min-

utes after the sample has been fixed and, preferably, within 30 minutes after the sample has been collected. The procedure for titration is as follows:

1. Cool the sample to below 70° F. If necessary, place the TIGHTLY STOPPERED sample bottle in cold water until the sample is cool enough to test. Caution: Do NOT submerge the bottle!

2. Using the aspirator, fill the sodium thiosulfate burette to its tip. Refill it, and allow it to drain down to zero.

3. Place the iron ring in the holder which is located between the sodium thiosulfate bottle and the rack for sampling bottles.

4. Place the clean casserole in the ring beneath the tip of the sodium thiosulfate burette.

5. Pour off any liquid above the stopper of the sample bottle. Remove the stopper, and pour the entire contents of the sample bottle into the casserole.

6. Add 10 drops of starch solution from the dropping-bottle to the sample in the casserole. (The starch solution must be made up fresh on the day of the test.)

7. If the solution does not turn blue, report the dissolved oxygen content of the sample as zero.

8. If the solution turns blue, add thiosulfate solution from the burette, drop by drop, stirring continuously with a glass rod or length of glass tubing, until the blue color just disappears.

9. Read the burette and multiply the reading by 0.2. The resulting figure is the dissolved-oxygen content of the sample in parts per million. For example:

If burette reading is 2.7 Dissolved oxygen=2.7x 0.2=0.54 ppm

Interpretation of the Result of o Test

A report of zero oxygen does not signify that dissolved oxygen is completely absent from the sample. With all the precautions previously outlined, the equipment of the dissolved-oxygen test cabinet is still incapable of detecting less than 0.02 ppm of dissolved oxygen in water.

Naval deaerators are guaranteed by their manufacturers to reduce the dissolved-oxygen concentration of feed water to 0.014 ppm or less. The occurrence of blue color at step 6, or the detection of dissolved oxygen in excess of 0.02 ppm, is evidence that the deaerator is not operating properly. There should be no relaxation of the precautions outlined previously (temperature of samples, careful handling of samples, frequent preparations of fresh reagents) because zero oxygen is recorded on successive readings.

STOCK SOLUTION FOR THE PREPARATION OF REAGENTS

Reagents for use in the various tests on water are supplied as standard stock solutions, either 10 or 20 times reagent strength. These stock solutions must be diluted accurately to reagent strength with the equipment provided in the boiler water testing cabinet. For diluting stock solutions, use double-distilled water or alcohol, as specified.

Requisitions for standard stock solutions of reagents should state that they are for use with the boiler water testing outfit. For information regarding stock catalog numbers, see chapter 56 of BuShips Manual.

SUMMARY OF TESTS AND MANDATORY REQUIREMENTS FOR BOILER FEED WATER

It might be well to stop here and emphasize the specific requirements in the preparation and chemical treatment of boiler water and feed water, and in the testing procedures. Remember that, unless proper care is exercised over the boiler water, there is likely to be a formation of scale and sludge, capable of reducing heat transfer and thus indirectly causing tube failure. Results of improper treatment of boiler water are illustrated in figure 6-3.

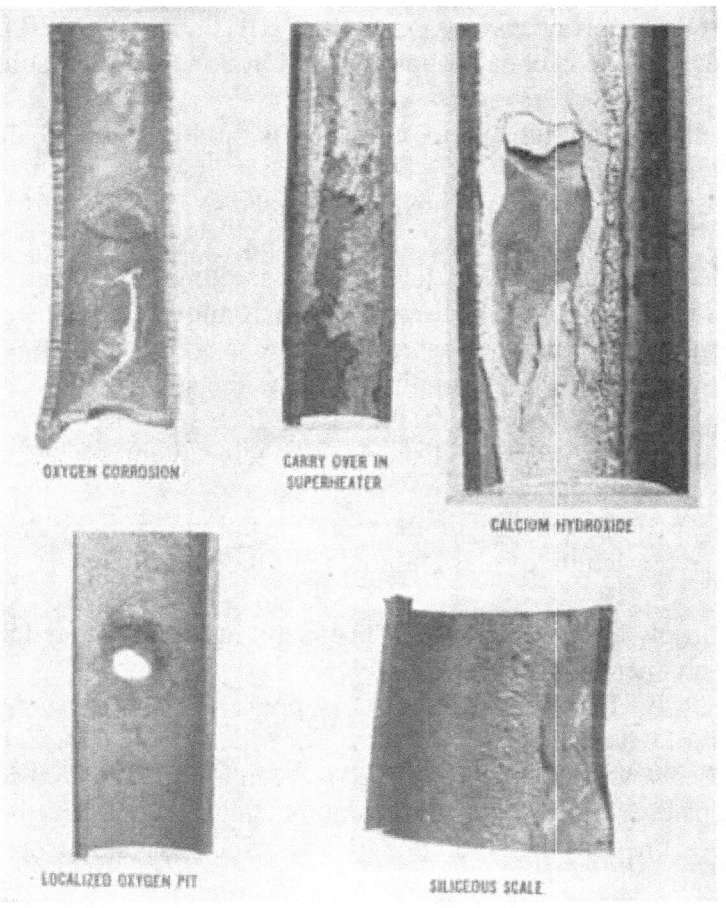

Figure 6-3.—Tube samples showing the results of improper treatment of boiler water.

OXYGEN CORROSION   CARRY OVER IN SUPERHEATER   CALCIUM HYDROXIDE   LOCALIZED OXYGEN PIT   SILICEOUS SCALE

Feed Water Requirements

The following provisions must be observed in the care of boiler feed water, being deviated from only in emergencies.

Make-up feed water having a chloride content in excess of 0.5 epm is not to be used. Make chloride test on reserve feed tanks weekly and just prior to use.

The hardness test must be made on feed tanks daily, on reserve feed tanks weekly, and on all tanks immediately prior to use of their contents as boiler feed water. The hardness of make-up feed water should not exceed 0.5 epm.

The evaporator-coil drains are to be tested when the evaporator plant is started. Whenever a measuring tank is filled, the distiller discharge to the reserve feed tanks should be tested for chloride, and must not exceed 0.065 epm.

Test the condensate from the main condensers for chloride every 15 minutes under way and every 30 minutes while standing by. Test auxiliary condensers every 30 minutes. Condensate should not exceed 0.1 epm. Electrical salinity indicator readings should be checked frequently by the chemical method.

Test the contents of the deaerating feed tank for chloride once each watch. The chloride content of the de-aerating feed tank should not exceed 0.15 epm and MUST NOT exceed 0.5 epm.

Aboard vessels provided with dissolved-oxygen testing kits and having sample coolers installed, the water discharged from the deaerators should be tested weekly. Other vessels should make these tests during the first quarter after commissioning and at any time when it is suspected that the deaerating feed tanks are not functioning properly. Ships not supplied with dissolved-oxygen testing kits will require shipyard or tender assistance to make these tests. Deaerated feed water should not con-

tain dissolved oxygen in excess of 0.014 ppm. Since the shipboard test cannot detect dissolved oxygen in concentrations of less than 0.02 ppm, it is obvious that any indication of dissolved oxygen must be taken as a warning that the feed water is not properly deaerated.

Boiler Water Requirements

Steaming boilers should be tested daily for alkalinity, chloride, and hardness. These same tests should be conducted weekly on idle boilers. Boiler water hardness is to be maintained at zero; alkalinity of boiler water should be maintained between 2.5 and 3.5 epm. The chloride limit for water tube boilers with 2-inch or larger tubes is 25 epm. The limit for boilers with tubes smaller than 2 inches is 15 epm.

The results of all boiler water tests must be entered in the boiler record sheets. Ships should make up such additional record sheets as are necessary for entering results of other tests.

EARLY TYPES OF FEED WATER SYSTEMS Open Feed System

In the open feed system, condensate from the main and auxiliary condensers was discharged directly to a feed and filter tank by a reciprocating "wet" air pump. Normally, the feed and filter tank received the drains, and was vented to the atmosphere. Make-up feed water was taken into the system through lines connecting the condensers with the reserve feed tanks.

By careful plant operation, the oxygen content of the feed water was to some extent reduced, but satisfactory service depended largely on the use of relatively low boiler pressures.

Semlciosed Feed Systems

The open feed system was replaced by the semiclosed feed system for naval vessels, with the increase in main steam pressures from 300 to 400 pounds. In the semi-

closed system finally developed, centrifugal main and auxiliary condenser condensate pumps discharged to the surge tank through air ejector inter- and after-condensers. The surge tank constituted the storage reservoir necessary for reliable plant operation under conditions of rapidly fluctuating load.

Reasonably good deaeration was secured under steady operating conditions if the temperature of the condensate could be kept relatively high, the condensate pump glands water sealed, and the feed heater drains discharged to condensers. Make-up feed water was taken into the condensers from reserve feed tanks.

Vacuum-Closed Feed System

In the vacuum-closed system, which was the next stage in the development of the feed water system, the surge tank was enclosed and vented to the condenser, so that a much lower pressure existed on the surface of the feed water in the surge tank. At the temperatures and pressures existing in the vacuum surge tank, the tank served as a deaerating agent, removing some of the air from solution in the water; the air passed over to the condenser, where it was removed by the air ejector.

The vacuum-closed system was capable of maintaining the oxygen content of the feed water at about one-half that secured with the semiclosed system.

PRESSURE-CLOSED FEED SYSTEM

The high pressure and temperature of the more modern plant (600 psi and 850° F.) caused too much corrosion even with the vacuum surge tank. In modern combatant naval vessels the open, the semiclosed, and the vacuum-closed feed systems have been superseded by the pressure-closed feed system. In this system, auxiliary exhaust steam boils the feed water (removing dissolved air) and scrubs it (removing suspended air). The steam is condensed by the cooler condensate, and the air is vented to the atmosphere.

The pressure-closed system eliminates the necessity for the shell-and-tube type of feed water heater and feed water heater drain booster pump. This has effected a substantial reduction in weight.

Arrangement of System

Figure 6-4 illustrates how the condensate from the main and dynamo condensers is discharged by the condensate pumps through the air-ejector condensers to the de-aerating feed tank. Feed booster pumps take suction

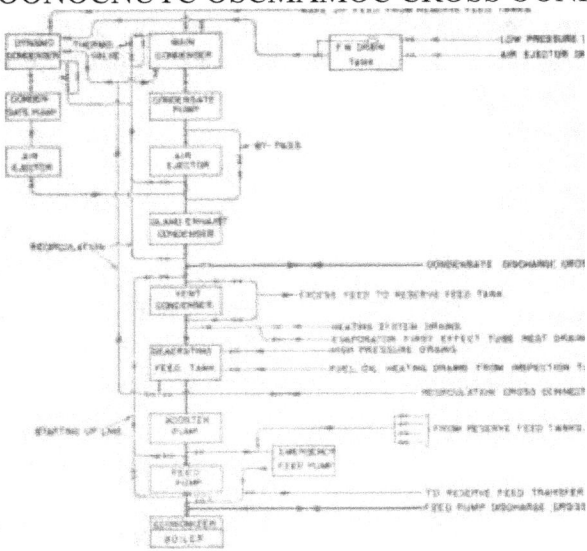

Figure 6~4. Pressure-closed feed system.

from the deaerating feed tank and discharge to the suction of the main or auxiliary feed pumps, which in turn discharge through the economizer to the boiler. Note that the entire main condensate system, with the exception of the condensate pump suction line, is under pressure.

Deaerating Feed Tank

In the deaerating feed tank the boiler feed water is heated and deaerated by direct contact with auxiliary exhaust steam. Figure 6-5 indicates diagrammatically the basic arrangement of the type of deaerating feed tanks commonly used in naval installations. (The flash type deaerating feed tank, very few of which remain in naval vessels, is not dealt with in this discussion.) Major functions of the deaerating feed tank include: 1. The provision of a storage reservoir in the feed system;

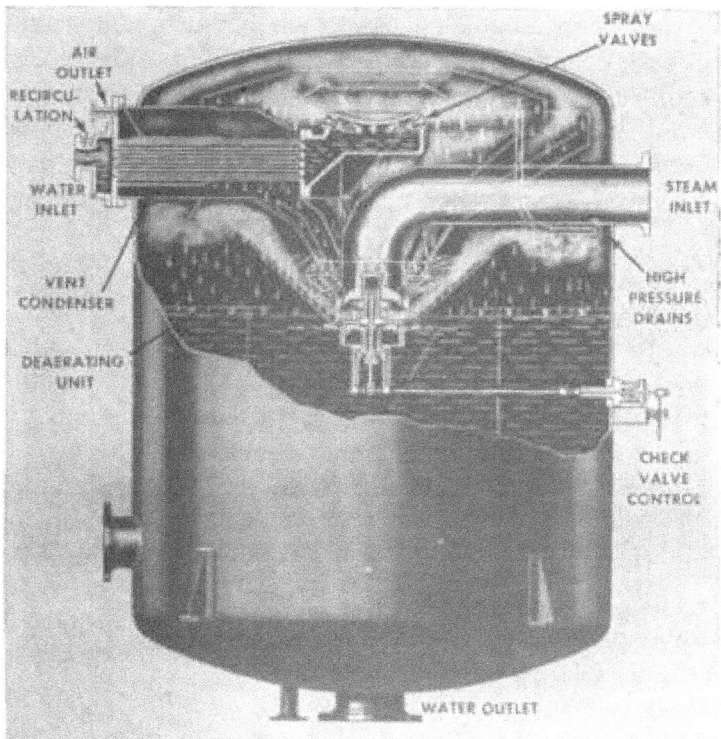

Figur* 6-5.—Deaerating feed tank.

2. Heating the feed water to a temperature closely approaching that of the auxiliary exhaust steam;

3. Deaeration of the heated feed water;

4. Maintenance of the reserve supply of feed water, stored in the lower part of the tank, in a thoroughly heated and deaerated condition.

Water flow. —The mixture of condensate, drains, and make-up feed water constituting the inlet water to the deaerating feed tank enters through the tubes of the vent condenser. By the pressure of the condensate pump discharge, this inlet water is forced through the spray valves of the spray head and discharged in a fine spray throughout the steam-filled top or preheater section of the deaerating feed tank. The tiny droplets of water are heated and scrubbed by the relatively air-free steam, and most of the dissolved air is released. Here the partially deaerated and heated water is picked up by the incoming exhaust steam and thrown radially outward and upward against the lower side of the conical baffle, in a finely atomized spray. The water then falls into the storage space at the bottom of the tank, where it remains under a blanket of air-free steam until needed for the boilers.

Steam flow. —Auxiliary exhaust steam flows directly into the spray head. A check valve is provided in the spray head, or in the line leading thereto; this valve operates to prevent a return flow of water into the auxiliary exhaust line in case the deaerating feed tank should be flooded. A portion of the incoming steam is condensed, the condensate collecting with the heated deaerated feed water in the bottom of the tank. The remaining steam flows into the shell of the vent condenser where it is further condensed as it heats the incoming water passing through the tubes. The condensate from the shell of the vent condenser drains into the deaerating feed tank. The steam not condensed in the vent condenser flows out through the vent line of the vent condenser, carrying

with it all the dissolved air which has been removed from the incoming feed water.

Venting. —To obtain effective deaeration, it is necessary to vent sufficient steam from the deaerating feed tank to sweep out all the air which has been separated from the feed water. The amount of the vented steam is controlled by an orifice installed in the vent condenser vent line, and the valves in the vent line should be kept wide open under all operating conditions. The vent condenser vent line is led to a gland exhaust condenser, or to the after condensers of the main and dynamo air ejectors, in order to recover the feed water content of the vapor and the heat contained in the vented steam.

Internal check valve. —Remote operating gear is provided for the check valve so that it may be operated manually in the event of derangement of the automatic feature. Some types of deaerating feed tanks are provided with an external automatic operating mechanism for adjustment of the internal check valve, in order to control the velocity of the steam issuing through this valve into the tank; the internal valve in this case is termed an atomizing valve. The check or atomizing valve should be kept in proper adjustment at all times, in accordance with the manufacturer's instruction book, in order to obtain the designed velocity of steam issuing through the valve into the tank.

Control of steam to deaerating feed tank. —Vessels equipped with deaerating feed tanks should always carry auxiliary exhaust pressure within the design range for the particular vessel (usually from 10 to 15 pounds). Automatic back-pressure regulating valves are ordinarily provided to bleed live steam into the auxiliary exhaust main whenever the pressure falls below the minimum auxiliary exhaust pressure (usually 8 pounds) necessary for proper operation of the deaerating feed tanks. These valves should be regarded as for emergency use only as it is uneconomical to use live steam to heat feed water. De-aerating feed tank supply valves should always be kept wide open when in use, in order to obtain proper deaera-tion.

Warming Up the Deaerating Feed Tank

It is important that a secured deaerating feed tank be kept isolated from the system and that its contained water be deaerated before the tank is cut into the system to supply boiler feed water. If the secured tank is empty, it may be filled by the emergency feed pump, taking a suction from a reserve feed bottom and discharging through a starting-up line to the main condensate line ahead of the vent condenser. On destroyers a 1-inch pipe leads overhead and, behind No. 3 boiler, joins the condensate cross-connection line. If pressure is placed upon the cross-connection line, water can be led through the 1-inch line into the condensate cross-connection for the purpose of filling either or both deaerating tanks. During this operation auxiliary exhaust should be supplied in order that the incoming water will be heated and deaerated. In warming up a cold deaerating tank the temperature should be brought up slowly in order to avoid sudden temperature changes within the tank. When the tank is filled to a normal operating level, a feed booster pump should be started to circulate the heated water from the tank back through the vent condenser for about 10 minutes, to ensure complete deaeration of the water.

If the secured deaerating feed tank is not empty, it may be warmed up by the use of a booster pump for recirculating the contained water. Auxiliary exhaust is supplied to the tank during recirculation and the water is gradually heated and deaerated. Recirculation should continue for about 10 minutes after the temperature of the water has reached the temperature corresponding to the pressure in the tank.

When the deaerating feed tank is fully warmed up, the starting-up line valve should be throttled before the tank is cut into the system. During normal operations the starting-up line should be secured and the feed pump recirculating line relied upon to protect the booster pump as well as the feed pump from overheating.

Condensate Recirculation

Under normal operating conditions recirculation to the main condenser at light loads is automatically controlled by the thermostatic recirculating valve. Water for normal recirculation is taken from the main condensate line either beyond or ahead of the vent condenser, and the branch of the recirculating line from the air ejector discharge should remain secured except v^hen warming up the plant. Recirculation of water taken from the condensate line beyond the vent condenser assures proper functioning of the deaerating feed tank at light loads, but involves the use of a recirculation cross-connection when enginerooms are running cross-connected.

Recirculation From the Deaerating Feed Tank

Provision is made for recirculation from the storage section of the deaerating feed tank to main and dynamo condensers. This line is necessary to supply water to the condenser hot wells prior to warming up the system, and under conditions when insufficient water is supplied to the gland exhauster or after condenser of the air ejector through normal functioning of the thermostatically controlled recirculating valve. Recirculation from the deaerating feed tank in normal operations should be avoided, as recirculating heated water through this line is not as economical as recirculating cool water from the condensate line ahead of the deaerating feed tank.

Make-Up Feed

Make-up feed is admitted to main and dynamo condensers as necessary through the make-up feed line connecting with the reserve feed tanks. Excess feed is normally discharged to reserve feed tanks from the main

condensate line beyond the air ejector discharge or, in the case of some installations, beyond the deaerating feed tank vent condenser. Deaerating feed tank water level may fluctuate rapidly during maneuvering conditions, but will not normally rise above nor fall below the high and low operating levels marked on the deaerating feed tank in the vicinity of the gage glass.

Transfer of Feed Water

When enginerooms are running cross-connected with more than one deaerating feed tank in operation, water may be transferred from one tank to another, if necessary, by use of the bypass around the thermostatic recirculating valves. Opening the recirculating line in one engineroom reduces the pressure in the main condensate line, thereby permitting some of the water delivered by the condensate pump in another engineroom to be discharged through the condensate cross-connecting line, and raising the deaerating feed tank level. This method avoids the necessity of discharging excess feed to a reserve feed tank in one engineroom and then admitting it in another engineroom to control deaerating feed tank levels.

Operation Under Way

At ship's speed up to about one-half power, one or more deaerating feed tanks are normally secured, condensate from the unit with the secured deaerating feed tank being transferred to another engineroom via the condensate discharge cross-connection.

In some installations, where condensate and feed booster pumps are combined, the booster pump cannot be secured when running cross-connected. The vent valve on the idle booster pump should be cracked open, so that vapor generated by recirculation from the booster pump may escape through the vent condenser.

Securing one or more deaerating feed tanks under way at low and medium speeds is desirable, as more effective deaeration is obtained in the operating tanks, control of make-up and excess feed is simplified, and some improvement is obtained in over-all operating economy.

When the vessel is under way at speeds of more than one-half power, all deaerating feed tanks should be put into service. Each engineroom then may be operated independently and cross-connecting lines between engine-rooms may be secured.

Operation in Port

Normal operation of the feed system in port with the main units secured involves securing all but one of the deaerating feed tanks and using the emergency feed pump for supplying water to the boilers in operation.

If an unusually large amount of feed water is required under port operating conditions for any purpose, such as filling boilers, the main propelling unit may be started.

QUIZ

1. What are the two units now used by the Navy for reporting the results of water analyses?

2. What unit is used for reporting the amount of dissolved oxygen?

3. What is the difference between boiler scale and baked sludge?

4. What is the most important scale-forming salt?

5. Why is boiler scale dangerous?

6. Would high or low alkalinity be the most likely to cause a general etching of the waterside surfaces, accompanied by local pitting?

?• At what point in the feed water system should the dissolved oxygen content of the feed water be reduced to its lowest possible level?

8. What is the solubility of oxygen at the boiling point of water?

9. What is the most likely cause of corrosion ocurring as scattered, localized pits, and without general corrosion in the areas between the pits?

10. How can foaming be controlled?

11. What is the most important reason why shore water should not be used for naval boilers?

12. What is used as an indication of the total amount of dissolved and suspended solids in the boiler water?

13. When the boiler water chloride content exceeds the limit, how is the excess chloride removed?

14. When the chloride content approaches the maximum limit, what method of reducing it will be most economical of fresh water?

15. How long a time may elapse between the collection of a dissolved-oxygen sample and the completion of titration?

16. How much dissolved oxygen must be present before it can be detected by shipboard test equipment?

17. How frequently should chloride and hardness tests be made on reserve feed tanks?

18. How frequently should steaming boilers be tested for alkalinity, chloride, and hardness?

19. Under normal operating conditions, why is it best to avoid recirculation from the deaerating feed tank?

CHAPTER

BOILER OPERATIONS

It is important to know the exact meaning of the standard terms used in connection with naval boilers. The following definitions have been established for the purpose of ensuring uniform use of terms throughout the service.

FiREROOM and boiler room. —A fireroom is a compartment which contains boilers and the station for operating them. A boiler room is a compartment which contains boilers but which does not contain the station for operating them.

Boiler operating station. —The station from which a boiler or boilers are operated is referred to as a boiler operating station. This term is most commonly used to describe the compartment from which bulkhead-enclosed boilers are operated.

Boiler emergency station. —The boiler emergency station is so located that, in the event of trouble, the Chief Boilerman on watch may proceed with minimum delay to any fireroom, boiler operating station, or boiler room.

Boiler full-power capacity. —The total quantity of steam required to develop contract shaft horsepower of the vessel, divided by the r.umber of boilers installed, gives boiler full-power capacity. The quantity of steam is given in pounds of water evaporated per hour. Full-power capacity is given in the manufacturer's instruction book for each boiler.

Boiler overload capacity. —Boiler overload capacity is specified in the design of the boiler. It is given in terms of steaming rate or firing rate, depending upon the individual installation. Boiler overload capacity is usually 120 percent of boiler full-power capacity.

Operating pressure. —Operating pressure is the pressure at the final outlet from a boiler, after steam has passed through all baffles, dry pipes, superheaters, etc., when the boiler is steaming at full-power capacity. Operating pressure is specified in the design of the boiler, and is given in the manufacturer's instruction book. Under actual operating conditions, when the boiler is steaming at less than full-power capacity, the pressure at the superheater outlet will vary from the specified operating pressure, provided a constant drum pressure is maintained.

Steam drum pressure. —Like operating pressure, steam drum pressure is specified in the design of a boiler and is given in the manufacturer's instruction book. Steam drum pressure is the pressure which must be carried in the boiler steam drum in order to obtain the required pressure at the turbine throttles, when steaming at full-power capacity. Ordinarily, the designed steam drum pressure is carried for all steaming conditions.

Designed pressure. —Designed pressure is usually 103 percent of steam drum pressure.

Total heating surface. —The total heating surface of any steam generating unit consists of that portion of the heat-transfer apparatus which is exposed on one side to the gases of combustion and on the other side to the water or steam being heated. Thus the total heating surface equals the sum of the generating surface, the superheater surface, and the economizer surface.

Generating surface. —The generating surface is that portion of the total heating surface in which the fluid being heated forms part of the circulating system. This surface is measured on the flue-gas side. The generating surface includes the boiler tube bank, water walls, water

screens, and water floor (when installed and not covered by refractory).

Superheater surface. —The superheater surface is that portion of the total heating surface where the steam is heated after leaving the boiler steam drum. This surface is measured on the flue-gas side.

Economizer surface. —The economizer surface is that portion of the total heating surface where the feeding fluid is heated before entering the generating system. This surface is measured on the flue-gas side.

SUPERHEATER DEFINITIONS

There are two basic types of modern naval superheater installations: (1) controlled, and (2) uncontrolled. In a boiler with controlled superheat, the degree of superheat can be changed by regulating the intensity of the heat passing through the superheater tube bank, without substantially changing the intensity of heat passing through the generating tube bank. This control is possible because the boiler has two furnaces, one for the saturated side and one for the superheat side. A boiler with UNCONTROLLED superheat, on the other hand, has only one furnace; and since the same furnace gases must be used for heating both the generating tubes and the superheater tubes, the degree of superheat cannot be controlled.

The superheater on a controlled superheat boiler is called an integral, separately-fired superheater; the other type, where the superheat is not controlled, is known as an integral, not separately-fired superheater. The term "integral" is used here to indicate that the superheater is installed as a part of the boiler itself. Practically all modern superheaters are integral with the boilers. In some older type ships, however, separate superheater boiler units were provided for superheating the steam from a number of saturated steam boilers.

Some controlled superheat boilers have radiant-type superheaters—that is, the superheater tubes are exposed to the radiant heat of the furnace. More commonly, how-

ever, the superheater tubes are protected from radiant heat by water screen tubes. The water screen tubes absorb the intense radiant heat of the furnace, and the superheater tubes are heated by convection currents passing out of the furnace. This type of interdeck or water-screened superheater is ofen referred to as a convection-type superheater. Superheaters on uncontrolled superheat boilers are heated by convection rather than by radiation. Superheaters on controlled superheat boilers may be heated by radiation, by convection, or by a combination of radiation and convection.

Figur* 7-1.—BoiUr with uncontrolled tuperhoat.
ST I AM OVUM
SCtUttell CLCMCMT
SURFACf BLOWri^E
CVCtONE STfAM SePAIIATO«
MUM &Af CTT VALVfS
CCONOMlXtl LIT
MTfRNAL ftiO^mi
ST Cam collecting
AW INLET TO DOUftLE CASINC
METVACTA»LF SOOT BLOWER
STUD VALL • ATCR COOLfC SIDE «ALL
FtAVTIC CMIIOMC OIIE
DO«NCOliE«
VATEV HALL MfADEK
IMPELLER ^LATE DffMSf FiREtKICK
WCM TEMP. INUILATIMG BRtCk
UNCALCIMtO DIATOMACEOUS EARTH BLOCKS
WATER SCREEN HEADER PROTECTION PLATES A BAPFLE MIX
SPECIAL OIL BURNER LIGHT INC PORT A4R LOCK TY^I
AIR VEMT CONNECTION DRY PIPE
ECOWOMIZfR ELEMENTS

STEAM DRUM PROTECTION PLATES

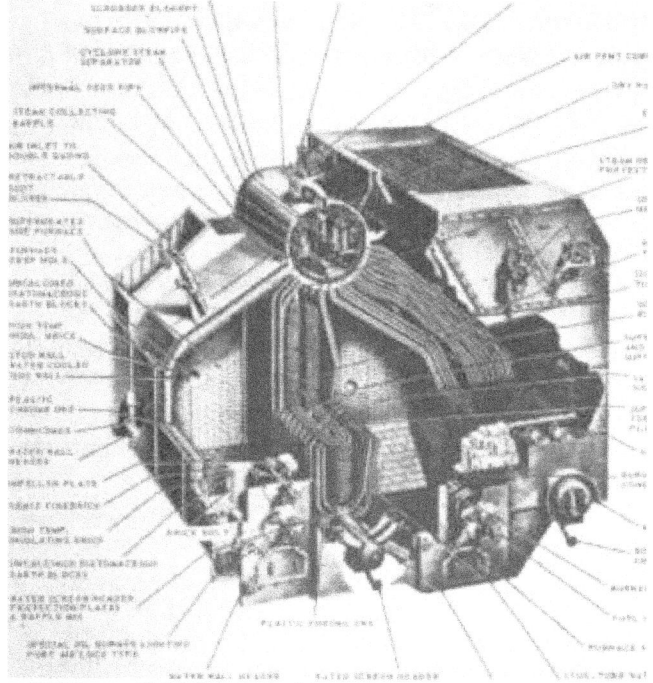

»aTBR wall NCAOt* OR AIM COMNBCTION
WATER SCREEN HEADER BOTTOM BLOW COMNBCTION
DOVMCOMERS
SOOT BLOWER HEADS
CEMERATIItC TUBES
SUPERHEATER TUBES
SOOT BLOWER ELEMENTS
SUPERHEATER AND DRUM SUPPORT TUBS^
SATURATED . SIDE FURNACE
SUPERHEATER TUBE SUPPORT PLATE
DOWWCOmERS
BURNER BLADEO CONE
BOTTOM BLOW CONNECTION
BURNER AIR OOORS
FUEL OIL BURNER
PURNACE ACCESS DOOR
STUO.TUBt WATER COOLIO DIVISION WALL

Figure 7-2.—Two furnoce tingle-uptake superheat control boiler.

Figure 7-1 shows a boiler with uncontrolled superheat Figure 7-2 shows a superheat control boiler.

LIGHTING OFF BOILERS

The following instructions for lighting off boilers apply, in general, to most modern express boilers. Special additional instructions for lighting off superheaters are giveh after the general procedure has been indicated. In all cases, this information should be supplemented by specific operating instructions issued by the Engineer Officer and by the manufacturer of the equipment.

In general, these are the steps to be followed in lighting off a boiler:

1. Remove the stack cover.

2. Inspect the bilges to be sure that they are free of oil. If necessary, wash and pump the bilges.

3. Inspect the bottoms and inner front casings of air-encased boilers, to see that they are free of oil accumulations. Be sure that the register drip holes are not plugged.

4. Check the fuel oil strainers to be sure that they are clean and in good condition.

5. Inspect all atomizer assemblies. Be sure that they are the correct size, and that they are clean and properly made up.

6. Move all air register doors to be sure that they operate freely.

7. Check individual atomizer (needle) valves and manifold valves. They should be CLOSED.

8. Wipe up all oil from floorplates, etc.

9. Run the forced draft blower, with the air register doors open; this will ventilate the furnace and clear it of accumulated gases.

10. Examine all casing doors to be sure that they are closed and that they are airtight.

11. Open the steam drum aircock and the superheater vents.

12. Check to see that the water gage cut-out valves are open and that the drain valves are closed.

13. Check to see that the surface blow valve and the sea valves are closed and that they are not leaking.

14. Run down or pump down the water in the boiler until it is just out of sight in the 10-inch water gage glass. To do this, you must open the bottom blow valve and open the drain valves to the bilges (including the superheater valves to the bilges). The line to the bilges has a hose connection so that the water can be pumped overboard if desired.

15. After the superheater has been emptied, close the superheater valves to the bilges and open the superheater gravity (open-funnel) drains.

16. Examine the hand gear for lifting safety valves, and operate this gear as far as this can be done WITHOUT lifting safety valves.

17. Open the auxiliary feed stop and check valve, start the emergency feed pump, and bring the water level to about 1 inch above the bottom of the glass in the 10-inch water gage. This procedure fills the economizer with water, tests for possible obstructions in the feed lines and water gages, and tests the operation of the emergency feed pump.

18. Open the main feed stop and check valve, start the main feed pump, and raise the water level in the boiler about >4 inch. This procedure tests the main feed pump and the lines.

19. Ease up on the boiler main steam stop valve stem, WITHOUT lifting the valve disk off its seat. This procedure will prevent the valve from sticking when it is heated.

20. Check to be sure that all cocks and valves in the line to the steam drum pressure gage are open.

21. Line up the fuel oil system. Open all necessary valves from the service tank to the service pump. Bypass the meter, and open all valves between the service pump and the burner manifold. Open the recirculating valves, and start the service pump.

22. If the oil is very cold and viscous, so that the service pump has difficulty taking suction, use the tank heating coils to warm the oil.

23. Cut in steam to the fuel oil heater.

24. Run the forced draft blower SLOWLY until the first burner is lighted off.

25. When the oil has reached atomizing temperature, partly close the recirculating valves to allow fuel oil pressure to build up to at least 200 psi.

26. Light off the center atomizer or the atomizer designated as No. 1. This atomizer must have a small or "port" size sprayer plate. Use a hand torch for lighting off. Stand clear to avoid injury from A flareback!

27. As soon as the burner has been lighted, close the recirculating valves and regulate the oil pressure with the micrometer valve.

28. Open the fuel oil meter inlet and outlet valves, and close the meter bypass valve.

29. Light off additional burners, as required. When burners are numbered, be sure that they are lighted off in the proper sequence. A hand torch must ALWAYS BE USED TO LIGHT THE FIRST BURNER, AND TO LIGHT ALL ADDITIONAL BURNERS UNTIL THE FURNACE BECOMES INTENSELY HOT.

30. Check the level in the water gages as the water in the boiler begins to heat up, and frequently after steam has been formed.

31. Close the steam drum aircock and the superheater vents, after steam has formed and has blown sufficiently to exclude all air from the boiler.

32. Check the steam drum pressure gage to see that it registers pressure, after the aircock and vents have been closed.

33. When the boiler pressure is about 150 to 200 psi below the safety valve reseating pressure, request permission from the officer of the deck to lift safety valves by hand. BEFORE lifting safety valves, open the superheater drain valves to the bilges and make sure that the superheater is free of condensate. When the boiler pressure is within 100 psi of the safety valve reseating pressure, lift the valves sufficiently to blow any dirt or other foreign matter from the valve seats.

34. Check the working of the water gages by opening the drains and blowing through.

35. See that the water in the boiler is kept at steaming level.

36. Before cutting the boiler in on the main or auxiliary steam lines, use the bypass valves (if fitted) or crack the boiler stop valve slightly to warm up the lines slowly and to equalize the pressure between the boiler and the line. Be sure that the lines are properly drained during the warming-up period.

37. Cut in the boiler when the proper pressure has been reached. Caution: A boiler must first be cut in on the auxiliary steam line before being cut in on the main steam line.

Lighting Off Uncontrolled Superheat Boilers

The following additional instructions for lighting off boilers with uncontrolled superheaters must be observed in order to protect the superheater while steam is being raised in the boiler:

1. On or before lighting the first burner, establish a positive flow of steam through the superheater. Open the superheater protection exhaust valve and the superheater protection steam inlet valve.

2. Be sure that the superheater is thoroughly drained at all times. When lighting off, open the gravity (open-funnel) drains and keep them open until the boiler has built up enough pressure (about 50 psi) to allow use of the high-pressure drains.

3. When the steam drum pressure reaches about 100 psi (or, in any case, by the time the steam drum pressure equals the pressure in the superheater protection steam supply line), close the superheater protection steam inlet valve.

4. While raising steam, be very careful not to exceed the allowable rate of combustion.

The superheater protection steam supply does not give adequate protection to the superheater at high rates of combustion.

5. Leave the superheater protection exhaust valve open until the auxiliary steam stop is opened and the boiler is furnishing steam to the auxiliary steam line. Then close the superheater protection exhaust valve.

Lighting Off Controlled Superheat Boilers

The following precautions must be observed in connection with boilers having controlled superheaters:

1. The superheater must be thoroughly drained at all times. When lighting off, open the low-pressure drains and keep them open until the boiler has built up enough pressure (about 50 psi) to allow the use of the high-pressure drains.

2. Never light burners on the superheater side until an adequate steam flow has been established through the superheater! Never light a burner on the superheater side unless one or more burners are in operation on the saturated side!

3. When operating with the saturated side only, keep the air registers tightly closed on the superheater side. This precaution is necessary in order to prevent air leakage through the superheater furnace, with consequent condensation in the superheater.

4. When there is little or no flow through the superheater, be careful that you do not over-fire the saturated side. Excessive heat from the saturated-side furnace can damage the superheater tubes.

Lighting Off Boilers on a Cold Ship

Under a cold ship condition, with no steam available for operating forced draft blowers or fuel oil pumps or for heating the fuel oil, particular care must be taken in lighting off the boilers. Insufficient draft, inadequate circulation of air between the casings, and poor atomiza-tion may result in the fire being extinguished, in flarebacks, in explosions between the casings of air-encased boilers, or in other casualties. The following instructions should be followed, insofar as they are applicable to the installation in question and the circumstances that prevail.

Diesel oil, if available, should be used for lighting off until sufficient steam has been raised to place the fuel oil heaters in service. Navy Special Grade fuel oil can be used, but poorer atomization will be obtained than with Diesel oil, and, because of the cold furnace, difficulty may be experienced in igniting the oil and in maintaining ignition. If Navy Special is used, a fire should be started on the furnace floor in way of the burner. Use oil-soaked rags, wooden boxes from which all metallic objects have been removed, or other combustible material, thereby warming the furnace and ensuring that the burner does not become extinguished.

To offset the difficulty of obtaining sufficient air for combustion, the sprayer plate used in lighting off must always be of the smallest capacity available. Fuel oil pressure should be maintained at 75 to 100 psi by means of a hand pump, if no motor-driven pump is available. Because of the difficulty in maintaining fuel oil pressure and in obtaining sufficient air for combustion, it is generally not practicable to light an additional burner until after steam has formed and the forced-draft blower and the fuel pump can be cut in.

Except in cases where motor-driven blowers are installed and power is available, natural draft must be used in lighting off. To light off a CLOSED FIREROOM by natural draft, open the fireroom hatches and all air registers to provide as much air as possible. When the furnace becomes thoroughly heated and considerable draft has been induced by the difference in temperature, it may be necessary to close the extra registers to prevent the formation of white smoke. The color of the smoke produced, if any, and the relative effect on the flame of opening and

closing the register doors should be used as a guide in adjusting the registers not in use.

With AIR-ENCASED BOILERS, care must be taken to prevent the accumulation of combustible gases between the casings after the boiler is lighted off. A flow of air must be established between the air casings from the back of the boiler to the burner front, to prevent the gases from backing up in the casings; or the outer casing access doors must be removed to permit the escape and dilution of such gases. If power is available from the Diesel generator set, it is possible to simplify the lighting-off procedure by placing the fireroom (or inner casing) under air pressure through use of the ventilating blowers. Establishing air pressure by means of ventilating blowers may be accomplished in two different ways. One method is to close the fireroom hatches and cover the inlet to the exhaust fans (with canvas or other means) and then to operate the ventilating supply blowers at full speed. This will put the fireroom under an air pressure of approximately 1 inch of water. Air should then be admitted to the boiler by removing the access doors in the air ducts between the boiler and the forced-draft blowers, or by removing a superheater or economizer outer casing access panel, thereby providing air flow between the boiler casings in the usual manner.

When the ventilation exhaust fans discharge into an uptake space or plenum chamber from which the forced-draft blowers takes suction, the following method of establishing air pressure in the inner casing may be used. Cover the outside air intake openings for the forced-draft blowers with canvas or special covers. The plenum chamber can then be placed under pressure by means of the ventilating exhaust fans. Air will be forced through the forced-draft blowers and between the casings to the boiler front in the usual manner. If the blower flaps do not open automatically it may be necessary to hold them open with a bar or by other suitable means. However, wVen air pressure is being established in this manner.

the fireroom hatches should remain open, otherwise a negative air pressure may be obtained in the fireroom. This would happen if the ventilating exhaust fans should operate at a higher speed, or have a greater capacity than the intake fans—thereby creating a false impression of positive air pressure in the boiler casing.

When either of these methods is used, doors of the forced-draft blower rooms should be closed, to prevent the escape of the air pressure. Following the establishing of an air-flow into the boiler by either of these methods, the boiler may be lighted off in the normal manner.

If no means of providing forced draft is available and it becomes necessary to light off an air-encased boiler by natural draft, the outer casing furnace access door on both the saturated and superheater side should be removed to permit entrance of air to the burners. In addition, the outer rear economizer doors, outer side casing access doors, and the access doors in the air ducts between the blowers and the boiler should be opened so as to minimize any tendency for a back draft into the casings and to permit the escape and dilution of any combustible gases which may be drawn into the casings. The boiler may then be lighted off in a manner similar to that described for lighting off a closed fireroom under natural draft. When suflRcient steam is formed to operate a forced-draft blower, fires should be secured, the boiler closed, and the blower operated to supply air in the usual manner.

BOILER OPERATION

As a First Class or Chief Boilerman, you will be required to direct and supervise the lighting off, operating, and securing of boilers, when ordered. It will be your responsibility to make sure that each watch station is properly manned, that all equipment is operating normally, that proper pressures and temperatures are developed and maintained, that the water in the boiler is at the proper level, and that combustion requirements are being met. You must be constantly alert to indications

of faulty operation of equipment. You must make sure that all safety precautions are being observed, and that unsafe operating conditions are not allowed to exist.

When getting under way and when coming into port, be sure that the boilers are operated in such a way that any steam demands up to boiler full-power capacity can be met. Before reporting the fireroom ready to answer all bells, you must make certain checks and inspections to be sure that everything is in order. It is your responsibility to KNOW that the boilers have been properly lighted off and that they have been cut in on the auxiliary and main steam lines. Be sure that all alarms and annunciators have been tested, and that they are in good working condition. Be sure that all forced draft blowers on steaming boilers are in operation.

Check the fuel oil system. The main fuel oil service pump should be in operation, and a standby should be warmed up and ready to take suction from the standby fuel oil service tank. The fuel oil heater must have a sufficient number of units in operation. For getting under way, large size sprayer plates must be used. Sets of standby atomizer assemblies with smaller sprayer plates should be made up and placed in the racks which are generally provided for this purpose.

Check the feed system. The main feed system and the main feed pump should be in use, and a standby main feed pump should be warmed up. The emergency feed pump will be operating slowly, in a standby condition. If there is an emergency feed system, be sure that it is properly lined up.

Safe Minimum Steam Flow

The safe minimum steam flow through the superheater when the superheat control boilers are operating at full steam temperature varies from approximately 5,000 to 9,000 pounds per hour, depending upon the type and size of boiler in question. The corresponding reading (pressure drop) on the superheater steam flow indicator is approximately 2 inches of water in case of convection-type superheaters, and 5 inches of water in case of radiant-type superheaters. The exact figure for each class of vessels is specified in the boiler instruction book or in special instructions issued by BuShips. On boilers with radiant-type superheaters, all burners on the superheater side must be secured when the steam flow indicator is in the red zone, or when the warning signal is lighted. Convection-type superheat control boilers are less critical and may be operated at reduced superheat (650° F. or below) with the pointer in the red zone, if the boiler is connected to its own turbogenerator on the ship's load, and if a positive steam flow is ensured. This latter exception is usually the case only on large combat installations and even then must conform with manufacturers* instructions as approved by BuShips.

Raising and Lowering the Temperature

The boiling point of water under a pressure of 600 psi (gage) is approximately 489° F. This also will be the temperature of the saturated steam rising from water boiled at that temperature. Modern naval boilers as installed on most combat vessels operate with a superheater outlet temperature as high as 850° F. The difference between the superheater outlet temperature and the saturated steam temperature is known as "the degree of superheat." There is usually a carry-over of heat from the saturated furnace to the superheater tubes even when the superheater burners are secured. This gives the steam at the superheater outlet a slight degree of superheat (the total steam temperature usually reading 500° F. or slightly higher) with the superheat side secured. Thus when operating at maximum superheater outlet temperature (850° F.) there is a degree of superheat of approximately 350° F. With this in mind it is easy to understand the importance of raising and lowering the outlet temperature slowly. Time and circumstances

PERMITTING, NEVER RAISE OR LOWER OUTLET TEMPERATURE

AT A GREATER RATE THAN 50° F. EVERY 5 MINUTES. NEVER fire the superheat side at a greater rate than the saturated side.

Keeping the Temperature Steady While Steaming

Variations in superheater outlet temperature must be kept at a minimum for the following reasons:

1. Wide or frequent fluctuations in the degree of superheat, above or below that for which the machinery is designed, would result in a considerable loss of overall efficiency of the engineering plant.

2. Excessively high temperatures would result in severe damage to superheaters, piping, and machinery.

3. Sudden large changes in superheated steam temperatures would result in sudden expansion or contraction of piping and machinery casings, resulting in possible damage to valves, piping joints, turbine casing joints, and superheater handhole plates.

Cutting In When Paralleling Superheat Control Boilers

Cutting in and paralleling boilers requires much the same procedure as putting the first boiler or boilers on the line. Cutting in and paralleling superheat control boilers when the superheaters of the boilers already on the line are in use requires additional instructions, as outlined in the following paragraphs:

1. Before any additional boiler is cut in on the main line, all the steam lines from the incoming boiler must be thoroughly drained and warmed, and the boiler then cut in on the AUXILIARY line.

2. Before an additional boiler is cut in, the outlet temperature of the steaming boilers

should be reduced to approximately 600° F. This reduction in temperature is made gradually, and as time and circumstances permit; preferably it should be made at a rate not exceeding 50° F. every 5 minutes.

3. The blow-down of the superheater drains to the bilge should be thoroughly checked, as a final precaution to make sure that all moisture is eliminated.

4. The main line slop valve should be opened, and the main boiler stop valve cracked and then slowly opened. The order in which these valves will be manipulated depends upon the ship's piping arrangement.

5. The superheater side of the incoming boiler should be lighted off as soon as the superheater protection device indicates a sufficient steam flow through the superheater.

6. The superheater outlet temperature is equalized with the line temperature, and the loads on all boilers equalized. The superheat should be raised evenly on all boilers, at the rate of 50° F. every 5 minutes.

Paralleling Superheat Control Boilers in Emergencies

Under battle conditions and in other emergencies, it may be necessary to bring in additional boilers on the line quickly without dropping superheat on the steaming boilers. In such a case, maximum protection against joint or gasket leaks in the main steam lines will be ensured by the following procedure:

Bring the boiler up to pressure and cut it in on the auxiliary line. Meanwhile, bleed the superheater of the boiler copiously to the bilge and to any other connection provided, such as the auxiliary exhaust line. Warm up the steam lines to the main line stop valve by means of the bypass valve or by cracking the boiler stop, and drain these lines copiously to the bilge. With the boiler pressure 5 to 10 pounds above the line pressure, establish flow from the incoming boiler to the main line carrying superheated steam, by opening the bypass valves. If the steam flow indicator shows that sufficient steam flow has been established to protect the superheater, light off the superheater and gradually bring up the temperature until

it equals that of the other boilers. Then open the main and line stops, and close the bypass valves on these stops.

If insufficient steam is obtained through the bypass, carefully crack the main stop and the line stop to give adequate steam flow, and then proceed as above. The rate of firing of the superheater side will depend on the adequacy of steam flow. Start with a small burner and watch the superheater outlet thermometer. If it shows a normal increase in temperature, bring the temperature up until it equals that of the other boilers, and then open the stops. If the thermometer does NOT show a normal increase in temperature within 5 minutes, secure the superheater fires, establish more flow, and light off again.

Experience has shown that greater precautions must be taken in cutting in and cutting out boilers that have radiant superheaters.

Backing With Superheat

Maximum steam temperature is seldom used in the astern turbines. In order to prevent damage to these turbines, superheater outlet temperatures are usually lowered from the maximum when a backing bell is received. Backing temperatures vary on different installations. For specific data on various installations, consult chapter 41 of BuShips Manual, and turbine manufacturers' instructions.

Control of Superheat Control Boilers

Superheat control boilers that are being operated cross-connected at a comparatively steady steaming rate should be operated with the same number of burners, the same size sprayer plates, and the same oil pressure. To take care of variations in load, one machinery space or fire-room should be designated to control each group of boilers. Minor adjustments to the number of burners and the oil pressure should be made on each boiler, to maintain the desired steam pressure and temperature. For the large

changes in load, all boilers should be operated individually, with the burners cut in or cut out as may be necessary to keep the pressure and temperature from rising or dropping. If practicable, each fireroom should be informed prior to any large change in load, so that any excessive rise or fall in pressure may be avoided.

Under ordinary conditions the saturated steam load of a vessel with superheat control boilers should be evenly divided among the boilers in use. As pointed out earlier, no boiler should be cut in on the main steam line until it has first been cut in on the auxiliary steam line; in addition, any combination of boilers connected to a common main steam line must also be connected to a common auxiliary steam line. That is, if two boilers in the same fireroom are both furnishing steam to the main steam line, the valve in the auxiliary steam line between those two boilers must be open; similarly, if the cross connection in the main steam line between the port and starboard engines and/or between the forward and after engine-rooms is open, the corresponding cross connection(s) in the auxiliary steam line must be open. Otherwise, an increase in the saturated steam load on the boiler (or group of boilers), without a corresponding increase in the firing rate on the saturated side of the boiler (s) concerned, will cause a drop in drum pressure and may cause stagnation or reversal of flow through the superheaters of those boilers, with consequent overheating of the superheaters. If a vessel is to be operated with the cross connections open, the cross connection in the auxiliary steam line should be opened before the cross connection in the main steam line is opened.

Control on Small Ships

On small ships, the burnerman at the boiler is largely dependent upon the boiler steam-pressure gage and the superheated steam thermometer for indications of changes in steam demand. As conditions change, he rapidly cuts in or cuts out burners so as to maintain as

steadily as possible the desired boiler steam pressure and temperature.

The efficient handling of large and frequent changes in rate of combustion requires careful training of the fire-room force. When steaming steady and with more than one, fireroom in operation (cross-connected), minor variations in steam demand are taken care of in one fireroom, designated the control fireroom.

Rotation in Use of Boilers

The use of the boilers should be such that the steaming hours will be, as nearly as practicable, the same for each boiler. By this it is meant that the total steaming hours per boiler since commissioning of the ship will remain nearly equal.

It is good engineering practice to stagger the steaming hours of the boilers, between cleaning periods, so that all boilers will not need cleaning at the same time. The only time when it is practical to clean all boilers at once is when the engineering plant is secured for a routine naval shipyard overhaul period.

SECURING BOILERS

The following instructions for securing boilers apply in general to most modern express boilers. Special additional instructions for securing superheaters are given after the general procedure has been indicated. The information given here must, of course, be supplemented by specific instructions issued by the Engineer Officer and the manufacturer of the equipment.

In general, these steps are to be followed in securing a boiler:

1. Blow tubes, after permission has been obtained.

2. Close the atomizer valves, one at a time, and at the same time close the air registers.

3. Slow down the fuel oil service pump, as the burners are being secured.

4. Secure the steam to the fuel oil heater.

5. When all burners have been secured, stop the fuel oil service pump.

6. Slow the forced draft blowers, but continue to run them until the furnace has been cleared of all gases of combustion.

7. Open the fuel oil recirculating valves, and leave the fuel oil system lined up as though for recirculation, until the fuel oil heaters have cooled. This procedure will allow the hot oil in the heaters to expand without lifting the relief valves.

8. Remove the atomizer assemblies, and clean them as soon as possible.

9. Close the furnace tightly, to keep cool air from flowing into the hot furnace. Sudden cooling causes serious damage to furnace refractories and to tubes.

10. When the boiler has cooled enough so that it is no longer generating steam, close the boiler steam stops and fill the boiler to the three-fourths level in the water gage glass.

11. Secure the feed pump and the feed system, when feed water is no longer needed. Caution: Do NOT secure the feed system while the boiler is still generating steam!

12. Clean out the accumulated oil in all drip pans, and wipe down all machinery and floor plates. Clean out any oil accumulations in bottoms of air casings, and wipe down any drips on the inner fronts of air-encased boilers.

13. When all fires are out in the boilers leading to one stack, and when the stack has cooled sufficiently, put on the stack cover.

Securing Uncontrolled Superheat Boilers

When securing a boiler with uncontrolled superheat, open the superheater protection exhaust valve BEFORE the steam flow through the superheater has stopped. Keep this valve open until the steam drum pressure has dropped to 100 psi or less.

Securing Superheat Control Boilers

When securing a superheat control boiler, follow these special instructions:

1. Drop the superheat temperature to about 600" F, at the rate of about 50° F every 5 minutes.

2. Secure the superheater fires.

3. Secure the saturated-side fires.

4. Close the boiler steam stop valves.

5. Bleed the superheater until the boiler has ceased generating steam.

IN-PORT STEAMING

Ship's service turbogenerators should, whenever possible, be operated on steam taken through the superheater of a boiler regardless of whether or not the superheater is being fired. This practice affords maximum protection to the superheater. In ships not fitted with separate turbogenerator steam lines direct from the superheater outlet, or with desuperheaters, the following procedures should be used in order to protect the propulsion turbines:

1. On vessels where the steam piping system is so arranged that steam can be prevented from entering the gland system of the propulsion turbine, a section of the main steam line must be used.

2. On vessels where the steam piping is not so arranged, the steam for the turbogenerators must be taken from the auxiliary steam line.

Avoid rapid raising of steam. Except in emergency it is advisable to take from 1 to 2 hours to raise steam in a water tube boiler, depending on the size and type of boiler. If the brickwork is new a longer period will be required. With express type boilers, rapid raising of steam from a cold boiler is very injurious to the brickwork, since rapid changes of temperature tend to spall and loosen the firebrick.

With two-furnace single-uptake superheat control boilers, hot gases from the saturated side penetrate the

superheater and make necessary certain restrictions on the rate of firing of the saturated side when steam is not flowing through the superheater. These restrictions have been pubhshed for each class of boiler. When steam is raised rapidly, extra precautions must be taken to safeguard the superheater. Circulation should be provided by continuously blowing superheater drains to the bilge, by using auxiliary exhaust or other connections if installed, and by operating a turbogenerator if possible.

CARE OF ECONOMIZERS

When boilers are equipped with economizers having aluminum gill-ring extended surfaces, special care must be taken to prevent torching, and melting of the aluminum. Soot fires are likely to occur if excessive amounts of soot are allowed to collect in the economizer tube nest. To reduce the danger of such soot fires the economizer should be maintained as free of soot as possible; in addition, sparky fires, most prevalent at low rates, and high quantities of excess air are to be avoided.

When a boiler is steaming, water must be fed continuously to the unit in order to avoid formation of steam in the economizer with subsequent overheating of the extended surfaces. During low-rate steaming in port, economizers should be vented as often as necessary to prevent steam pockets from forming in the upper tubes.

When raising steam rapidly during emergency lighting-off procedures, admit water to the economizer even though the water level in the steam drum becomes so high as to necessitate surface blows. It is better to overfeed in this manner than to risk overheating and damaging of the economizer because of lack of circulation before the boiler is placed on the line.

BURNING OTHER THAN NAVY SPECIAL GRADE FUEL OIL

It sometimes becomes necessary to burn other than Navy Special fuel oil, either because of inability to obtain Navy Special fuel oil or because of the necessity of dis-

posing of contaminated Diesel or other light fuel oil. With other than Navy Special fuel oil, tank heating may be required to permit pumping, and more frequent strainer cleaning may be necessary. Light fuel oil will require little or no heating. Precautions should be taken during all operations to avoid excessive fuel oil temperatures, because high temperatures are conducive to heater fouling, especially if the fuel oil is unstable. Present fuel oil specifications, however, minimize the possibility of naval ships being furnished with unstable fuel oils.

If it becomes necessary to burn DISTILLATE or Diesel fuel oils, changes in operating procedure and special precautions are necessary to obtain optimum results. The following instructions should, therefore, be observed:

!• Under normal conditions distillate fuel will not require preheating prior to combustion, the fluidity of such oil being adequate to ensure efficient atomiza-tion at temperatures above approximately 35° F.

2. A slightly wider flame angle may be expected when burning distillate fuel oil because of the more rapid burning characteristics of the fuel. The position of the atomizers and the register door adjustment must, therefore, be checked and reset as necessary to prevent the spray striking the furnace opening ring and to secure the best operating conditions.

3. Atomizer assemblies should be examined to ensure the integrity of metal joints, as the lower viscosity of distillate fuel increases the danger of leakage.

4. Because of reduced atomizer capacity attendant with the use of distillate fuel oil, it may be necessary to use a larger size sprayer plate or a greater number of burners for a given rating than was necessary with Navy Special Grade fuel oil. Where special sizes of sprayer plates have been furnished to enable full-power operation with distillate fuel oil, such plates must not be used under any other condition of operation.

5. Because there is an increase in fuel pump slippage when distillate fuel oil is used, it may be necessary to use a higher pump speed to maintain the desired fuel oil pressure. The same result may be accomplished by using booster pumps to maintain a positive pressure on the service pump suction. Complete isolation of Navy fuel oil and Diesel oil should be effected insofar as practical. However, in cases where mixtures of these fluids are encountered or where it becomes necessary to dispose of contaminated Diesel or other light fuel, their handling and combustion may be accomplished without great difficulty if the following precautions are taken:

1. Isolate these mixtures to prevent contamination of other oil.

2. Schedule such mixtures for immediate use.

3. Avoid long storage in fuel oil tanks.

4. Empty the tank completely of the mixture.

5. Upon securing a burner, immediately remove and clean the atomizer. Shift atomizers more frequently where carbon formations are interfering with the quality of the atomization.

BOILER OVERLOADING

Three principal features limit the capacity for any given boiler operation—the "end points" for WATER circulation, for MOISTURE CARRY-OVER, and for COMBUSTION.

The end point for water circulation is affected by the following factors:

1. Location of burners, whether in the end or side of the boiler

2. Arrangement of baffles in the tube banks

3. Arrangement and size of downcomers

4. Arrangement of tubes in the tube banks

5. Steaming rate—that is, the rate of steam generation (pounds per hour) or of fuel consumption (expressed usually in pounds per hour per square foot of generating surface)

The end point for MOISTURE carry-over in steam is affected by special arrangements of baffle separators and screens which are installed in modem boilers to separate moisture from the steam before it enters the dry pipe.

The end point for COMBUSTION depends upon the maximum amount of fuel that can be burned properly and efficiently in a given boiler; this in turn depends upon the amount of air which can be forced into the furnace, the ability of the burner apparatus to mix this air with the fuel, and the volume and shape of the furnace itself. In a properly designed boiler, the end point for combustion should occur at a lower rate of steam generation than the end point for moisture carry-over, and the latter should occur before the end point for circulation is reached.

Because of the possible adverse effect on the circulation of water in a boiler, with subsequent warping or failure of tubes, DO NOT fx)rce a boiler, except in an EMERGENCY, to a rating higher than that required to obtain the full power of the ship with all boilers in use. However, in cases where specific "assigned maximum firing rates" have been authorized, these latter rates must govern.

Be particularly careful to guard against overloading a boiler when the ship is steaming on less than a full complement of boilers.

For each boiler installation, BuShips (usually upon the recommendation of the Naval Boiler and Turbine Laboratory) has established an allowance of ratios and sizes of sprayer plates to be used. Included in this list of sprayer plates is a size of sufficient capacity for boiler overload steaming rates. Sizes of sprayer plates larger than those on the allowance list should not be used.

Importance of Boiler Steaming Rates

A Boilerman 1 or C should have a good understanding of the amount of load that can safely be placed on the boiler under various conditions of operation.

Great emphasis is placed on the full-power speed of a ship. Although full-power speed depends upon a number of factors, the important one, as far as the boilerman is concerned, is the full-power rating of the boiler. When full-power runs are held, the designed full-power rating should not be exceeded, merely to increase the speed of the ship. In addition to being dangerous engineering practice, exceeding the full-power rating in this manner will give false information as to full-power speed. This false information could be passed on and on, and later result in over-firing and subsequent casualties.

Checking on Boiler Steaming Rates

For practical everyday shipboard use, the best method of determining the load on a boiler is to check on fuel oil consumption. The number of gallons of oil per hour can be read from the fuel oil meter, and under steady steaming conditions the meter reading can be used to determine the load. But there are times when the boilers are steamed at a very high rate for short periods of 15 or 20 minutes. In this case the fuel oil meter reading would not give an accurate estimate of the maximum boiler steaming rate during the hour period.

The amount of oil that should be used by the boiler at a specified firing rate for a specified period can be computed from the sprayer plate capacity curves, knowing the size and number of sprayer plates in use, and the fuel oil pressure. Knowing the amount of oil actually used, you can compute the percentage of load on the boiler. You can also figure out in advance the steaming rate and the boiler load that will be obtained under any firing conditions. You will then know the number and size of sprayer plates, and the oil pressures, that should be used for any specified firing rate. From a number of such computations you could make up a table that would show, in percent of full load, the load on the boiler for any firing set-up.

Before attempting to compute the amount of oil that will be fired in a boiler under specific conditions, it would be well to do a little research work and further study. Such good reference material is available in the manufacturer's instruction books, fuel oil sprayer plate curves, and the Navy Boiler and Turbine Laboratory report on test runs for your type of boiler. The manufacturer's instruction book will give you detailed information on the designed rate of combustion for your boiler, and will also give you the oil rate, oil pressure, air pressure, and burner adjustment for various speeds.

There are usually blueprints which give sprayer plate capacity curves for the allowance of sprayer plates carried aboard your ship. Curves on sprayer plate capacities are included in manufacturer's instruction books. For a given oil pressure at the burners, the capacity curves will give you the amount of oil burned per hour for each size of sprayer plate.

The Navy Boiler and Turbine Laboratory test report should be in every engineer's log room. This report consists of data compiled from the various tests made on your type of boiler before it was accepted by the Navy. The book also contains the results of a large number of test runs made on your type of boiler under various conditions of load. This information will be useful to you when checking up on your boiler operating loads and steaming rates.

Boiler Full-Power and Overload Ratings

Each boiler used for propulsion in the Navy has a full-power rating and an overload rating. These ratings can be found in your manufacturer's instruction book. For example, the information given for a superheat control boiler on a destroyer is indicated in table 1.

Similar data for your own type of boiler will give you detailed set-up information on firing your boiler under full-power and overload steaming conditions.

> The numbers under caption "Air Door Adjustment" designate number of notches closed from wide open (W. O.) position.

The important consideration is that a boiler should not be steamed beyond the overload limit. A Boilerman 1 or C in charge of a fireroom should have a good understanding of the relationship between the combustion or firing rate of a boiler and the load on the boiler.

Checking on Boiler Loads

It's a good idea to check on the performance of your boiler, especially the full-power trial runs of your ship. By obtaining the necessary figures from the full-power trial form or the fireroom operating log you can determine the actual loads on your boilers. There may be a possibility that there is a discrepancy between the full-power rating of the ship and that of an individual boiler. Normally, the requirements for the full-power speed of

207

the ship should check with the full-power rating of your boiler.

Take for example a ship with the same type of boilers shown in table 1. Suppose that the ship has just completed a full-power run and you want to check on the performance of number 1 boiler. The fireroom operating log showed that fuel oil consumption during the period of 1 hour was 1,260 gallons.

A quick check can be made on the percentage of full load of the boiler. The weight of the fuel oil is approximately 8 pounds per gallon. This figure varies according to different conditions (see chapter 40 of BuShips Mamial), but 8 lb/gal is sufficiently accurate for practical purposes. The figure 1,260 gal/hr is multiplied by 8 lb/gal; this gives us 10,080 Ib/hr. The full-power fuel oil combustion rate in table 1 is 9,220 Ib/hr. To obtain the percentage of full-power load, the following computation is used:

A ^ 1M80 10,080X100 ^
100 9,220 9,220

This 109.3 percent indicates that the boiler is being steamed at approximately 10 percent overload in steaming the ship at full power.

In checking the possible causes of this 10-percent overload, there are several things to take into consideration. The firesides and the watersides of the boiler may be dirty. The boiler may not have been fired properly, to obtain the most efficient combustion in the furnace. The sprayer plates may be worn or damaged, and are giving unsatisfactory service. The fuel oil may have contained sludge, or water in excess of specifications. The fuel oil meter may be in need of an overhaul, or the tank soundings may have been inaccurate. These and other items should be considered and investigated, if practical. In any case, you should inform your division officer or the engineer officer of any discrepancy between the actual and designed full-power rating of the boiler.

A further detailed check on the boiler performance can be made by using the following table:

Table 2.— Boiler Operation Data —/
Boiler No. 1
(Known data from Oper. log)

Fuel oil reading (gal/hr)
No. of burners
Size of sprayer plates
Fuel oil pres.sure (psi)
(Computed data)
Lb of oil per burner (sprayer plate)..
Lb of oil per side
Lb of oil for boiler Cial of oil for boiler.
Percent of full power of boiler.
8. H. side
3
2712-D 264
1, 590 4, 770
Sat. side
4
2912-D 247
1, 330 5, 320
> Full power is 9,220 lb of oil per br.

From the fireroom operating log you can copy the "known data" (see table 2). Then you can compute the additional data as shown under the heading "computed data." In this work you will need the Sprayer Plate Capacity Curves (fig. 7-3) for the boiler installation on your ship.

In table 2, the fuel oil pressure on the superheater (S. H.) side of the boiler was 264 psi. In figure 7-3, locate the line which gives you the reading of 264 psi on either side of the graph. With a straight edge, follow this horizontal line until it intersects with the sprayer plate curve (which is in this case plate number 2712-D). From this point of intersection follow the vertical line to the bottom of the graph. This will give you a figure of 1,590 pounds of oil per hour. This value is for one

OIL PRESSURE AT BURNERS-LB. PER SQUARE INCH

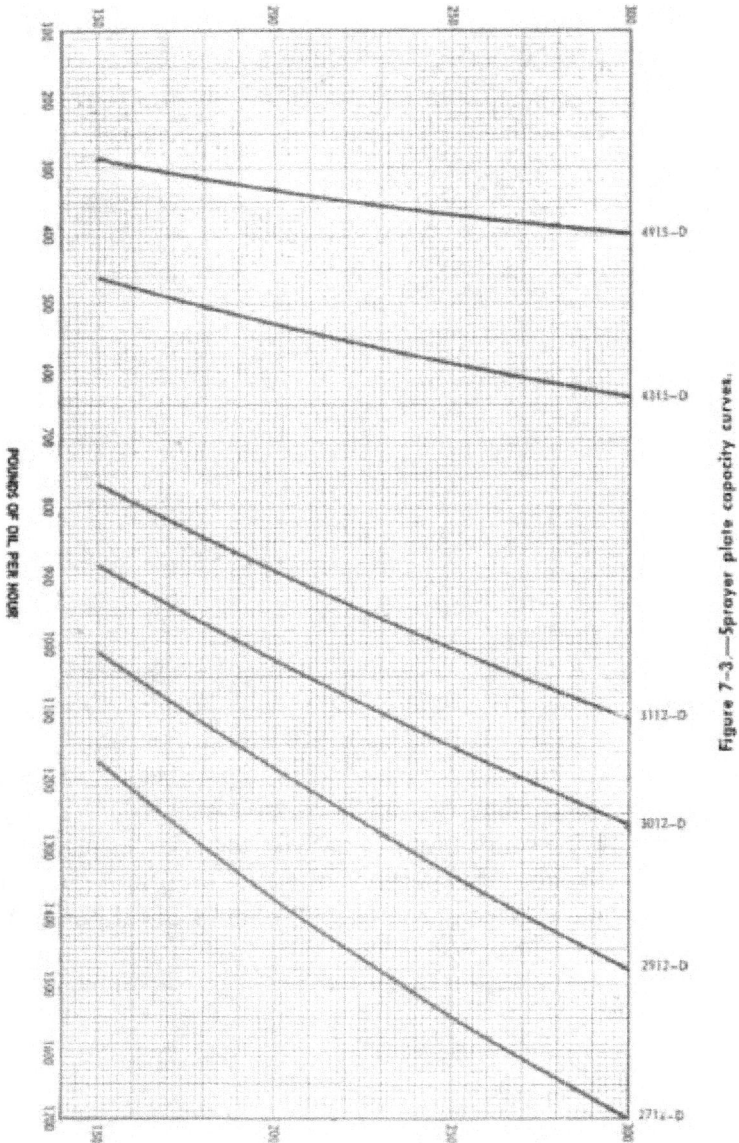

Figure 7-3.—Sprayer plate capacity curves.

burner or sprayer plate. Since 3 S. H. burners were used, multiply 1,590 by 3; this gives an oil consumption figure of 4,770 pounds for the superheater side of the boiler.

The same procedure is used for obtaining the figures for the saturated (Sat.) side of the boiler. The two figures, 4,770 and 5,320, added together give the sum of 10,090 pounds of oil per hour. Dividing 10,090 by 8 (lb/gal) you obtain the figure of 1,261 gallons of fuel oil per hour, which checks very closely with the figure of 1,260 gallons per hour copied from the fireroom operating log. The correspondence between the figures will give you an idea of the accuracy of the fuel oil meter and fuel oil tank soundings.

To obtain the percentage of full-power load you proceed in the same manner as in the previous example of checking. The figures of 10,090 and 9,220 pounds of oil per hour should give you 109.4 percent of full power. This checks with the figure of 109.3 obtained in the previous example of checking the percentage of boiler full-load power.

Steaming With Reduced Boiler Power

During most of the time when the ship is under way the engineering plant is operated at reduced power with one or more boilers secured. When steaming under this condition it is easily possible for the main turbines to consume more steam than the boilers in use can safely supply.

When the ship is maneuvering, and especially during high rates of acceleration in speed, the boilerman in charge of the fireroom must be careful to see that the temporary load on the boiler does not exceed the allowed limits as specified in Engineering Department Orders. Sometimes these instructions are not specific in detail, and it is up to the boilerman in charge of the fireroom to determine what constitutes the allowed limits.

Checking the amount of fuel oil used per hour does not take care of cases where the boiler is steamed at high

rates for 10 or 15 minutes during the hour period. The manufacturer's instruction book will give data on full power and overload set-ups. For short periods of time the load on the boiler has to be checked by the oil pressure and the sprayer plates in use. From the oil pressures and the number and size of sprayer plates you can roughly compute the load on the boiler for any given instant.

For the purpose of explanation we will take the same type of boiler used in previous examples. Assume that a destroyer operating on 2 boilers answers an emergency bell for a period of 5 minutes. In the engineroom is a new throttleman, who does not follow the standard acceleration table. He opens the throttle valve to give the ship full power ahead. In order to keep the steam pressure up, the fireroom force cuts in all large burners and brings up the oil pressure.

In this assumed case, let us take a check to see what is happening. The most important question is—What is the steaming load on the individual boilers? To find this out we must check on the number and size of sprayer plates in use and at the same time the oil pressure. This data is jotted down on a piece of paper; then at a later time the details can be figured out, and a boiler operation data sheet (table 3) can be drawn up.

The final figure in table 3 shows that the boiler was operated, during the 5-minute period, at 126.8 percent of full-load power—that is, the boiler was steamed 6.8 percent beyond the maximum allowed limit.

The Chief Boilerman in charge of a fireroom should make sure that the maximum firing data is posted on or adjacent to the boiler. This information should include the data shown in the manufacturer's instruction book (120 percent), for the overload operating condition. Personnel in charge of a fireroom watch should make sure that the maximum operating data for the firing (combustion) rate is not exceeded. Fireroom personnel should

Oil pressure.
(Computed information) Lb of oil per sprayer plate.
Lb of oil per side
Lb of oil per boiler
Gal of oil per boiler
Percent of full power of boiler.
290
1, 670 5, 010
290
1, 670 6, 680
11, 690 1, 461
126. 8

be instructed in the dangers of overloading the boiler by using an excessive boiler firing rate.

PROCEDURE IN CASE OF LOW STEAM PRESSURE

When running at high speeds, particularly when attempting smokeless operation, you may have a drop in steam pressure. If this occurs, demands for steam must be cut down decidedly and without hesitation, until the pressure begins to build up; otherwise, the operation of blowers, pumps, and other auxiliaries which control the supply of air and fuel will become slower and slower. Ships have been brought to a full stop with all fires out as the final result of bleeding the steam from the system by keeping the main engine throttles open too wide. So unless a boiler is to be secured or cut out, do not allow the steam ressure at any time to fall to less than 85 percent ol ne authorized boiler operating pressure. The rise in spe Jfic volume of the steam at a reduced pressure will be likely to interfere with the circulation in the boiler, with consequent danger of overheating the boiler at high firing rates. If special circumstances should make it necessary for a boiler to continue to furnish steam after the pressure has dropped to below 85 percent of the

213

authorized boiler operating pressure, the boiler must not be operated at a rating of more than one-half of full-power capacity.

CARBON FORMATIONS

Carbon on Atomizer Tips

Carbon on atomizer tips may be a sign of improper location of the tips. If the atomizer is inserted too far in, an eddy is formed in the entering air currents. A fine fog of oil is drawn back behind the face of the tip and is deposited back on the atomizer pipe. This latter is an extreme case, where the atomizer has been inserted several inches beyond the proper point. In extreme cases jets of ragged flame are also drawn back by this eddy, sometimes completely hiding the tip from view.

Carbon on Furnace Openings

Carbon formation on the edges of the furnace openings indicates that atomizers have been drawn back in the registers too far or—as is often the case—that because of patching of the brickwork, the furnace opening is irregular. Nearly closing air doors while maintaining relatively high or double casing pressures is a cause of carbon deposition. Care should be taken to avoid undue throttling of the register air doors, especially during in-port operation. In extreme cases carbon deposition may be sufficient to intercept the particles of atomized oil and cause them to pass backward and downward inside the double casings as an oil drip, with consequent fire hazard.

Carbon Deposits on Tubes and Furnace Bottom

Deposits of hard carbon on tubes and on the furnace sides and bottom are due to the oil striking the tubes or the brick before it has had time to burn. This is frequently experienced even with the light grades of oil, and is caused by two defects—one a defect in design, the other in operation.

In older boilers, the wing and bottom burners are sometimes placed too near the tubes and the furnace bottom. As a result the oil strikes these surfaces and is carbonized before it has time to form an intimate mixture with the air and to become completely consumed. Because of the necessity for cutting down to a minimum the furnace volume in naval boilers, this trouble has frequently been encountered. However, burners in recent boilers are so designed and installed that difficulty from this cause will be minimized provided the atomizers are properly adjusted.

Under operation the causes that produce carbon deposits are the same as those that produce excessive soot and smoke—namely, the failure to obtain an intimate mixture of the entering air and the oil at the register. Carbon deposit is more often caused by the improper location of the atomizing tips than by any other factor. When excess air enters around the cone of oil the flame cools down so that combustion is not obtained until the spray has passed some distance back in the furnace. As a result the particles of oil forming the outer surface of the oil cone strike the furnace bottom, the side walls, or the tubes near the front of the furnace, and the oil cools below the ignition temperature. The heat causes a coking of the hydrocarbons and coke adheres to the surfaces as a solid mass. A similar action occurs when there is insufficient air for complete combustion, but this need not be considered, as the tendency aboard ship is toward a great excess of air. If an excess of carbon forms on the tubes or the furnace walls, the burners should be carefully checked to ensure that the atomizers are in the proper position as prescribed in the instruction book for the installation in question.

POSITIONING OF THE ATOMIZER BARREL IN THE DISTANCE PIECE

The position of the atomizer barrel in the distance piece should be checked and adjusted, if necessary, by screwing

the distance piece (jacket tube) inward or outward until the proper relationship exists between the face of the atomizer tip or nut and the diffuser. (See fig. 7-4.)

The distance pieces of registers of air-encased boilers are marked so that the position of the atomizer-distance piece assembly with reference to the burner throat can be determined from the fireroom. Operating instructions

jfL vtt-cti MTUHII

1> r~ no* *ro>iii(»

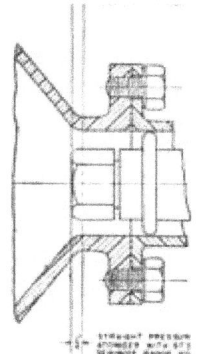

«I(-MI WruMi
>iM «ronin
ATOWXCn *|TM IT'D »
TOOO 'MEX-FStSS* REOISTEI*

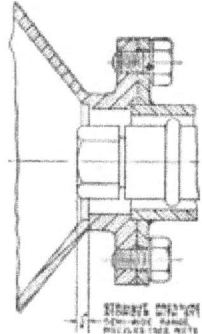

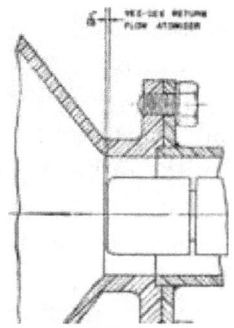

,^ mm VM'm, o.

hOZUl% (MC NOTE II

TOOO BULKHEAD ANO AIR ENCASED* 'RACIAL-CYLINOeRiCALr REGISTER
uncii M

TOOO COMMERCIAL TYPE 'DUO-PRESS' REGISTER

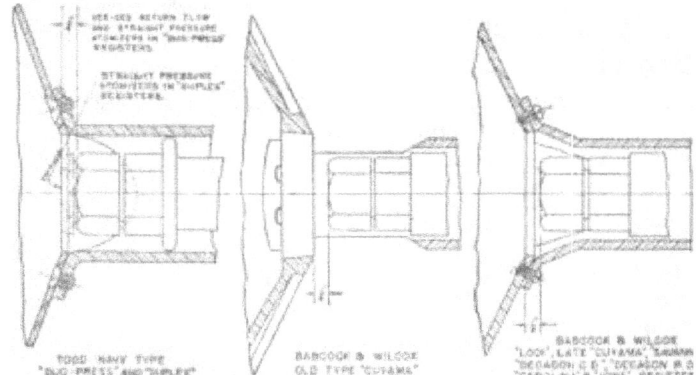

TOOO NAVY TYPC •OUO-PRESS-ANO-OUPtCK* lieSlSTCRS

BA8C0CK a WILCOX OLD TYPE 'CUYAMA* REGISTER

BABCOCK a WILCOX "LOD<".LATE "CUYAMA'.'^SiWANNAM; "DECAGON C O'.'OeCAGON R 0', 'CAROLINA'a'IOWA' REGISTERS

•OiLCat MCMIASt TMt

0iMCNt»oii ro S/«T

Figure 7-4.—Atomizer adjuttment.

covering the individual boiler-burner installations should be studied and closely followed.

LAYING A SMOKE SCREEN

On combatant ships, each boiler is provided with a special smoke-screen burner. Regular burners must not be used for making smoke if smoke-screen burners are fitted. A smoke-screen atomizer is similar to a standard atomizer, but it has a longer barrel and a special nozzle. A large-capacity sprayer plate is used with the smokescreen burner.

A special opening in the boiler front admits the end of the atomizer assembly; the assembly is fitted so as to prevent the entrance of air. Thus the oil is forced into the firebox without being mixed with air. Since the oil cannot burn completely, smoke is formed.

When a smoke-screen burner is being used, excess oil tends to accumulate on the furnace floor. In order to help prevent the accumulation of unburned oil, the regular burners adjacent to the smoke-making burner should be in use at the same time so that the heat from the regular burners will vaporize the oil from the smoke-making burner.

When the smoke-screen burner is used, soot tends to accumulate in the tube banks, economizer, uptake, and stack. Soot accumulations interfere with heat transfer and, in addition, may constitute a serious fire hazard. If the soot becomes ignited, great damage may be done to tube surfaces, economizer rings, and uptake casings. If at all possible, soot blowers should be used immediately after making smoke. When it is necessary to make smoke for an extended period of time, tubes should be blown at least once each hour, and more often if practicable.

FIREROOM SAFETY PRECAUTIONS

The importance of fireroom safety precautions cannot be overemphasized. As a First Class or Chief Boilerman,

you have a particular responsibility in the matter of safety. You must be entirely familiar with the precautions to be observed in operating boilers and all fireroom auxiliary machinery, and you must be constantly alert to prevent any violation of these safety precautions.

Safety precautions for various auxiliary units are discussed in the appropriate chapters throughout this training course. A few of the most essential safety precautions pertaining to boilers are given below. These precautions must, of course, be supplemented by safety precautions prescribed by the Engineer Officer on your own ship, and by precautions listed in chapter 51, BuShips Manual.

The following precautions must be observed PRIOR TO

PLACING A BOILER IN SERVICE:

1. Make all prescribed tests of safety valves.

2. Test all automatic safety devices to make sure that they are operating properly.

3. In lighting off, run the water just out of sight in the bottom of the water gage glass. Then use the emergency feed pump to bring the water up to about 1 inch in the glass.

4. Always blow through the furnace with air or steam before lighting off. Always blow through the furnace before relighting burners, when all atomizers have been extinguished.

5. On or before lighting the first burner in a boiler with uncontrolled superheat, open the superheater protection exhaust valve and the superheater protection steam inlet valve, to establish a positive flow of steam through the superheater.

6. Never light a burner on the superheater side of a superheat control boiler until one or more burners are in operation on the saturated side. Never light A burner on the superheater side until you are

SURE THAT A POSITIVE FLOW OF STEAM HAS BEEN established THROUGH THE SUPERHEATER.

The following precautions must be observed while the

BOILER IS BEING OPERATED:

1. Do not exceed the authorized maximum steam drum pressure.

2. Do not leave disconnected atomizers in place.

3. At least once each hour, test the drains from the fuel oil heater.

4. As long as a boiler is furnishing steam, do NOT shut off the feed supply even for a short time.

5. Blow through the water gage glasses before the boiler is cut in on the line; at the end of each watch; and whenever there is any question as to the water level in the boiler.

6. Observe the following precautions to reduce the danger of flarebacks:

a. Do not allow oil to accumulate in the furnace. Keep atomizer valves tight.

b. If burners are suddenly extinguished, shut off the oil supply and blow through the furnace with steam or air.

c. Do not use large amounts of excess air.

d. Always use a torch to light off the first burner! Always use a torch to light off additional burners, until the furnace has become very hot! Never attempt to relight burners from hot brick work !

e. When lighting off a burner, stand clear to avoid injury from a flareback.

7. Never blow down the division wall headers, the water screen headers, or the water wall headers until after all burners have been secured.

8. When securing a boiler with uncontrolled superheat, open the discharge connection to the auxiliary exhaust before the steam flow through the superheater has stopped. Keep this connection open until the steam drum pressure has dropped to 100 psi or less.

9. When securing a superheat control boiler, drop the superheat temperature slowly — about 50° F every 5 minutes. Always secure the burners on the superheater side BEFORE securing those on the saturated side.

The following precautions must be observed after the

BOILER HAS BEEN SECURED:

1. Remove atomizer assemblies as soon as possible after securing.

2. Close all openings to the furnace after all burners have been extinguished.

3. After the boiler has ceased generating steam, fill the boiler to the three-fourths level in the water gage glass and close the feed check and stop valves and the steam stop valves.

4. Before removing any fittings or parts subject to pressure, and before loosening a manhole or hand-hole plate fitting, take steps to ensure the complete absence of pressure. Open the aircock and the superheater vents, and test the blow down from the upper water gage cut-out valve.

QUIZ

1. On combatant installations, what percent of boiler full-power capacity is referred to as "boiler overload capacity"?

2. What is "designed pressure"?

3. What term is used to indicate the pressure at the final outlet from a boiler?

4. What three surfaces must be added together to give the total heating surface of any steam generating unit?

5. When the superheater on a superheat control boiler is to be lighted off, how many burners must be in use on the saturated side before the burners on the superheater side are lighted oflF?

6. When a superheat control boiler is operating at full steam temperature, what steam flow through the superheater is considered to be a safe minimum?

7. What term is used to describe the difference between the superheater outlet temperature and the saturated steam temperature?

8. What is the most rapid rate at which superheater outlet temperatures should be raised or lowered?

9. To what temperature should the outlet temperature of the steaming boilers be reduced before an additional boiler is cut in?

10. Which type of superheater requires the most careful handling, particularly with regard to cutting in or securing?

11. What type of oil is best used for lighting off under cold ship conditions?

12. What capacity sprayer plate should be used when lighting off under cold ship conditions?

13. What causes distillate fuel to have a slightly wider flame angle than Navy Special Grade fuel oil?

14. What are the three "end points" which limit boiler capacity?

15. In a properly designed boiler, which of the three "end points" should occur at the lowest rate of steam generation?

16. Under steady steaming conditions, what is the most practical method of determining the load on a boiler?

17. From what factors can you compute the load on a boiler at any given moment?

18. What is likely to result from improper location of atomizer tips?

19. What causes deposits of hard carbon on tubes and on furnace walls?

BOILER EFFICIENCY

SAVING OF FUEL

The military value of a naval vessel depends in large measure on her cruising radius. This in turn depends on the efficiency with which her engineering plant is operated. As a Boilerman First or Chief, you are responsible not only for maintaining the boilers in operating condition, but also for operating them at peak efficiency.

To do this job, you and your fireroom force must have a sound knowledge of the principles of combustion as applied to a boiler furnace. Greater savings in fuel, with resulting increase in steaming radius, can be made in the fireroom than in all the rest of the Engineering Department combined.

The complex design of modern naval boilers necessitates a high degree of technical knowledge on the part of the fireroom personnel. Some of this knowledge is obtained through actual practice in the operation of a fireroom, but the practical experience must be supplemented by a knowledge of the principles of combustion.

COMBUSTION OF FUEL

The rapid chemical union of fuel and oxygen, accompanied by the evolution of heat and light, is called combustion. The combustible components of fuels are mainly carbon and hydrogen, largely in the form of hydrocarbons.

Sulfur is also present in small quantities. The noncom-bustible components of fuels are chiefly oxygen and nitrogen.

In almost all burning processes, the principal reaction is the combination of carbon and hydrogen in the fuel with the oxygen of the air to form carbon dioxide and water vapor. To a lesser extent, sulfur dioxide is also formed from the sulfur in the fuel. Combustion may therefore be considered a chemical process. When a fuel burns, the chemical reaction between the combustible elements in the fuel and the oxygen in the air results in ijew compounds.

In the combustion of fuel oil—that is, the chemical combination of the combustible elements of the fuel with the oxygen in the air entering the furnace—certain chemical reactions must be considered; these reactions are shown in equation form in table 1.

Table 1.— Chemical Reactions in the Combustion of Fuel Oil

In the equations an element is designated by a symbol, and a compound is designated by a combination of the symbols of the elements forming the compound. For example, the equation 2C-f2C02=202 indicates that two parts of carbon combine with two parts of oxygen to produce two parts of carbon dioxide. In the absence of sufficient air to form carbon dioxide, carbon monoxide will be formed in accordance with the equation, 2C +0^=200.

The substances that enter into combustion reactions, and the symbols and molecular weights of these substances, are shown in table 2.

Substance

Oxygen

Hydrogen
Carbon
Sulfur.. _ _
Nitrogen
Carbon monoxide Carbon dioxide _.
Sulfur dioxide
Water vapor
Molecular symbol
O2
c S2
CO CO2 SO, H,0
Atomic weight (approx.)
16 1
12
32 14
Molecular
weight (approx.)
* Carbon exists in various forms.

It must be remembered that the source of oxygen for the combustion reactions listed in table 1 is atmospheric air, and that air is a mixture of oxygen, nitrogen, and small amounts of carbon dioxide, water vapor, and inert gases. The approximate composition of air is shown in table 3.

Table 3.— Approximate Composition of Air

At the proper temperature, the oxygen in the air separates from the nitrogen and combines chemically with the combustible substances in the fuel, producing the reactions indicated in table 1. The nitrogen, which is 76.85 percent by weight of all air entering the furnace, serves no useful purpose in combustion. It is a source of direct loss, since it absorbs heat during its passage through the furnace, and carries off a portion of the heat when leaving the boiler.

224

Temperature and Heat of Combustion

To insure complete and continued combustion and thereby to avoid losses from incomplete combustion, the temperature of the fuel must be kept at the ignition temperature. When combustion reactions take place, a definite amount of heat is produced. The heat of combustion of the fuel is the sum of the heat of combustion of each element in the fuel.

Table 4.— Heat of Combustion of Fuel Oil Elements
Element
Hydrogen
Carbon (to CO). Carbon (to CO2) Sulfur (to S(h)..

Table 4 shows the large difference in the amount of heat of combustion evolved when carbon is burned to carbon monoxide (4,440 Btu), as compared with the heat of combustion evolved when it is burned to carbon dioxide (14,540 Btu). You will find the difference to be 10,100 Btu. In burning to carbon monoxide, the carbon is NOT completely oxidized, whereas in burning to carbon dioxide is has combined with all the oxygen possible, and thus oxidation is complete. This latter fact is most important to you, since incomplete combustion of the carbon in the fuel is a source of considerable loss in the boiler furnace.

Air and Combustion

In the previous discussion we have assumed that the oxygen necessary for combustion was present in the exact amount required to ensure complete combustion of all the combustible elements in the fuel. It must be remembered.

Symbol

Heat of combustion (Btu per pound)

c c

62,000 4, 440

14,540 4,050

however, that the physical introduction of oxygen into the furnace in a manner to ensure complete combustion of all the combustible elements in the fuel is a most difficult problem. All the factors involved in this problem must be understood by operating personnel, if the best results are to be achieved. You have already learned that the source of the oxygen for combustion is atmospheric air, and that this air is composed of approximately 23.15 percent oxygen by weight and 76.85 percent nitrogen by weight. It will be apparent to you, therefore, that to supply 1 pound of oxygen for combustion it will be necessary to supply $^{\wedge}$ 2315 4.320 pounds of air.

The nitrogen in the 4.320 pounds of air will be $0.7685 \times 4.320 = 3.32$ pounds. As stated before, the nitrogen serves no useful purpose in combustion and is a source of direct loss.

Now consider the reactions (table 1) that take place in the combustion of fuel. Notice, for example, that, in the combustion of carbon, $2C + 2O_2 = 2O_2$;.—that is, two molecules of carbon combine with two molecules of oxygen to form two molecules of carbon dioxide. If the molecular weights as given in table 2 are now substituted for the appropriate molecular symbols in the equation above, the equation becomes: $(2 \times 12) + (2 \times 32) = 2 \times 44$. This means that 24 parts by weight of carbon combine with 64 parts by weight of oxygen to produce 88 parts by weight of carbon dioxide. Any weight of carbon dioxide will be composed of 24/88 or 27.27 percent of carbon and 64/88 or 72.73 percent of oxygen, by weight; that is, 1 pound of CO_2 contains 0.2727 pound of carbon and 0.7273 pound of oxygen. From the above it follows that in the

0 7273

combustion of 1 pound of carbon to carbon dioxide q 2727

or 2.667 pounds of oxygen will be required. Since air contains only 23.15 percent by weight of oxygen, 1 pound of oxygen will be obtained from 4.320 pounds of air; the complete combustion of 1 pound of carbon will require

2.667 X 4.320 or 11.52 pounds of air. Since each pound of air contains 76.85 percent of nitrogen, the 11.52 pounds of air will contain 11.52x0.7685 or 8.85 pounds of nitrogen, which will pass up the smoke pipe without serving any useful purpose in combustion. To summarize, the products in the complete combustion of 1 pound of carbon are:

1 pound C + 2.667 pounds O. = 3.557 pounds CO2 2.667 X 3.32 pounds N2 = 8.85 pounds N.

The reactions involved in the complete combustion of fuel oil are analyzed in a similar manner in table 5.

Table 5.— Analysis of Chemical Reactions in the Complete Combustion of Fuel Oil

The amounts of air theoretically necessary for the perfect combustion of the various elements in the fuel are shown in table 5. In practice, however, the amount of air necessary to ensure complete combustion must be in excess of that theoretically required. However, too much excess air will serve no useful purpose, but will absorb and carry off heat.

Combustion and Heat Losses

When fuel is burned in the boiler furnace, the difference between the heat input and the heat absorbed represents the heat losses. These losses may be unavoidable,

avoidable, or—in some cases—avoidable only to a certain extent. Some losses, also, are considered to be normally unaccounted for.

The following heat losses are important in determining boiler efficiency:

1. Loss DUE TO THE MOISTURE CONTAINED IN THE FUEL.

Some fuels contain a certain amount of moisture which must be evaporated and superheated to the flue-gas temperature. To keep this loss to a minimum, you should use every precaution to keep the moisture in the fuel to a minimum.

2. Loss DUE TO THE MOISTURE FORMED IN THE BURNING

OF HYDROGEN IN THE FUEL. The hydrogen in the fuel will unite with the oxygen in the air and form water vapor, which must be evaporated and superheated as in the case of moisture in the fuel. This process is explained in the preceding paragraph.

3. Loss DUE TO THE MOISTURE IN THE AIR SUPPLIED FOR

COMBUSTION. Any moisture in the air must be heated in the manner just stated.

4. Loss DUE TO HEAT CARRIED AWAY IN THE SMOKE PIPE

GASES. This is the greatest of the heat losses. The factors involved in its determination are: (a) the weight of the dry gas, (b) the difference between the temperature of the atmosphere and the lowest theoretical stack temperature (unavoidable loss), and (c) the difference between the temperature of the exit stack gases and the lowest theoretical stack temperature. This type of heat loss may be minimized by keeping the heat transfer surfaces of the boiler clean and by using no more excess air than is actually required for combustion.

5. Loss DUE TO THE INCOMPLETE COMBUSTION OF CARBON.

When the carbon in the fuel is burned to carbon dioxide, the heat of combustion is 14,540 Btu per pound; but when it is burned only to carbon monoxide, the heat of combustion is only 4,440 Btu per pound. In-

complete combustion of the carbon, therefore, may result in a loss of 10,100 Btu per pound. This high loss may be prevented by the admission of air in the proper way and in sufficient quantity to make complete combustion possible.

6, Loss DUE TO UNBURNED HYDROCARBONS, TO RADIATION,

AND TO CAUSES NOT ACCOUNTED FOR. Losses that cannot be measured or are impracticable to measure are: (a) unburned hydrocarbon losses due either to the gaseous or to the solid unconsumed contents of the flue gases, (b) radiation, which will vary with the boiler design, steaming rate, ambient air temperature, and the nature of the installation in the ship, and (c) other losses not normally accounted for.

Boiler Test Data

Tests conducted by the Naval Boiler and Turbine Laboratory showed that increasing the excess air from 10 to 90 percent (at two constant firing rates) had important effects upon boiler characteristics. The effects of increasing excess air percentages may be summarized as follows:

1. Over-all and fireroom efficiencies dropped markedly (about 10 percent).

2. Losses chargeable to the furnace and burners were trebled.

3. Steam consumption of the blowers was doubled.

4. Equivalent evaporation decreased almost 10 percent.

5. CO_2 percentage dropped from 14.5 to 8.5.

It is clear, therefore, that serious heat losses will result from the use of 90 percent excess air. When you consider the fact that, in actual practice, 200 percent excess air is a matter of common occurrence, you can understand that the use of high excess air is the common cause of low operating economy. Efficient fireroom operation requires the reduction of excess air to the minimum which is consistent with complete combustion.

PRESSURE ATOMIZATION

From the inlormation in the preceding paragraphs, it is apparent that boiler efficiency depends largely on the completeness of combustion. Complete combustion is not possible, however, without proper atomization. To accomplish the latter, sprayer plates, nozzles, and registers must be maintained in the best possible condition. Furthermore, the correct range of oil pressure must be maintained and the oil must be heated to the viscosity that will ensure proper atomization. The latter two factors are highly important and will be discussed more thoroughly later in this chapter.

Principle of Pressure Atomization

The pressure atomizers most commonly used in the naval service are provided with passages through which the oil is forced under high pressure. In order that the action of centrifugal force may break up the oil, the passages are grooved (either spirally or tangentially) to give the oil a high rotational velocity. Through these grooves the oil discharges into a small cylindrical chamber, coned out at the tip end, and with an orifice at the apex of the cone.

As the oil leaves the cylindrical chamber by way of this orifice, it is acted upon by three forces—centrifugal force due to rotation, translation along the axis of the atomizer, and gravity. As a result, the oil breaks up into a cone of very fine foglike particles. A strong blast of air, which has been given a whirling motion in passing through the register, catches the oil fog and mixes with it, and this mixture enters the furnace, where combustion takes place.

Atomizing Pressure Range

When the smaller sizes of sprayer plates are used, oil pressures should be maintained between 125 and 300 psi, depending upon the firing rate required. The lower pressure of 125 psi should not be used except to take care of

temporary maneuvering conditions. When the larger sizes of plates are used, every effort should be made to keep the oil pressure between 200 and 300 psi. The higher oil pressures are desirable because they ensure better atomization of the oil.

Atomizing Temperatures

In order to obtain the optimum atomization for efficient burning, fuel oil at the burner manifold must be maintained at a temperature that will give a viscosity of 135 Seconds Saybolt Universal (or 17.2 Seconds Saybolt Furol). Since fuel oils differ in viscosity, the temperature required to provide the prescribed viscosity will also differ for various fuel oils.

Figure 8-1 shows the relationship between viscosity and temperature for Navy fuel oils. As you can see, there is a straight-line relationship between the temperature and the viscosity of oil, when the temperature-viscosity points are plotted on this special type of graph paper. Thus, if you know the viscosity of an oil at any two temperatures, you can find its viscosity at all other temperatures by plotting the two known points on the chart and drawing a straight line which passes through both points. If you know only one temperature-viscosity point for an oil, you can get a close approximation of its other temperature-viscosity points by plotting the one known point and drawing a straight line which passes through the point and is parallel to the other fuel oil lines shown on the chart.

On figure 8-1, notice the line marked optimum burning VISCOSITY. Find the points at which this line intersects the two lines which indicate the range of Navy Special fuel oils, and drop vertical lines from these intersections to the baseline. This procedure gives you the range of atomizing temperatures for Navy Special fuel oils (115° to 145° F). To find the best atomizing temperature for any particular shipment of oil, you must know where the oil falls within the Navy Special range shown on the chart. Find the intersection of the line marked OPTIMUM BURNING VISCOSITY and the line indicating the character-

istics of your particular oil. Then drop a vertical line from this intersection to the baseline, to find the best atomizing temperature.

Figure 8-1 also shows the best viscosity for pumping and transferring fuel oil. In the manner just described, you can drop vertical lines from the appropriate intersections to find the optimum pumping and transfer temperatures for fuel oils. For Navy Special fuel oils, the range of optimum pumping and transfer temperatures is from 75° to 100° F. To find the best pumping and transfer temperature for any given shipment of Navy Special fuel oil, you must know where the oil falls within the Navy Special range shown on the chart. Check the temperature of the oil with a thermometer, and then consult the chart to see whether or not it would be advantageous to heat the oil before pumping or transferring it.

Figure 8-2 is a simplified chart which may be used to determine the correct atomizing temperature for an oil if you know either the Saybolt Universal viscosity or the Saybolt Furol viscosity at 122° F. If the viscosity at 122° F is 200 Seconds Saybolt Universal, or 23 Seconds

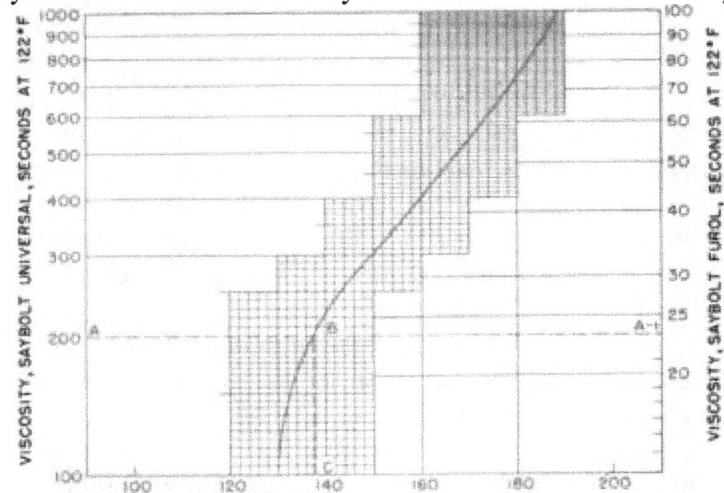

Figur* 8-2.—Atomizing temperature (*F).

.T.M. STANDARD VISCOSITY-TEMPERATURE CHARTS OR LIQUID PETROLEUM PRODUCTS (D 341-39) CHART a: SAYBOLT UNIVERSAL VISCOSITY 30ILER a DIESEL FUEL OILS FOR NAVAL VESSELS

Saybolt Furol, draw a horizontal line from A or A-1 until it intersects the curve at B. Then drop a vertical line from B to the baseline, and take the point of intersection, C, as the atomizing temperature—in this case, 138.5° F.

If you have no chart to work from, and if you know only the Furol viscosity of the oil at 122° F, it is still possible to make a fairly accurate estimate of the correct atomizing temperature for a shipment of Navy Special fuel oil. Consider that the temperature required for best atomization will be equal to 115° F plus 1° F for each second of Furol viscosity at 122° F. For example, if you know that your fuel oil has a viscosity of 24 SSF at 122° F, the best atomizing temperature will be 115 plus 24, or 139° F. This method of obtaining the atomizing temperature does not give as accurate a figure as you can get from using the charts, but it is accurate enough for most practical purposes.

MAINTAINING STEADY CONDITIONS

When your ship is cruising, and often when it is operating in port, the rate of combustion is fairly constant, since the variations in steam demands are very slight under these circumstances. Every effort should be made to attain steady conditions on as many boilers as possible of those in use. The fireroom force should observe the following procedures:

1. Keep the oil temperature as specified.

2. Regulate the blowers to the proper air pressure and keep this pressure steady as long as no change is made in the number of burners or oil pressure.

3. Carry a steady water level.

4. Adjust registers carefully so that all are giving the best results.

5. Keep atomizers clean and adjusted in the proper positions.

6. Make symmetrical distribution across the front of the boiler of the atomizers specified for use, in

accordance with the lighting-off order specified in the boiler manufacturer's operating instructions, or the specifications of BuShips.

7. Maintain a constant steam pressure and temperature at the superheater outlet.

8. Either the engineroom or the fireroom force, as applicable, should maintain constant feed pressure and feed water temperature.

9. When more than one boiler is in use, as many of them as possible should operate at the same rate of combustion; and the same number of burners should be used on each boiler.

10. To ensure best results, use the same size sprayer plates in both furnaces.

QUIZ

1. In what part of the engineering department can the greatest saving of fuel be made?

2. What are the main combustible components of fuels?

3. What are the chief noncombustible components of fuels?

4. In most burning processes, what is the principal reaction?

5. What is the source of oxygen for the combustion reactions which take place in the furnace?

6. What gases are present in atmospheric air?

7. In what manner does nitrogen constitute a direct loss in combustion?

8. What percent of oxygen, by volume, is found in atmospheric air?

9. If you know the heat of combustion of each element in a certain fuel, how can you determine the heat of combustion of the fuel?

10* Oxidation is complete when carbon is burned to what?

11. How many pounds of air are necessary for the complete combustion of 1 pound of carbon?

12. Why is it more desirable to burn carbon to carbon dioxide rather than to carbon monoxide?

13. In practice, is the amount of air necessary to ensure complete combustion greater or less than that theoretically required?

14. When fuel is burned in the boiler furnace, what is meant by heat loss?

15. Which of the heat losses is greatest?

16. Excess air has what effect on the percentage of CO_2?

17. What is the most common cause of low operating economy?

18. What forces act upon the oil as it leaves the orifice of the atomizer?

19. Why are the higher oil pressures desirable?

20. What oil pressure range should be maintained when full power and overload sprayer plates are used?

21. What should be the temperature of the fuel oil in the burner manifold?

CHAPTER

BOILER RETUBING

USING THE RIGHT TOOL

Boiler tube replacement is a most important phase of a Boilerman's work; still more important is the supervision and training of your men to use the correct tools and to do the job properly. As a Boilerman First or Chief you have an opportunity to lay a foundation for future engineering efficiency by training your men in the methods of boiler retubing. With training in mind, this chapter deals with the tools and procedures used in the replacement of boiler tubes, as prescribed in chapter 51 of BuShips Manual, and by the Naval Boiler and Turbine Laboratory, Philadelphia, Pa.

REMOVING TUBES FOR INSPECTION

Tubes that are badly warped, bulged, or in which the belling is likely to give trouble, or that leak after having been rerolled, should be replaced. Never straighten tubes in place, as the joints are likely to become strained and ultimately leak. Possible permanent injury to other parts of the boiler also may result from attempts to straighten tubes in place.

Do not undertake general cutting out or renewal of tubes without first obtaining approval from BuShips. Whenever it comes to the attention of the commanding officer of a ship that a general renewal of tubes in a boiler is required, a report to that effect is made, via

force commanders, to BuShips. A representative specimen of the faulty tubes accompanies this report, except in cases where renewal is made necessary by external corrosion resulting from a known cause.

TUBE FAILURES

Determining the cause of the tube failure is often a difficult task. The most common reasons for the failures are discussed here, to assist you in determining, by visual inspection, the probable cause of failure.

Each type of failure has distinguishing physical characteristics which will, in practically every case, identify the cause. Such evidence includes the nature of the fracture, the size and shape of the opening, the appearance of the edges, and the condition of the tube surfaces. Since there may be more than one cause involved in a specific failure, the visual evidence may be complex. In order to determine the cause of the failure, tubes in the immediate area of the defective tube may also have to be examined.

Oxygen Pitting

The basic cause of oxygen pitting is dissolved oxygen in the feed water. You can prevent or eliminate oxygen pitting by correcting all possible sources of air leakage into the feed system and by maintaining the efficiency of the equipment used for removing air from the feed water. Oxygen attack is less severe when the alkalinity of the boiler water is kept within the prescribed limits both in the steaming and in the idle boilers. Oxygen pitting of the small scattered type is illustrated in figure 9-1.

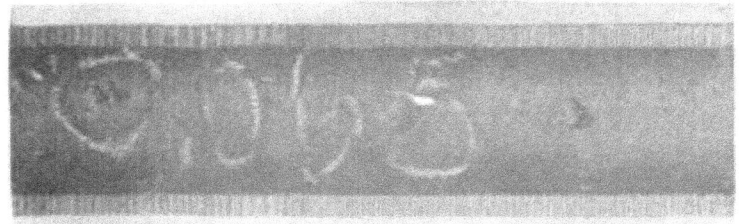

Figure 9-1.—Oxyg«n pitting.

237

Bulging

Bulging results when the tube metal attains a temperature high enough to permit what is known as plastic FLOW. It occurs on the hottest side of the tube; thus, prolonged heating will cause a large bulge and eventual tube failure. When bulging occurs, the tube wall at the bulges is severely thinned. The most common cause of bulging is interior scale, which slows the heat transfer from the tube surface to the water or steam inside the tube.

Forwarding Samples for Examination

Whenever you have extensive tube failures—as, for example, when a condition of low water has existed in a boiler—only a few representative tubes that failed need be forwarded to the U. S. Naval Engineering Experiment Station at Annapolis, Md. Each tube submitted may be cut into convenient lengths for shipping, but each length should be marked so that the sections can be readily reassembled.

Submitting Scale Samples

If unaccounted-for scale is encountered during a boiler inspection, a sample of BOILER WATER and SCALE should be forwarded to the Engineering Experiment Station, Annapolis, Md. The sample should be accompanied by a letter containing complete information on the boiler conditions and the samples submitted. The method used in examining the boilers should be described and any unusual results observed in the water analysis in the period preceding the discovery of the scale should be noted.

When tubes have been removed and split for examination, you can sometimes obtain a large enough sample by removing the scale with a sharp instrument. Otherwise, remove large pieces of scale by clamping the tube piece in a vise and bending the tube back and forth until the scale breaks loose. Be certain that the material removed includes the body of the scale from the tube surface and

not surface scale only. It is preferable to collect scale from tubes taken from the fireside row, where the most severe conditions will be encountered. When submitting samples be sure that you send at least 2 ounces of scale.

In addition to the scale samples, always include one or more tubes representative of the worst conditions found. Each tube submitted should be cut into convenient lengths for shipping, marked as previously described, and identified as to boiler-tube bank, row, and side nearest the fire.

A 1-gallon sample of water taken while the boiler is being emptied, or just prior to its being emptied, should also be submitted with the scale and tube samples. See that the bottle containing this sample is almost full and that it is tightly stoppered. Label the bottle, giving full details on the source of the water and the ship's data on chemical tests made on the water.

Cutting Exploring Blocks

You may find it necessary to remove some boiler tubes in order to determine the condition of the remaining tubes and the necessity for major tube renewal. If you have to do this, remove the tubes in such a way that the resultant space will be in the shape of a rectangular block not more than ten tubes wide, but deep enough to include the center row of tubes.

Whenever you renew the tubes in the A row, you will also have to renew the corresponding tubes in the B row, REGARDLESS of the condition of the latter.

Tube failures in other than the A and B rows generally occur in the outer half of the tube bank, and are the result of external corrosion just above the mud drums. Where such failures have occurred, either during operation or under a hydrostatic test, or when an examination of the tubes removed by the cutting of exploring blocks shows that the tube thickness is less than half the original thickness, you should renew all the tubes, regardless of their condition, from the center row to the outer row inclusive, over a suflftcient length of the tube bank to ensure a thorough coverage of the affected area.

All tube renewals should be recorded on boiler tube data sheets similar to those shown in chapter 18 of this training course. The record should show all tube renewals, the dates of renewal, the tubes that have actually failed during the operation, and the causes of such failures. It should be filed with the boiler record sheets. Either a copy of this chart, or a written summary covering all failures and renewals since the last report was submitted, should accompany all requests for authority to retube or to cut "exploring blocks." An accurate record of tube failures and renewals is essential in carrying out a tube renewal policy that is progressive and economical, and that gives reasonable assurance of the reliability of a boiler.

PREPARING FOR TUBE RENEWAL

As in any repair job, a most important factor in a boiler retubing job is the preparation for the actual repair operation. Your preparation for this task should include:

1. Checking to see that the proper SPARE tubes are available

2. Checking and sometimes changing the watch list to assure that adequate personnel are available for the job

3. Selecting the proper equipment

4. Examining the equipment to see that it is in good working condition

For this job the cutters must be sharp, the air hose and fittings for pneumatic tools in good condition, the electric extension cords properly fitted, the junction boxes and switches in order, and the staging of proper length and condition.

Steam Drum

The steam drum will have to be opened and a sufficient number of steam baffles and fittings removed to permit access to the affected tube ends.

Precautions: Make a personal check to see that all cross-connection stops between the boiler being repaired

and any steaming boiler are locked or wired in closed position and tagged: danger — do not open. Remember that your men must not enter the secured boiler until all rules of venting have been complied with. The portable blowers should be kept running at all times while men are working in the boiler.

Removing Defective Tubes

Using an air-powered side cutter, you should cut the defective tube flush with the water drum or header, as illustrated in figure 9-2. Be careful not to damage the

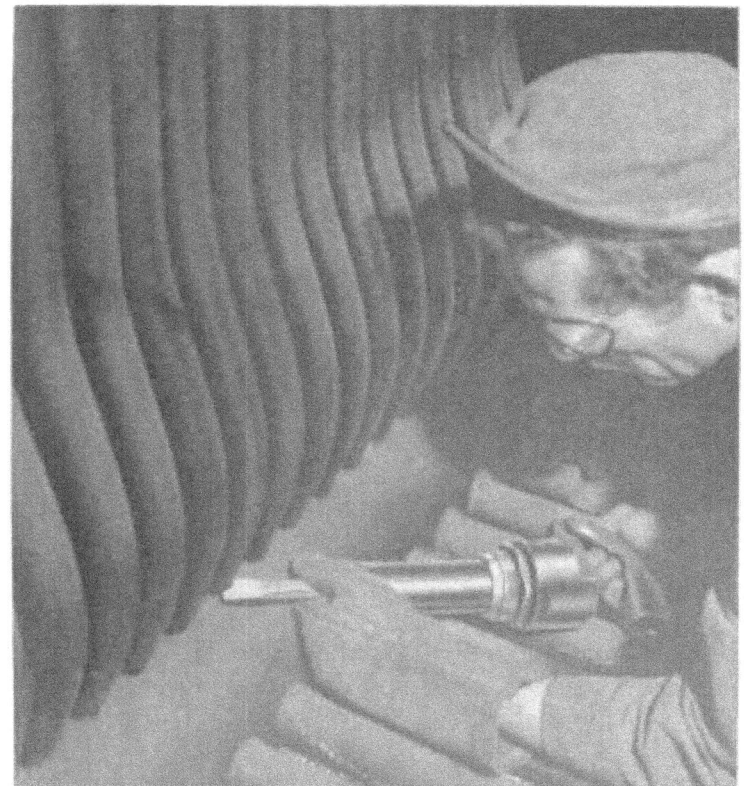

Figure 9-2.—Cutting the tube with an air-driven side cutter.

surface of the header or the water drum. Practice will teach you to RIDE the bottom level of the side cutter on the drum or header surface. By doing this you will not cut into or gouge these surfaces.

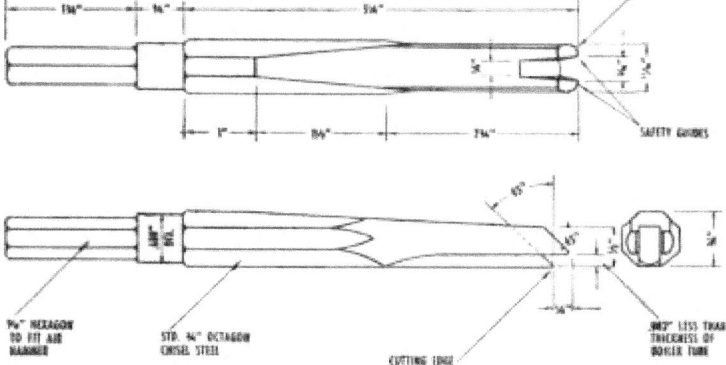

Figur* 9-3.—Safety beiUr tub* ripping chis*l.

Figure 9-4 —Bocking-out tool.

After removing the defective tube from the boiler, you next make a cut on the inside of the remaining portion of the tube with a "safety ripping chisel" (fig. 9-3). This tool is so designed that it cannot cut through the tube and score the tube sheet.

After the tube has been cut approximately three-fourths of the way along the tube sheet, crimp the edges of the tube, and drive out the stub with a backing-out tool. (See figs. 9-4 and 9-5.) No harm will come to the tube-sheet hole, as the diameter of the backing-out tool is slightly less than the diameter of the tube-sheet hole.

Figurs 9-5.—Using the backing-out tool.

If a safety ripping chisel is not available, remove the tube by means of an air-driven round-nose gouge; as illustrated in figure 9-6, you can lift a section of the tube with the sharp end of the gouge riding almost parallel with the tube being cut. If the gouge is used in this manner, all the metal making up the wall thickness of the tube will not be cut with the tool. Many times when the gouge has cut to about two-thirds of the thickness of the tube sheet, the tube will ride out through the hole in the sheet

Figure 9-6.—Using the round-nose gouge.

without further operations being necessary. But if the tube does not loosen after one or two cuts have been made, drive it out with the air-driven backing-out tool. If a backing-out tool is not available you can crimp the tube, and then drive it from the tube sheet with a blunt-nose tool or a tool ground to conform to the curvature of the tube-sheet hole.

CLEANING AND INSTALLING REPLACEMENT TUBES

Before replacement tubes are set into place, clean them thoroughly to remove preservatives and foreign matter. Best results can be obtained by immersing the entire tube in a standard cleaning solution, A satisfactory cleaning

solution can be made by mixing boiler compound to saturation with hot water, and then adding a small portion of kerosene. If a tank is not available or material to build a tank is not to be had, you may clean tubes by improvising a steam jet siphon, using small-size pipe fittings and a flexible steam hose that can draw from a drum or bucket containing either of the solutions above.

Another method used on some ships and stations is to immerse the tubes in a bath of Diesel oil and swab them with rags. Regardless of the method used, be certain that the tubes are really cleaned, both inside and out.

Figure 9-7 shows a furnace view of the warped generating tubes of a boiler that has been retubed and then steamed at a fairly high rate, without the heavy preservative on the watersides of the tubes being removed.

After the tubes have been thoroughly cleaned, prepare the tube ends by further cleaning the outside with medium fine grit emery cloth to a distance of not less than 3 times the thickness of the tube sheet. The tube end should be

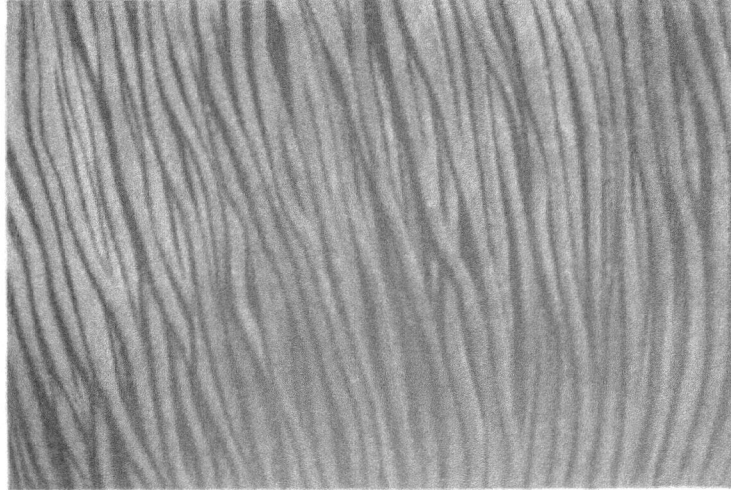

Figur* 9-7.—Warping due to failure to romovo procervativos.

rounded sufficiently with a file so that no square or sharp ends are left. This is done to prevent the tube from splitting when belled.

Preparation of the Tube Sheet

Prepare the tube-sheet hole prior to inserting a replacement tube. This preparation is best accomplished by using a piece of hardwood turned to a diameter slightly less than the hole and covered with medium fine grit emery cloth. This wood jig, when passed back and forth through the tube sheet or header hole, will thoroughly remove foreign matter and irregularities. You can then finish the job by using fine emery cloth wrapped around your finger. But don't give the tube-sheet or header hole **a lick and a promise" and call it good enough. See that the hole is clean and smooth before proceeding with the job. After you are certain the emery has done its work properly, remove any remaining foreign matter by wiping the hole with a cloth dipped in Diesel oil.

The tube may now be worked into place. Be certain that the tube extends through the tube-sheet or header holes at least inch. If you fail to meet this requirement, the result will be a poor belling job.

Expanding a Tube

With the tube in place you are ready to expand the ends. Expanding should be done with the expander supplied by the manufacturer for tubes in the particular part of the boiler in which you are working.

There are two types of expanders—the roller type and the BALL-DRIFT type. Roller-type expanders are furnished for use by the ship's personnel for emergency tube replacements. This type of expander gives satisfactory results, but expanding by this method is much slower than with the ball-drift expanders. Ball-drift-type expanders are used by a naval shipyard in the assembly of new. boilers but are not too practical for shipboard use, since

WITH BELLING ROLL

WITHOUT BELLING ROLL

Figure 9-6.—RelltMyp* tubs •xpand«ra.

their operation demands a highly skilled operator. Figure 9-8 illustrates the roller type of tube expander.

Where tubes holes are not easily accessible, as in headers, a series of adapters are furnished. These adapters (fig. 9-9) when rigged to their best advantage will make possible the expanding of any tube.

Carbon-steel tubes should be expanded just enough to secure tightness, in order that ample life may be left in the metal for future expanding should it be found neces-

L0N6 EXTENSION

SHORT EXTENSION

UNIVERSAL JOINT

SQUARE END SLEEVE

Figure 9-9.—^Tube-expqnder adapters.

sary. Figure 9-10 shows how to expand a tube with an air-driven tube expander. The rolling should continue only until the rollers appear to operate evenly all around the tube, with sufficient pressure on the tool.

Figura 9-10.—Expanding a tub*.

You will find that alloy tubes such as chrome molybdenum alloy superheater tubes are relatively hard in comparison with carbon steel tubes, and require much heavier rolling. Roll tubes of these types until they are tight under hydrostatic tests, but do not roll them to the point where they become distorted or damaged. While rolling tubes of this type, you'll find that the rolled portion may begin to FLAKE. But this is no cause for alarm. Leading boiler manufacturers at the present time are of the opinion that tubes made of character 4-6 chrome molybdenum alloy must be rolled almost, or in some cases fully, to the point where flaking occurs. The best criterion of sufficient rolling, however, is tightness under hydrostatic test.

The rolled portions of alloy tubes should be thoroughly cleaned of any flakes that may have developed during the

tube rolling operation. This is best accomplished by using a piece of hardwood turned to a diameter slightly less than the tube opening and covered with medium grit emery. Take care that these flakes do not enter the steam circuit, where they may be deposited on valve seats and damage the valves.

Belling a Tube

As a safeguard against pulling out, bell all tube ends after expanding with a roller- or drift-type belling tool. The drift-type tool is illustrated in figure 9-11. While belling, be careful not to overdo the operation. Regard-

Figur* 9-11.—Belling tool.

less of the size of tube, the tube ends should not be belled more than 14 inch over the outside diameter nor less than Vie inch over the outside diameter. Figure 9-12 illustrates belling a tube with an air-powered belling tool. Bear in mind, however, that some expanders are fitted with belling rolls, as illustrated in figure 9-8, thus making belling as a separate operation unnecessary.

Figur* 9-12.—Belling a tub*.
I

PLUGGING BOILER TUBES

As an emergency measure, it is occasionally necessary to plug a defective boiler tube until such time as it can be replaced. Tube plugs are carried as spare parts aboard ship. The plugs are tapered to the proper shape, and are generally drilled and threaded at the larger end so that they may be removed with a pulling tool.

Before driving the tube plugs into place, be sure that the plugs and the inside of the tube ends are entirely clean so that the plugs will make metal-to-metal contact with the tube. Drive the plugs in far enough to ensure their holding, but do not drive them in so far that they will damage the tube sheet.

The defective tube should be punctured in order to prevent the build-up of pressure in the tube, since high pressure in the tube would tend to blow the plugs out. If the tube contains a brush or some other obstruction article, be sure to puncture the tube both above and below the obstruction.

USING JIGS TO TEACH

Cutting, expanding, and belling tubes are best learned by actually doing the jobs. It is suggested that you make up a jig, similar to that shown in figures 9-5, 9-10, and 9-12, to teach your men the proper methods and procedures in boiler tube replacement. For example, you can set up the jig in the fireroom and conduct a class for men off watch. It is a good idea for you to expand and BELL a scrap tube before you conduct the class. In this way, you can start the instruction in tube removal with a tube ready in the jig.

A good way to conduct a class is to have one man remove a tube, while you explain to him and the class each step of operation in detail, and then to have him clean the end of a specimen tube, and insert the specimen tube in the representative hole in the jig. Next have him expand and bell this tube. A tube is now available in the jig for the next man to remove. After each man has completed the procedure in this manner, it is well to have him go through the operation again without your help and have him explain each detail to the class. The members of the class can act as critics and offer suggestions. By conducting the class in this manner you not only teach proper procedure but also give the student a chance to "get the feel of the operation" and to learn how to teach others.

After the class has secured, leave the jig in place with ample specimen tubes at hand. You'll find to your surprise that men will practice the operation on their own time without supervision, and they will get a big kick out of doing it.

ESTIMATING A REPAIR JOB

At one time or another you may be called upon to furnish the engineer oflftcer with an estimate of the time and material necessary to accomplish a repair job. When this occasion arises you are indeed "on the spot/' But, if you will take the sensible and practical approach to the problem, such an estimate can be quite accurate.

No set rules or regulations on how to estimate any particular job are given here, but suggestions are offeree on how to arrive at a basic method of estimating any job, whether it is a minor repair, a complete overhaul of a refractory lining, or a boiler retubing job.

The basis for any estimate is PAST experience in domg exactly the type of job you wish to estimate. By the time you have made Boilerman Chief you should have had enough experience to have a general idea of the operations involved in each job.

Jt is suggested that you analyze this past experience and note the time you spent in aoing eacn separate operation. This time should be set down in man-hours. For example, a job that required you and 4 Fir emen to work 7 hours a day for 3 days would be computed as follows:

7 hours X 3=21 hours 5 men X 21 hours=105 man-hours

By adding together the man-hours for each operation, you obtain a total of man-hours required to accomplish the task. This figure may be used as a basis for planning or estimating future work. You must realize, though, that this total of man-hours consumed in the past is but ONE FACTOR in making an estimate for a similar job.

Let's consider some of the other important factors necessary to arrive at an accurate estimate for completing a particular job:

1. Number of men available to do the work, and number of hours a day they are available.

2. Experience of the men available, particularly in the task to be performed.

3. Amount of work that will have to be done in preparing for the job (For example, will several hours be consumed getting to the part or parts to be replaced or repaired, will you have to move or work around other machinery, or will you have to rig staging?)

4. Availability of material you are going to use (Is it readily available or must all or part of it be fabricated, cut to size, or prepared for use in any other way?)

5. Difficulty of job (Under no circumstances should you estimate a job that you feel you cannot do; it is better to admit you don't know how to accomplish a task than to give an estimate and then not be able to live up to your commitment.)

How to Make an Estimate

For study purposes, let's examine the factors to be considered in making a theoretical estimate for replacing 30 boiler tubes.

1. Are the replacement tubes available and are they the correct size for the job? How much time will it take to get them to the job? (Note this in man-hours.)

2. Are the PROPER tools available and in condition to do the job? Is staging required? Is it available? How much time will be spent in rigging this staging? (Note this in man-hours.)

3. Will the boiler be cool enough to work on or will you lose time securing it? (If the

latter, note the time in man-hours.)

4. Are the available men experienced? Or will you have to give them detailed supervision? (If the men are not experienced, decide on a percentage to use as a basis for the 'time-lost" element. For example, if you figure that the men have only a reasonable amount of experience, set your lost-time factor at 25 percent. In your final analysis you'll have to add this loss to the over-all time estimate.)

5. Now use YOUR past experience and figure "how long it took the last time." (Note this in man-hours.)

6. Check the plan of the day for scheduled drills, liberty, or any items that will affect the continuance of the

work. Make another allowance for loss of time. Add this to item 4. 7. Consider the amount of time necessary to clean up after the work is completed and to test and get a satisfactory OK on the completed job. (Note this in man-hours.)

How to Figure

1. Add the total of man-hours noted for each item.

2. Add your lost-time factor (25 percent more or less) to this total.

3. Allowing 8 hours for a man-day, divide the total man-hours by 8 to arrive at the number of man-days.

4. Divide the number of man-days by the number of men available. This will give you the number of calendar days necessary to complete the work.

When estimating the time it will take to do a job, be very cautious in the use of the unit of man-hours. The man-hour is a unit used to measure the amount of work done—it it not a measurement of the actual number of hours required to do a job.

When you have arrived at a final figure, go back over your figures and check each item for error or omission. In many cases this checking will save you untold embarrassment during the course of the job.

Points to Consider

1. Do not make an estimate or start a job until you know that the right type and amount of material is readily available and in good condition.

2. Never underestimate in order to impress the engineer officer with your skill. If you underestimate and then cannot produce the required results, you will "lose face" before your men and your superiors.

3. Never make an estimate on information given to you by some other person. He may be wrong!

4. Never assume that the proper tools and equipment to do a job are available and in good condition. See for yourself!

5. Never GUESS at what is to be done. Know in detail what the job will require. Study your blueprints and manufacturer's instruction book.

6. Never RUSH through a job in order to make the time limit that you have set. If necessary, get more help.

7. Always make allowances for the unexpected.

8. Always use a minimum number of men. There is less chance of things going wrong with a small, efficient crew than with a crew that is loaded with excess personnel.

9. Allow enough time to do the job the safe way.

10. Allow sufficient time for final tests and for any corrective work that may be necessary.

11. Do not slow a job down just to complete the operation on schedule. If you have overestimated, note the item responsible for the miscalculation and remember it the next time.

12. Keep a record of each job completed. This record will serve as a valuable guide for future estimates.

Estimating Material Required

Make a list of the individual items needed to do the job; estimate the amount of each material required, and make allowance for waste, depending on the nature of the material and its condition.

Listed below are suggestions of items you may overlook on a material estimate:

1. Lubricating oil (for tools).
2. Emery cloth.
3. Waste (for wiping down).
4. Bits for electric or pneumatic drills.
5. Extension cords and extra light bulbs.
6. Diesel oil for cleaning purposes.
7. Boiler compound.
8. Tite Seal.
9. Gaskets.
10, Antiseize compound.
!!• Valve-grinding compound.

Ordinarily not all of the above will be used or even requisitioned for any one occasion. But you must consider them, because if such items are not readily available the job may be delayed until they can be obtained.

QUIZ

1. Why should boiler tubes not be straightened in place?

2. From whom must approval be obtained before a general cutting out or renewal of tubes is undertaken?

3. What is the basic cause of oxygen pitting?

4. What is the most common cause of bulging?

5. To whom are representative tubes forwarded when extensive tube failures make necessary a general renewal?

6. When scale samples are submitted, what quantity of scale should be forwarded for analysis?

7. How much boiler water should be sent with the scale and tube samples?

8. What is the first step taken to actually remove a defective tube?

9. After the defective tube is cut flush with the water drum or header, what tool should be used to make the cut on the inside
of the remaining portion of the tube?

10. What tool should be used to drive the stub from the drum or header after the tube is cut and crimped?

11. Why should the replacement tube ends be rounded by a file so no square or sharp ends are left?

12. What is the best method of preparing tube sheet holes prior to the insertion of replacement tubes?

13. How far should the tube extend through the tube sheet, for a good belling job?

14. To what extent should carbon-steel tubes be expanded?

15. To what extent should chrome molybdenum tubes be expanded?

16. After expanding a tube, what is done to safeguard against the tube pulling out?

17. What is the basis for an estimate?

18. What unit of time should be used in figuring an estimate?

19. How should you estimate a job you are not sure you can accomplish?

20. In making an estimate of a repair job, what is the **time loss'* factor?

CHAPTER

BOILER REFRACTORIES

The refractory materials which line the boiler furnace retain heat for a relatively long period of time, and so help to maintain the high furnace temperatures which are necessary for complete and efficient combustion of the fuel. Refractories are also used to form baffles which direct the flow of combustion gases and protect drums, headers, and tubes from flame and excessive heat.

CARE OF BRICKWORK

Under normal conditions, the boiler brickwork should last for a number of years without complete renewal, provided it is given the proper care. Boiler furnaces should be inspected regularly to determine the condition of the refractories, and minor repairs should be made as required. In particular, the burner cones must be inspected frequently, and should be rebuilt whenever they differ by more than 5"^ from the specifications.

All holes should be patched as soon as practicable, and broken or loose brick should be repaired with plastic refractory material or else replaced. Figure 10-1 illustrates the proper method of patching with plastic refractory material. When a furnace wall is partially rebricked, the grade of brick used should be the same as that already installed in the remainder of the wall.

It is important to know something about the factors that are most likely to contribute to the deterioration of

UNOMCUT FIMtmCK TO PMVENT PATCH PWM FALLINO INTO FUKNACE

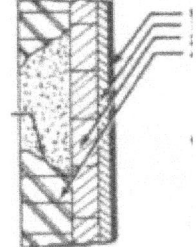

CASINC ———

1" INSULATING BLOCK 2h" INSULATING BRICK 4h" FIREBRICK

Figur* 10-1.—Hew to undsrcut firebrick wqII for patching.

furnace refractories. Slagging is probably the greatest cause of failure of the firebrick used in naval boilers. Slag is the fused (glassy) or unfused (dry) substance that forms on the firebrick walls and floors as the result of a chemical reaction between the firebrick and the ash or unburned particles of fuel oil present in the furnace. Slag accumulations cannot be entirely avoided, since all fuel oils contain some slag-forming elements.

In itself, slag would not be particularly harmful to a boiler furnace. As the slag penetrates the firebrick, however, it creates a surface layer of brick that has a different composition and a different coefficient of expansion from the original brick. When temperature changes occur, a cleavage plane is formed and the slagged layer cracks off. This cracking is known as SPALLING.

You should never make any attempt to remove slag from furnace walls or floors, except when rebricking the furnace. It is not possible to remove even a thin layer of slag without also removing some firebrick, thus exposing a new surface of the firebrick for the slag to act upon. The spalling caused by slag action can continue for some time without endangering the walls or the floor, provided the anchor bolts are not affected. Since the slag has a tendency to flow downward to the furnace floor, damage to the walls is likely to be greater than damage to the floor; but slag accumulations will probably be thicker on the floor than on the walls. After a period of years, the

slag accumulation on the floor may build up to such an extent that it will be struck by flame and unburned oil from the burners. When this occurs, the furnace floor will require rebricking. Occasionally one or more firebricks should be removed from the furnace floor in order to determine the degree of slag penetration.

Spalling from thermal changes alone was formerly a major cause of refractory failure. However, the firebrick now used in naval boilers is of very high quality and is highly resistant to spalling from thermal causes alone. Figure 10-2 shows a furnace wall in which the brick to the left of the expansion joint has been destroyed by true thermal spalling; to the right of the expansion joint, some slag penetration has occurred, and some spalling from this cause has begun.

Figure 10-2.—Thermal ^palling (left), and flag action (right).

Since firebrick expands when heated and contracts when cooled, it is obvious that rapid temperature changes in a boiler furnace are likely to cause damage to the firebrick. Rapid raising of steam in a boiler is probably more injurious to the refractories than to any other part of the boiler. Emergencies may arise which require the rapid raising or lowering of furnace temperatures; but it is important to remember that the refractories cannot stand such treatment very often. As a rule, raising the furnace temperature too rapidly is likely to cause breakage of the firebrick at the anchor bolt; lowering the furnace temperature too rapidly is likely to cause deep fractures in the firebrick.

Refractories are also damaged by any type of mechanical strain. Continued panting or vibration of a boiler may cause a weakened area of wall to be dislocated so that the bricks will fall out onto the furnace floor. Proper operation of the boiler, with particular attention to the correct use of burners and forced draft blowers, will generally prevent panting or vibration.

TYPES OF REFRACTORIES

Many different materials are used as refractories, and these diverse materials react quite differently under different conditions. Refractories, therefore, are classified into three groups—acid, basic, and neutral—according to their behavior with acid and basic reagents.

Acid refractories are those which contain an appreciable amount of uncombined silica. They generally have good resistance to acid slag attack, but poor resistance to basic slag attack.

Basic refractories are those which are composed largely of the chemically basic oxides, lime and magnesia. They generally have good resistance to basic slag attack but poor resistance to acid slag attack. Magnesite brick is one type of basic refractory. Basic refractories are not used in naval boilers and will not be discussed further.

Neutral refractories are those which are neither definitely acid nor definitely basic. They are not affected appreciably by the corrosive action of either acid or basic slags.

Acid Refractories

The following refractories are included in the acid group:

Fireclay brick is made principally or entirely of fireclay, which is an earthy or stony mineral aggregate, essentially hydrous silicates of aluminum with or without free silica, of suitable refractoriness for use in refractories. This type of brick is used in direct contact with flame in the furnace. Fireclay brick is divided into three grades.

1. Grade A brick is 60 percent alumina brick with a softening point equivalent to 3245° F and with a high resistance to spalling. For furnace walls that must frequently be replaced because of severe operating conditions, Grade A brick should be used in preference to Grade B brick. However, under normal conditions it is more economical to use the less expensive Grade B brick. In cases where it is desired to use Grade A brick, BuShips authorization is required. Grade A brick and Grade B brick must not be used in the same furnace except upon authorization from BuShips. Where their use in the same furnace is authorized, the Grade B brick must always be used at a lower level than the Grade A brick.

2. Grade B brick is a superduty brick having a softening point equivalent to 3170° F and a high resistance to spalling.

3. Grade C brick is a high-heat brick suitable for SHORE installation ONLY. It has a softening point equivalent to 3092° F and has less resistance to spalling than Grades A or B.

HiGH-TEMPERATURE INSULATING BRICKS are lightweight bricks (approximately 2i^ lb), possessing insulating properties and resistance to temperatures up to 2500° F. These bricks are used as back-up insulation for refractory furnace linings, and are not exposed directly to flame. There are two types. One type is composed essentially of fireclay; the other is composed of diatomaceous silica.

Insulating block. —This is composed of uncalcined diatomaceous earth mixed with asbestos fiber. It should not be used where the temperature will exceed 1500° F nor in any place where it may come in direct contact with flames.

Fireclay refractory plastic mix. —This is also known as plastic firebrick. It is a mixture of raw plastic clay and calcined clay tempered with water. Its particular usefulness lies in the fact that, because of its plastic nature, it can be pounded into places where otherwise a tile of special shape would be required. The refractoriness of plastic firebrick is practically equal to standard brick, but its volume stability and bond strength are somewhat inferior. This precludes its general use for side walls; but it is a suitable material for repairing brickwork, topping off side and back walls, repairing and constructing burner openings, and in general for use in any part of the furnace not exposed to temperatures in excess of 3000° F.

Baffle mix or castable mix (baffle material).—This is a mixture of refractory material and a heat-resistant hydraulic cement which will acquire its strength without being heated. Class A material is limited in use to service temperatures below 2500° F, and contains hard, dense grains as the refractory ingredient. Class B material is limited in use to service temperatures below 2200° F, and contains crushed insulating brick as the refractory ingredient. Use of these materials is limited to the cooler portions of a boiler.

High temperature refractory castable or 3000° F. castable refractory. —This is a mixture of hard, dense, refractory grains and hydraulic cement which will develop good strength in 24 to 48 hours without heating. The

mixture is shipped dry and must be mixed with water for use. This material may be used instead of plastic firebrick around burner openings and peepholes.

Air-setting mortar. —This is a finely-ground fireclay material that develops strength at room temperature and maintains this strength without fusion or shrinkage up to 3000° F.

Neutral Refractories

Plastic chrome ore (PCO), or boiler water-wall refractory plastic mix, is a moldable or ramming material composed of ground refractory chrome ore and liquid sodium silicate. It is used on the stud-tube dividing wall of naval boilers, at temperatures up to 2800° F.

Chrome castable refractory. —This is a mixture of ground refractory chrome ore and hydraulic cement which will develop good strength in 24 to 48 hours without heating. The mixture is shipped dry and must be mixed with water for use. This material may be used instead of PCO on studded tubes, and for fillets or wall copings. It should not be installed around burner openings.

INSTALLING THE FURNACE LINING

A general knowledge of refractories, their use, composition, and capabilities is a good foundation for a practical approach to the actual building of a furnace lining. The first step should be to consult the refractory drawing. From this drawing the amount of materials and the proper sequence for the construction of the wall and deck can be determined.

Construction of Rear and Side Walls

The hooked-end anchor bolts are secured by dropping the hooked ends into pads that are spot-welded to the corrosion-resistant steel casing. The pads extend % inch from the casing, and are inches high by 1 inch wide. Figure 10-3 shows an anchor bolt and pad.

Insulating block is used for the layer immediately inside the casing. Insulating block is 1 x 6 inches, and may be either 18 or 36 inches in length. It is laid with the length of the block in a horizontal position and the side which is 6" x 18" (or 6" x 36") against the casing; thus, the layer of insulating block is 1 inch thick.

4h" FIREBRICK 1
2)4" IMSULATIHC BRICK 1
1" INSULATING BLOCK ,

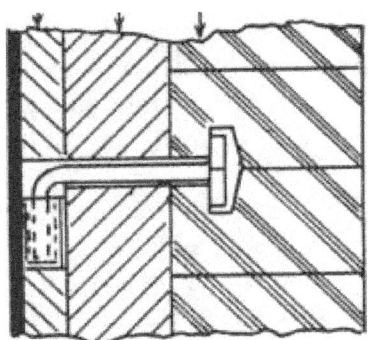

Figure 10-3.—Hooked-end anchor bolt and pad.

The insulating block must be cut out to accommodate the anchor bolt pads and the anchor bolts. It is necessary to allow about inch above the anchor bolt, so that the anchor bolt can be raised when the firebrick is being installed.

It is not necessary to use insulating block mortar for laying up the block. However, some mortar of this type should be available for filling in the spaces left by broken corners or edges, and the spaces around the anchor bolt pads. If insulating block mortar is not available, broken pieces or ends of insulating block should be soaked in water and used to fill these spaces.

Insulating brick is laid inboard from the insulating block. Insulating brick is 21/2" x 4>/2" x 9". It is laid with the length of the brick in a vertical position and the side which is 4i\'7b/' x 9" against the previously-installed insulating block; thus, the layer of insulating brick is 2V2 inches thick.

Insulating brick should be laid out from the vertical expansion joints, so that the necessary cutting for anchor bolts will be along the 21/2" x side of the insulating brick. As the insulating brick is laid, you will have to cut notches for the anchor bolts. The notches should be large enough to allow movement of the anchor bolt when the firebrick is being installed; as a rule, a notch about % ii^ch high and i/> inch wide is sufficient. Set the anchor bolts in their pads as you install the insulating brick.

Two vertical rows of anchor bolt pads, 4^^ inches apart instead of the customary 9 inches, indicate the location of a vertical expansion joint. At the expansion joint, the 2V-> inch thickness of insulating brick is replaced by a split insulating brick and a split firebrick; the firebrick is on the inboard (hotter) side.

If a corner of an insulating brick breaks off, trim the broken end and use the remainder of the brick. When laying insulating brick, take care to see that there are no openings between the bricks. If an insulating brick is warped or twisted, it should be rubbed flat. Both insulating block and insulating brick are very easily worked. They can be cut with » hacksaw blade or shaped with a very coarse file (preferably a wood file). They may also be shaped by rubbing them against a firebrick.

FiRERBiCK is used for the innermost layer of the furnace. The standard size for firebrick is the same as the standard size for insulating brick—that is, 2%"x4y/'x9". ANCHOR BOLT FIREBRICK is the same material as the regular firebrick, but has two spaces left in one of the 9" x 4i^" faces for inserting the head of the anchor bolt. In addition to the standard size, a 2\:/' x 414" x 13anchor bolt firebrick has recently been made available; this "brick-and-a-half" size firebrick should be used to border the expansion joints, thus eliminating the need for half bricks in these positions.

Firebrick is laid with the length of the brick in a horizontal position and the 9" x ly^' face against the insulating brick; thus, the layer of firebrick is 4^^ inches thick. A cardboard or wooden batten is used at each expansion joint; the battens are left in place and are burned out when fires are lighted in the furnace. Figure 10-4 shows an expansion joint at the end of a wall.

Fire brick is laid out from the expansion joints, and must be laid in such a way that it will break joints with the insulating brick layer, both vertically and horizontally. The refractory drawings will indicate whether the first course of firebrick should begin with a full brick or with a half brick (or a "brick-and-a-half") at the expansion

Figure 10-4.—Expansion joint at end of wall.

joint. If the refractory drawing is not available, you can determine this by laying up, dry, a sufficient number of bricks to locate the anchor bolts from the holes in the insulating brick.

Firebrick must be laid with air-setting mortar. Mortar should NOT be used between the floor pan and the first course of firebrick. If the floor pan is slightly warped, the best procedure is to cut the firebrick so that it fits the contour of the pan. Mortar should be used between the ends of the bricks in the first course.

Always complete a full course before starting to lay the next course. When the first row of anchor bolts is reached, the top edge of the firebrick should be about % inch above the top edge of the anchor bolt pads. Adjust the anchor bolts in their proper position as you go along.

As you put the anchor bolts in place in the firebrick, fill in the opening in the insulating brick with small pieces of insulating brick mixed with the air-setting mortar used for the firebrick. This fill-in should not grip the anchor bolt tightly, but rather should allow for the movement of the anchor bolt which must occur when the wall expands. If the anchor bolt is held so firmly in place that it cannot move in accommodation to the wall's expansion, it will put a very great strain on the wall and may even cause the wall to buckle. The anchor bolt space in the firebrick must NOT be filled in with anything.

When using air-setting mortar to lay firebrick, it is not necessary to first dip the brick in water. Dip one end of the brick into the mortar and then, with a wiping motion, dip the bottom of the brick. Lift the brick from the mortar and allow the excess mortar to drip off, as shown in figure 10-5. Do not spread mortar on the wall or on the brick. Quickly place the brick in position in the wall, and push it hard against the next brick in the course. Then tap it firmly, as shown in figure 10-6; tap it both sidewise and backwards, so that it will be forced into position. When no more mortar can be forced out of the joints, strike off the excess mortar even with the face of

Figure 10-5.—Dipping a brick.

Figure 10-6.—Tapping a brick into position.

268

the brick. The thickness of the joints may vary somewhat, according to the smoothness and uniformity of the bricks, but is should NEVER exceed inch.

The consistency of the mortar is extremely important. A full bag of mortar should be mixed at once, if possible. If only a small amount is needed, the dry mortar should be thoroughly mixed before removing the required amount. This is necessary because mortars are often composed of materials of different densities, and separation takes place in transit and in storage. Use sufficient clean water with the mortar to make a thick, soupy mix. The mortar should adhere uniformly over the surface of the brick. When the brick is set in place, the mortar should be easily squeezed from the joint, but should completely fill the joint. If the brick were to be lifted, mortar should be found over the entire surface of both firebricks, with part remaining on each.

The proper consistency must be maintained by the addition of more dry mortar or water as necessary, and by thorough stirring of the batch at frequent intervals.

At the ends of the courses, plastic may be used if the space to be filled is less than a quarter brick. On some boilers the front and rear furnace walls continue behind the boiler tubes. These corbel bricks behind the tubes are at an angle with the horizontal rows of brick of the front and rear walls. To offer adequate support to the corbel, the firebricks of the horizontal course can be cut to the length required to bring the top edge in contact with the corbel, and the small triangular space remaining can be filled with plastic.

Construction of the Front Wall

The front wall of the furnace is built in the same manner and of the same materials as the side and rear walls, except that the total thickness of the front wall may vary from one installation to another, depending upon the type and arrangement of the burners. In many installations,

split insulating brick V/^ inches thick is used instead of the straight 214-inch insulating brick. Be sure to consult the refractory drawings to determine the thickness of the front wall.

Burner openings are left in the front wall, and plastic firebrick or high-temperature castable hydraulic refractory is used to form the burner cones. The position of the brickwork near the plastic is shown on the refractory drawings; in general, one or two inches should be allowed between the burner cone and the closest brickwork. The weight of the brickwork over the plastic must always be supported by angle irons, so that the weight of the brickwork will not be borne by the plastic. The brickwork at each side of the plastic material surrounding the burner openings should be cut back 1 or 2 inches toward the cold side, so as to aid in holding the plastic in position. The older construction, in which staggered bricks extended into the plastic for the length of half a brick, is no longer used. Instead, the brickwork is arranged as shown in figure 10-7.

Figure 10-7.—Front wall of furnoco.

Construction of Furnace Floor

After the walls have been installed, the surface of the brickwork should be cleaned. A stiff brush may be used for this purpose, provided you are careful not to undercut the mortar in the joints. All dried material clinging to the brickwork of the walls must be removed before you begin to install the floor.

Insulating block is used for the first course of the furnace floor. The block is laid so that it is 1 inch high, and the 18-inch or 36-inch length is parallel to the side wall.

Insulating brick, inches high, forms the next course. The insulating brick is laid at right angles to the insulating block. Air-setting mortar should be used to lay the insulating brick, and the face of the brick should be painted with a light wash of the air-setting mortar.

Both the insulating block and the insulating brick are soft and easily damaged. When constructing a furnace floor, avoid walking on the insulating block or the insulating brick.

The last course is of firebrick, inches high, laid at right angles to the insulating brick. When installing sloping floors, the firebrick should be laid so that slag will not flow across the 9-inch length. Firebrick is also laid with air-setting mortar.

Until recently, furnaces were constructed with a i/4-inch expansion joint between the floor and the vertical walls. However, this construction allowed slag that penetrated the expansion joint to cause erosion or undercutting of the walls. When furnaces are rebricked, or when the base of a vertical wall has been seriously eroded or undercut, a fillet of plastic chrome ore or chrome castable refractory may be installed at the angle between the vertical wall and the floor. This type of installation is shown in figure 10-8. The plastic chrome ore or the chrome castable refractory should be installed with V^-inch expansion spaces at approximately 24-inch intervals. As you can see in

I INSUL. BLOCK
|. FIREBRICK
CASING

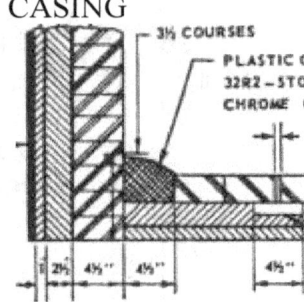

PLASTIC CHROME ORE SPEC 32R2-STOCK NO. 32R3S0 OR CHROME CASTABLE REFRACTORY

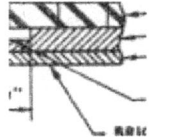

!4*' BATTEN BOARD
BRICK PAN
FIREBRICK 2H" INSUL. BRICK _ r* INSUL. BLOCK
1>^"x4h"K9" FIREBRICK

Figure 10-8.—Fillet ond •xpansion joint at junction of vertical wall and floor.

figure 10-8, a i^-inch expansion joint is also allowed in the floor, at some distance from the actual junction of the floor and the vertical walL

The plastic refractory material used around the burner openings should be secured by at least one anchor bolt per 100 square inches of surface area. At least two different lengths of anchor bolts must be used in the plastic front, in order to avoid creating a cleavage plane in the plastic. The heads of the longer anchor bolts must always be at least 2 inches from the furnace. Either hooked-end or straight anchor bolts may be used for the plastic front. When straight anchor bolts are used, they must be rigidly attached to the casing by a nut welded to the inside of the casing, as shown in figure 10-9.

Plastic firebrick is now being shipped in resealable all-metal drums. The drums should never be opened until the material is required; and if less than a drum is required, the remaining material should be covered with a damp cloth and the drum resealed. To prevent drying out, do not expose to the air more of the material than can be used in a short time.

Installing the Plastic Front

272

Drums that have been in storage may not be uniform in workability throughout. The contents of the entire drum should be dumped on a clean floor plate, cut into small lumps, and mixed with a shovel. When a large quantity is required, as for a burner installation, split the drums in two places with an axe. The material can then be dumped. If the material is a little too dry, sprinkle water over it while it is being mixed, and then cover the mass with wet cloths.

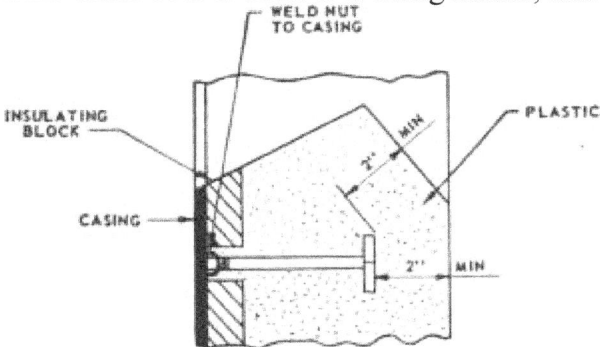

Figure 10-9.—Straight anchor bolt in plastic front.

Plastic that is very hard when it is removed from the container should be cut into small pieces and spread out on the deck—or better still, on wet rags or wet burlap bags. It is then sprayed with clean water, covered with wet bags, and allowed to stand for 4 or 5 hours, or until the material reaches a working consistency.

Plastic that is too soft for installation when it is removed from the container should be cut into small pieces and allowed to dry slowly and without any application of heat, until the material reaches a working consistency.

Plastic firebrick should be passed into the furnace in pans covered with damp cloths. If it is shoveled into the furnace, it can be kept moist by covering it with damp cloths. Be careful to prevent any of the material from

remaining on the floor for any considerable period while new material is passed in and used. You should also guard against picking up any slag, insulation, or other material with the plastic.

The plastic should be built up in such a way as to avoid the formation of vertical cleavage planes—that is, it should be built UP from horizontal layers, rather than be built OUT from vertical layers. Beginning at the bottom, install a horizontal layer of plastic on the firebrick. Use fist-sized lumps of plastic, and ram each lump into place before putting in the next lump. The plastic should be rammed by hand with a wooden mallet, or by means of a pneumatic rammer. The pounding should be done at a slight angle toward the casing. The edge of the mallet should be allowed to strike the surface of the plastic so as to roughen the surface; this roughness will help to bind the next layer. After one horizontal layer of plastic has been installed and pounded, begin on the next horizontal layer.

The plastic should be built up, layer by layer, to the centerline of the burners, allowing space for the forms. Slightly more plastic than necessary should be pounded, and then it must be cut out to accommodate the forms. A form smaller than the burner cone angle is desirable, as this allows room for ramming and helps prevent voids in the plastic forming the burner cone. The pounding must be continued after the lower burner forms are in place. The service life of the wall will depend on this pounding. If the pounding is done with a slight angle toward the casing, and with the layers at the same time kept level, little plastic will need to be cut off the face. Be sure to pound tightly around the anchor bolts. The direction of pounding is stressed because, if the material is built in vertical layers from the casing, cleavage planes parallel to the front casing will develop, and sections of the wall will drop off.

When the entire wall has been pounded, cut or scrape it to the proper thickness. If the forms are made to

extend into the furnace for the thickness of the wall, this will aid in cutting the entire wall to the same thickness. Next the forms are removed and the excess plastic cut from the burner cones with a sweep.

Caution: If the plastic seems too soft and begins to slump while the burner opening is being swept, allow the plastic to dry out for a few hours, and then complete the sweeping process.

After all the burners have been swept, use the sweep to test the angle of all the burners. Stipple the entire surface of the plastic, except the burner cone proper, with a wire brush. Smooth surfaces would tend to prevent the escape of moisture. Vent the plastic on 1 X'-inch centers by plunging a yig-inch rod, tapered to a blunt point in the last half inch, through to the casing. In the burner openings, the vent holes should be in the same direction as for the remainder of the wall; that is, at right angles to the casing.

The plastic installation is vented to allow for the rapid loss of moisture, since this type of material contains by weight about 10 percent of water when installed. An average front wall installation uses about 1 ton of plastic firebrick; therefore, approximately 200 pounds of water must be removed. If for any reason this water were trapped within the plastic itself, or behind it, steam would be generated as the furnace temperature increased; and an explosion might take place, causing the plastic to be deposited on the furnace floor. Even if the action were not rapid enough to cause an explosion, severe cracking might be caused; and the plastic would soon fall from position. Venting also allows the heat to penetrate deeper into the body of the plastic, and thus increases its strength. Since this material is basically a mixture of raw clay and calcined clays, a ceramic bond in this type of material does not start to form until a temperature of 1800° F. is reached.

Prolonged air-drying increases the amount of shrinkage in the plastic; therefore, immediately after the plas-

tic installation is completed, a fire should be started in the boiler. A laboratory test on plastic firebrick that has been thoroughly air-dried shows about 2 percent shrinkage, whereas a test on plastic firebrick that is fired immediately upon removal from the container will show only about 1 percent shrinkage. If it is impossible to start firing immediately, keep the plastic moist by covering it with a damp cloth.

When lighting off, first light off the center burner with a small-size sprayer plate and keep it in operation for about 15 minutes. Then secure the center burner and light off another burner for the same period of time. Continue this method of using the burners in rotation for a 6-hour period. After this initial firing, the furnace temperature should be gradually increased over a period of 6 hours to the maximum that can be attained under the particular steaming conditions. (The maximum attainable temperature will depend, of course, upon the purpose for which the boiler is being used. In port, it is likely that the boiler will be used only to supply auxiliary steam.

If possible, however, this final firing should be at the boiler full-power rating.) Final firing at the maximum practicable temperature should be maintained for a period of 6 hours. The plastic cannot develop its full strength unless it is fired at a high temperature for at least 4 hours.

On all types of boilers, be sure that the superheater is properly protected during the baking-out period. Special care must be taken to protect the superheater when baking out the superheater side of a superheat control boiler. Light off the saturated side and use steam taken through the superheater to operate a generator. Obtain additional flow through the superheater by blowing down to the bilges and by using the recirculating line to the auxiliary exhaust system (if fitted). When a sufficient flow of steam has been established through the superheater, the burners on the superheater side may be lighted off, in rotation, in the manner previously described. You must be especially careful not to overfire the superheater side, particularly when there is a relatively small flow of steam through the superheater. Use small-size sprayer plates and, if necessary, reduce the oil pressure to the minimum in order to prevent overheating of the superheater during the baking-out period.

When it is known that the boiler cannot be fired or that for other reasons the fronts cannot be baked out properly, high-temperature castable refractory may be used.

When baking-out procedures have been completed, the boiler should be secured and allowed to cool as slowly as possible. If time is available after completion of the baking out, it is advisable to open the furnace, as soon as the brickwork has cooled sufficiently, and inspect the settings. Concentricity of the burner openings should be checked, and any cracks of excessive size filled with plastic.

HIGH-TEMPERATURE CASTABLE HYDRAULIC REFRACTORY

High-temperature castable refractory is a hydraulic-setting material which acquires its strength without being fired. This material is received dry in 100-pound bags, and must be mixed with water for use. If possible, mix one full bag at a time. Printed directions for mixing are usually on the bag or on a tag attached to the bag.

Three degrees of consistency are recognized—pouring, tamping, and ramming. Pouring consistency has enough water added to cause the material to flow and level out. Tamping consistency has just the necessary water added to allow the material to be puddled into place. Ramming consistency has just sufficient water added to hold together when packed solid with a mallet. This amount of water, however, may not be sufficient to set all the hydraulic cement; and the finished material may consequently lack strength.

When installing a SOOO"" F. grade of castable refractory around the burner openings in a boiler front, use the same number of staggered square-headed anchor bolts as you would use for installing plastic firebrick. To compensate for their expansion, dip the head and shank in liquid paraffin or asphaltum. In service, the paraffin or asphal-tum will burn out, leaving an expansion space around each anchor bolt. Wood or preferably metal forms should be prepared for the burner openings, and these are best secured through the distance pieces. Be sure that these forms are of the exact size and angle required for the particular burners, since the castable refractory, unlike plastic firebrick, should not be scraped or trimmed after forming. The form for the furnace face consists of two or more vertical supports shored against the rear wall or the tube banks and held a sufficient distance from the front wall to permit slipping boards (approximately 6 inches wide and with the 6-inch dimension vertical) between the vertical supports and the front wall. Paint the faces of the boards and of the burner forms with oil. Soak the wooden plugs for peepholes, light-off hole, and oil drip holes in water before use, to prevent absorption of water from the castable and subsequent swelling of the wood plugs. Allow the plugs for oil drip holes to remain in place and burn out, but plugs for peepholes and for the lighting-off hole are most easily removed by hitting them inward from outside the boiler.

Each batch of castable refractory should consist of a full bag of material uniformly mixed with clean water to a consistency that will ensure flow around the burner forms and achor bolts when puddled with the hand or a long stick. Take care to prevent disturbing the anchor bolts. Install the material within 30 minutes after mixing. Any that cannot be installed promptly and that partially sets should be discarded, since to mix this partially set material with additional water will weaken it structurally. When the coated anchor bolts and forms for the burner openings, oil-drip holes, light-off hole, and peepholes are in place, and the vertical supports secured

against the front wall, place one strip of wood (approximately 6 inches wide) at the lower part of the setting. Then pour the mixed castable refractory over this strip and puddle around the burner cones, the anchor bolts, and the brickwork until all voids are eliminated.

When the top of the strip is reached, add another strip and repeat the procedure until the entire front is installed. A convenient method of installing the uppermost portion is to cut a wood strip to end approximately 2 inches below the bottom of the brickwork above the burner. Castable is installed to the top of this strip, and then a strip is installed with the width, making approximately a 45° angle with the vertical front; the top of the strip should extend into the furnace. Castable is poured over this strip and puddled until the top of the castable is above the bottom of the brickwork. The castable lip extending into the furnace may be removed after the forms are taken off.

Do not remove the forms for approximately 24 hours. If any voids are found after removing the forms, roughen the surface to be patched and fill the void with the cast-able refractory mixed to a slightly heavier consistency. Castable refractory will increase in strength with prolonged air-drying before heating, and should not be subjected to heat for at least 48 hours.

PLASTIC CHROME ORE (PCO)

PCO, the water-wall refractory used in naval boilers, is quite different from the fireclay base materials used in the construction of furnace walls. It is a mined chrome ore called chromite. . The desirable parts of the ore, from a refractories standpoint, are chromic oxide and iron oxide, which are united chemically to form the mineral chromite. It is almost impossible to separate this mineral from its impurities, so the material must be used as it is mined, even though the unwanted elements tend to lower the fusion point of the combination.

In order to take advantage of the important character-

istics of PCO, it is necessary to prevent it from softening and sagging of its own weight while in service; the stud-tube boiler wall has been developed for this purpose. The studs, welded securely to the comparatively cool tube, project through the PCO and not only support it, but also tend to keep it sufficiently cool so that it does not develop into a semisoft state. If the material were merely supported mechanically, without also being cooled, its surface would soften and erode rapidly away as a result of the action of the flame and gases in the boiler. If it were not supported mechanically by the studs, it would become overheated and would gradually sag or flow off the wall.

The terms flowing and softening do not refer to an extreme condition in which PCO would be as soft as molasses or even comparable to a piece of pitch at room temperature. The softening characteristic of PCO is merely a gradual decrease in its physical strength as the temperature rises above 2300° F.

PCO also has the peculiarity of bonding very strongly to the steel studs and tubes in a boiler wall, and it withstands a tremendous drop in temperature from its hot side to its cold side without shattering. For instance, the temperature of the surface exposed *o the flame may be 3000° F., and the temperature at the point of contact with the tubes only a few hundred degrees. This is a drop in temperature of about 2500° F. in a thickness of about V'j inch, and it is doubtful that any clay or other type of material would withstand this treatment.

The final and essential reason for the use of PCO is its HIGH resistance to the action of furnace ash, slags, sulfur, and other impurities present in naval boilers. At the temperatures present in naval boilers, these products of the combustion of the fuel tend to attack chemically and to eat away firebrick materials; this action increases as the temperature rises. Because PCO is a neutral refractory, it is highly resistant to all these furnace and fuel conditions, and the products of combustion simply

do not react chemically to any great degree, with this material.

PCO is shipped in airtight containers in a ready-to-use condition, never add Portland cement, water, or any other material. A small amount of liquid will, however, sometimes separate during shipment and all such liquid within the container must be thoroughly mixed with the plastic contents before using. One drum of PCO will be needed to fill a space of 1 cubic foot when it is rammed into place.

In severe weather, PCO may freeze in transit or while in storage. This freezing will do no harm, provided the ore is thoroughly thawed before the drums are opened. Allow it to stand at least 24 hours in a warm place to thaw. Whenever possible, of course, stow PCO where the temperature is well above freezing. When PCO must be installed during freezing weather, keep the temperature in the boiler furnace above freezing until the finished work has completely airset. To do this it may be necessary to fill the tubes with warm water or to rig up a steam jet that will maintain a tube temperature above 40^ F.

To mix the PCO, remove the lid from one container, turn the drum on its side, and roll it back and forth several times. Then turn it upside down on a clean surface and lift it off its contents. Using a sharp trowel, cut the PCO into small pieces of about 1 or 2 inches. Spread out the chopped material if it is exceptionally moist, and expose it to the air for an hour or so (depending on dryness of atmosphere), until sufficiently dried out. Often all that is necessary is to select lumps which appear to be of right consistency for use. Damp pieces are set aside for later use. Do not use or work PCO that shows any sign of having a hardened crust. No attempt should be made to salvage such material by the addition of water or other liquid. Cut the hard part off and throw it away. Any excess PCO opened but not used promptly

should be returned to the drum; damp bags should then be added to the container, and the lid replaced.

Before starting the installation, make sure that the stud tubes are dry, clean, and free of all loose or foreign particles. No preliminary coating of any other material should be used; however, the coating of chrome paint given tubes for protection in transit may be left on the tubes. Work is started at the bottom and carried upward. Place the PCO in position by hand and ram it solidly, so that the finished surface is FLUSH WITH THE TIPS OP THE STUDS. The ramming must produce a dense structure around the tubes and around the studs from base to tip. Try to regulate the amount of material so

that HARD BLOWS WILL GIVE EXACTLY THE DESIRED FINAL LEVEL. If too little has been applied, do not add more until the old surface has been made very rough. If too much has been added, do not continue blows with the mallet, but cut off excess with a sharp trowel and refinish with a mallet.

For the ramming process, use wooden mallets of a size as large as the size of the tubes and the space between them will permit. One face of the mallet head should be flat, while the other face should be dressed to form the desired contour of the finished wall surface between the tubes. The latter face will usually be shaped like a very blunt-nosed chisel. Figure 10-10 shows the mallet and indicates the method of ramming. See that the working faces of the mallet head are wire-brushed frequently, to prevent the refractory material from adhering to it. When ramming, use a slight wiping action of the mallet. This will facilitate bringing the finished surface of the refractory flush with the stud tips, as well as permit you to work excess plastic into position as required.

Install PCO to its full thickness initially; never apply it in layers. When the PCO is applied it should be rammed from one side of the water wall between two adjacent tubes, to a thickness of slightly more than half the thickness of the wall, and to a height of approxi-

Figur* 10-10.—Ramming malltt.

mately 18 to 24 inches. This portion should then be backed by a suitably braced, heavy, hardwood form having the exact shape required for the wall. The plastic is then rammed from the opposite side in such a fashion that the material thus forced in from both sides meets, and is solidly bonded. The PCO should be solidly rammed to produce a dense fill, but should not be overrammed. Overramming is indicated by a tendency to surface sponginess, and by the binder of the material rising to the surface.

After ramming is completed, never "slick" the surface smooth because this will produce blisters, excessive cracking, and other harmful effects. The best final sur-

Figurs 10-11.—Stippling plastic chrom* or*.

face is obtained by stippling with a stiff wire brush, after completion. (See fig. 10-11.) The direction of the brush is to and from the refractory face, not along the face. The average workman will want to slick the surface for appearance sake after the ramming, so care must be taken to prevent this.

Plastic chrome ore contains about 6 percent added water. This evaporates as the material dries, leaving enough internal voids to take care of any expansion that may occur when the material is heated to a high temperature in service. For this reason it is not necessary to leave expansion joints when installing PCO, but it is recommended that trowel cuts be made in chrome ore when it is used as water drum or header protection or as fillets at the base of a vertical wall.

The boundaries of any unfinished work must be kept from drying out until work is resumed. The best method
is to thoroughly soak burlap bags in water, then wring them and hang the damp bags in such a position that the lower parts of the bags (the sections that will remain damp longest) cover the edges of the work.

PCO may be applied at any time after the tubes are installed. After installation of the PCO, air-dry for 24 hours and then use a light fire for 4 or 5 hours.

If the stud-tube wall requires patching after a period of service, carefully pick out all PCO on both sides of the wall and between the tubes as well. Then, after wire-brushing the studs and tubes clean, ram in fresh PCO as directed above. It is not necessary to air-dry patchwork. Use a light fire for the first 4 or 5 hours if possible. In patching or making extensive repairs to the walls, it is not necessary to apply chrome paint to the tubes.

CHROME CASTABLE REFRACTORY

Chrome castable refractory is shipped in 100-pound bags and requires the addition of water before use. Dump the entire contents of the bag on a clean, dry surface, thoroughly dry mix to ensure uniform distribution of all ingredients, select the amount desired and return the remainder to the bag. Only that portion of the contents which can be installed within 1 hour should be mixed with water. Thoroughly mix the selected batch while adding cool fresh water in the amount necessary to secure the consistency of plaster. When a trial application on a vertical surface indicates stickiness without slipping, the material is ready for use and work should be continued immediately to completion. If the material tends to slump, the mixture is too wet. A small amount of dry chrome castable refractory added to the batch will correct this. The material contains a hydraulic binder; therefore, the batch should not be allowed to stand. If it begins to set before application, full, final strength cannot be obtained. The material is best applied by quickly tossing handfuls against the stud tubes until the required thickness has been built up. If it is of the proper con-

sistency it will thus be forced in around the studs, filling all spaces- Use of a small wedge-shaped piece of wood (tapered to fit between studs) will assist in assuring this. The surface should then be smoothed with a small piece of w^ood to remove excess material and to make the finished surface flush with the stud tips or flush with the furnace side of the first row of studs on partially studded tubes. Excessive smoothing should be avoided. The surface should not be stippled.

Chrome castable is also suitable for use when constructing fillets at the base of vertical walls or as copings on side walls to protect drums or headers. The use of forms is not considered necessary. Material should be mixed with water as described above, but should be slightly stiffer than that at which the fillet would deform or sag. It should not be vented or stippled. Expansion joints wide) should be formed on approximately 24" centers by use of paper or trowel cuts. Prolonged storage after installation without firing is not harmful.

Should the stud-tube wall require patching after a period of service, any loose or soft material should be removed from the hot side. In the case of full studded tubes the entire thickness of chrome castable should be removed even if hard. In the case of partially studded tubes, sufficient material must be removed, even if hard, to completely expose one row of studs on each side of each tube to be repaired. The exposed surfaces of materials should be moistened before installing PCO (by ramming) or chrome castable (by a plastering action as previously described).

BAFFLE MIX

Baffle mix, which is also referred to as refractory cast-able mix (baffle material), is a granular material com-posed of aggregate (fireclay grog, crushed firebrick, crushed insulating brick, or crushed pottery saggers), a hydraulic-setting binder, and, in some products, raw clay. The types of aggregates and raw clay used will

determine the temperature limits to which the material may be used. Refractory castable mixes have a low strength at a temperature of 1500 to 1800° F. where the binder is dehydrated. As the temperature is increased above this range, a ceramic bond is developed and the material increases in strength.

Use of this material is limited at present, but will probably increase in the future. It eliminates the use of special shapes and is especially adaptable where an odd thickness of brick is required or where it would be difficult to install firebrick. It also eliminates joints and makes a smooth-surfaced wall which has less resistance than firebrick to the flow of gases. The material can be easily handled and stored.

Baffle mix is always received in moisture-proof 50-pound or 100-pound bags, and must be mixed with fresh, clean, cold water for use. Mixing is generally accomplished in a large, flat, watertight container, such as a mortar box. If possible, one full bag at a time should be used. This should be thoroughly dry-mixed before water is added. If less than a bag is required (you will need about 120 pounds to fill a space of 1 cubic foot), be sure to mix the entire contents of the bag while dry; then remove the required amount and mix this with fresh water. The coarse clay grog and fine particles of clay and cement in this type of material tend to become segregated in transit or upon excessive handling.

Proper directions for the use of the material are usually printed on the bag or on an attached tag. If directions are not available, add only sufficient water to allow the material to reach all spaces to be filled by puddling the mix into place. As with high-temperature cast-able hydraulic refractory, three degrees of consistency are recognized: ramming, pouring, and tamping. Ramming consistency has just sufficient moisture to hold together when packed solid with a mallet. If the amount of water is not sufficient to set all the cement, the finished

piece will lack strength. Pouring consistency has excess water added to cause the material to flow and level out. This excess water will cause decreased strength and increased shrinkage. The TAMPING consistency has just the necessary water added to allow the material to be puddled into place. This consistency will give the maximum strength of the material at all temperatures, and the shrinkage will not be excessive. The material will not flow around restrictions, so sticks or other means of puddling must be employed to fill all spaces and to eliminate air bubbles and voids. When the material is mixed to this consistency, it will not level ofl in the mixing container.

When mixing, add the water slowly, even if the desired amount is known, and mix the material steadily, as it will quickly change from the loose, crumbly state to the proper consistency, with slight additional water. The material should never be mixed ahead of time, as the initial set of the material takes place very soon after water is added. The proper consistency can be judged by taking a handful of the material, and with the hand open and the fingers spread slightly apart, tossing it around and trying to ball it. If it is too dry, it will not cling together; if it is too wet, it will run between the fingers. Baffle mix of the proper consistency will soon form a ball which can be changed from hand to hand.

Prepare the forms before the material is mixed. Wooden forms are satisfactory, but they should be painted with a light oil to prevent their absorbing moisture. For a large section, build up the form about 12 inches at a time, if there are no anchor bolts or other restrictions. This will allow the puddling of the material to be uniform. If metal forms are used, oil them to prevent the adherence of the material. When the first part of the form is in place, mix the baffle mix to the proper consistency, and pour it into the form until it is about 3 inches deep. Puddle with sticks until all voids have been filled and the air bubbles eliminated. Add another

section of the form and continue in the same manner until the job is completed.

Leave the forms in place for 24 hours to give the material time to set thoroughly. If the boiler is needed immediately, the form can be removed 4 or 5 hours after casting, and the boiler fired immediately if a slow fire is used. The longer the drying period, however, the stronger the installation will be. As the baffle mix will contain perhaps as much as 20 percent water, it should be carefully fired. Too fast firing will cause the formation of steam within the material and large parts might be blown off.

Baffle mix should not be installed in a warm furnace. It should not be installed when the temperature in the furnace is below freezing.

QUIZ

1. What grade of firebrick should be used for partial rebricking of a furnace wall?

2. What is considered to be the greatest cause of firebrick failure?

3. Why is it impossible to entirely avoid slag accumulation in boiler furnaces?

4. In what manner does slag cause damage to firebrick?

5. Why must you not attempt to remove slag from furnace walls or floors except when rebricking the furnace?

6. When may Grade A and Grade B firebrick be used in the same furnace?

7. Which grade of firebrick is suitable for shore installations only?

8. How can you determine the extent to which slag has penetrated the furnace floor?

9. From what source can you obtain information on the proper construction of a furnace lining?

10. When the front wall of a furnace lining is being installed, how much space should be allowed between the burner cone and the closest brickwork?

11. How many anchor bolts are needed to secure plastic refractory material around the burners?

12. What treatment can be used to soften up plastic firebrick which has become too hard?

13. Why is the direction of pounding important in the installation of the plastic front?

14. If moisture is left within or behind the plastic firebrick, what is likely to happen when the furnace is fired?

15. What method of drying causes the greatest amount of shrinkage of plastic firebrick?

16. At what temperature does plastic firebrick begin to form a ceramic bond?

17. How soon after the installation of plastic firebrick should a fire be started in the boiler?

18. What type of drying is best for castable refractory?

19. How soon after mixing should castable refractory be used?

20. Why must you not use partially-set castable refractory which has been softened with additional water?

21. What should be done with PCO which has acquired a hardened crust?

22. How should th^ final surface of a PCO installation be obtained?

CHAPTER

11

BOILER CLEANING AND MAINTENANCE
BOILER WATERSIDES

It is of the utmost importance that the watersides of the boilers be kept free of scale or accumulations of sediment. Failure to do so results in marked decreases in the efficiency of the boiler because of poor heat transfer, and, in aggravated cases, is the direct cause of tube failure. Experience has shown that tube failures resulting from defective materials or fabrication are rare.

The great majority of tube failures (other than those resulting from water-level casualties) are due to the presence of hard scale or, more often, of an accumulation of relatively soft materials such as iron, copper and zinc oxides, dirt, conglomerates formed from impurities, the solids formed from scale dissolved by boiler compound, and the residue of boiler compound itself. Hard scale will not form if all preservatives are thoroughly removed from new or laid-up boilers, pure distilled water is always used, oils, and greases are not introduced into the feed water, and boiler compound is employed to maintain boiler water alkalinity and hardness at the prescribed levels.

Whenever there is reason to believe that deposits may be present in the boilers, they should be cleaned out at the first available opportunity. It must be borne in mind that boilers that contain deposits on their heating sur-

faces, or that have grease or other foreign matter in suspension in the water, are liable to over-heating and serious damage and their evaporative efficiency is reduced. It is particularly important that in such circumstances the boilers should not be steamed at high firing rates except in case of great emergency. Grease and other foreign matter in suspension in the water tend to produce priming and to aid the various processes of corrosion to which the interiors are particularly susceptible.

In the case of reactivated vessels, sludge originating from preservatives in the system (sometimes not entirely removed prior to placing the vessel in service) may collect in the boilers; particular care must be taken to detect and remove such deposits.

At all times, the exact condition of the watersides of a boiler should be known. Thorough inspections must be made at every overhaul and whenever undesirable conditions have occurred.

Inspection and Cleaning of Watersides

Each boiler should be inspected prior to and after each regular boiler cleaning, as well as at any other time it is deemed necessary to ensure the thorough cleanliness of the watersides. Should unusual cases of damage or deterioration be discovered at any time, a special report must be made to BuShips, stating in detail the extent of injury sustained, the remedies applied, and, as far as can be determined, the causes. Results of all inspections should be entered in detail in the boiler record sheet and the engineering log.

At regular overhaul periods, all drums, headers, and economizers must be completely opened. Internal fittings should be removed from the steam drum. At least one handhole plate should be opened on each pass of the superheater, except at each alternate cleaning period, when the entire superheater is to be opened. Cyclone steam separators are not to be removed from their mountings for cleaning. However, in order to facilitate cleaning of the tubes from the steam drums, the curved apron plates which follow the contour of the lower half of the drum must be removed. This will permit cleaning of all tubes from the steam drum except the last few rows located just below the horizontal centerline of the steam drum. Access to these tubes must be gained through the water drums.

Before the curved plates are removed from the drum they should be stenciled or otherwise identified as to location along the curved baffle. If this is not done it will invariably be found most difficult, in replacing the plates, to make them tight.

Since a wet boiler gives a false indication of its true condition, the boiler should be completely dried out prior to starting the inspection. The general appearance of all visible parts of the boiler should be noted and recorded, with particular attention being given to accumulations of sludge, the presence of oil or grease marks, the presence of general scale or rust, and any other items of importance.

Proceed with a thorough wire-brushing of all tubes, drums, and headers. The superheater tubes made accessible for inspection by removal of the handhole plates should also be brushed, and, if found dirty, the entire superheater must be opened for cleaning. Brushes must
BE PASSED THROUGH EACH TUBE A SUFFICIENT NUMBER OF TIMES TO ENSURE THAT ALL SCALE, SEDIMENT, OR RUST SUSCEPTIBLE TO REMOVAL BY A POWER WIRE BRUSH IS REMOVED. Care should be taken to use the correct size brush for the tube being cleaned and to see that the bristles of the brush are in good condition. Personnel entering steam or water drums of boilers for any reason should not carry in their clothing or otherwise on their person any material or tools that could possibly become lodged in a boiler tube. Changing of boiler tube cleaning brushes should be made by passing the flexible shaft or air hose of the boiler tube cleaner outside of the drum. Only experienced men should be assigned to work in the watersides of a boiler. Inexperienced men being trained for waterside work should be under the close supervision of an experienced man.

After cleaning, the boiler should be blown out thoroughly with air, followed by washing out with water. Air and water nozzles fitted with spring type shut-off valves should be used to blow and wash down each tube. When washing superheaters and economizers, care must be taken to prevent water entering the firesides of the boiler.

Dry the boiler out and explore representative tubes in all parts of the boiler with a power wire brush. Unless the watersides are thoroughly clean, the cleaning process must be repeated. It is to be noted that removal of all the fine dust from a boiler is rarely possible, but this should not be construed as permitting the acceptance of tubes that have not been thoroughly cleaned.

After the boiler has been cleaned, it should be inspected for corrosion of the metal parts and a record made of the conditions found. All accessible pits should be probed with a sharp tool to remove remaining corrosion products down to bare metal. Visible parts of tubes should be carefully observed to detect the presence of hard scale, which often is the same color as the tube. Rolled tube ends, especially in the superheater, the steaming level waterline in the steam drum, and fittings exposed in the steam space are the usual locations of pitting. Pits should not be filled with cement or "smooth-on" paste, as such filling does not prevent further corrosion in the pits, while it does prevent the reduction of metal being readily judged on subsequent inspections of the boiler.

All openings in the steam and water drums, especially those serving the water-gage glasses and the boiler blow valves, should be cleaned out carefully.

Frequency of Mechanical Cleaning and Boiler Inspection

Boilers should not be steamed for more than 1800 (or at most 2000) hours between successive mechanical cleaning of watersides. They should be opened, inspected, and, if necessary, cleaned whenever untoward steaming conditions have occurred. Whenever a boiler is opened for any reason, it is good engineering practice to hose down accessible parts. However, water washing should not be substituted for the mechanical cleaning period specified above.

Make a notation in the boiler record sheet each time the boiler is opened. The notation should include the following:

!• Reason opened;
2. Whether washed down;
3. Whether mechanically cleaned and, if so, to what extent*

Cleaning of Superheaters and Economizers

The interiors of superheaters should be cleaned mechanically at each alternate cleaning period except when unfavorable conditions exist in the superheaters. In such cases the superheaters must be cleaned at each boiler cleaning or oftener, as may be required. Particular attention should be paid to the condition of the superheaters when the boilers have been contaminated with salt water or when carry-over has occurred.

Clean the interiors of economizers at each boiler cleaning except when unfavorable conditions exist in the economizers, in which case they should be cleaned oftener as may be required.

Care of Watersides After Cleaning

Whenever the boilers are open for cleaning and overhauling, their interiors must not be allowed to remain in a damp condition longer than necessary to accomplish the necessary cleaning. The cleaning and washing out of the interiors must be completed as soon as possible after opening, and the boilers closed and filled. The hottest water available should be used for filling the boiler. If practicable, the boiler should be lighted off* for steam test within 4 hours after filling. Steam-testing a boiler soon after its overhaul is good engineering practice.

Safety Precoutions

1. To prevent accidental opening, close and secure by locking or wiring all connecting valves to the boilers in which men are working.

2. There is always danger that steam and/or hot water may leak to an idle or open boiler through a leaky bottom blow valve when the bottom blow valve of another boiler connected to the common blow line is opened. This same danger exists with respect to stop valves, feed valves, superheater high pressure drain valves, etc. Open the superheater bilge drain valves to permit drainage of any water leaking into the headers. If pressure is to be applied to any valve on an open boiler, no personnel must be allowed in the boiler until pressure has been applied and the tightness of the valve positively assured.

3. Lash in closed position the control valves of steam-smothering systems while men are working in the vicinity thereof, and remove the lashings when the work is finished.

4. The use of naked lights in an open boiler is prohibited. Electric leads of portable lights should be thoroughly insulated, and the portable lighting fixture must be of the watertight type.

5. While men are working within the interior of boilers, a man should be stationed outside to render assistance in case of accident.

6. Do not make repairs to the power-driven tube cleaner while it is inside the steam drum.

7. Men entering the steam drum should remove all articles from their pockets before entering.

8. Do not close a boiler unless the watersides have been carefully examined for dirt and other foreign matter.

9. Before closing a boiler, inspect the interior to see that no person is therein and that the boiler, including the economizer and superheater headers, is free from tools, loose material, and foreign matter.

BOILER FIRESIDES

Firesides of boilers should be kept scrupulously clean; a clean tube will not accumulate deposits as rapidly as will a dirty tube. Deposits of soot, scale, or slag on the tubes seriously reduce the efficiency of the boiler. Slag in fluid form contributes heavily to failures of such parts as superheater support plates, baffle and protection plates, and soot blowers. The deposits act as insulating material and prevent heat from the furnace and the combustion gases being conducted through the tubes to the water. Blocking of the gas passages through the tube banks requires excess blower pressures to force the gases through the boiler, and this represents an additional loss of economy. Such blocking of normal passages causes large quantities of uncooled gases to flow over protection plates, baffles, and seal plates, resulting in early failure of these materials. The presence of soot not only represents a corrosion hazard—from the possibility of becoming wet and forming sulfuric acid to attack tubes and drums—but it also represents a fire hazard. Keeping the firesides clean saves fuel, material, and over-all upkeep effort.

Frequency and Procedures for Cleaning Firesides

Boiler firesides must be cleaned after each 600 hours of steaming. In most cases, however, it is not considered good engineering practice to wait this long between cleanings. As a general rule, the furnace should be entered after each steaming period at sea, and the firesides cleaned as necessary to prevent the accumulation of deposits.

Whenever a boiler is secured, take that opportunity to remove the soot, slag, and fireside scale that has been deposited on the tubes. Work from within the furnace, from the outer sides of the banks, and from the ends of the boiler through access doors. Special tools for reaching between the lanes of tubes may be made from flat bar, from sheet-metal strips cut with a saw-toothed edge, from rods, or from similar stock. Aggravated cases of slag and scale accumulation may require the removal of bolted sections of boiler casings to improve accessibility for cleaning, but every effort should be made to accomplish the work without resorting to this procedure.

Superheater tubes, because of their relative inaccessibility, are more difficult to clean than the boiler tubes. Special tools and procedures should be devised for cleaning-superheaters. In aggravated cases of superheater fouling" and when mechanical procedures are ineffective, water-washing or the "sweat method" may be used. These methods will be discussed later in the chapter.

Ordinarily economizer tubes do not foul as readily as either the generating tubes or the superheater tubes. The deposits are usually soft and may be removed with probes and air lances, although in aggravated cases water-washing may be necessary.

Examining the Firesides

Tubes should be carefully inspected for evidence of bulging, sagging, formation of circumferential cracks (usually confined to radiant-type superheaters using chromium alloy tubes), pitting, scaling, reduction of thickness on the furnace sides (resulting from gas erosion) , and acid corrosion (especially adjacent to the water drums).

Bulging is caused by overheating of the tube because of dirty or scaled watersides, or by lack of sufficient steam flow through the superheater. When bulging is found on boiler and superheater tubes, immediate steps must be taken to clean the watersides. Bulging or failure of superheater tubes on superheat control boilers may be also caused by firing the superheater furnace with insufficient steam flow through the superheater or by firing the saturated furnace at high rates without steam flow through the superheater.

Sagging is caused by low water or by a general condition of overheating due to dirty or scaled watersides. As in the case of bulging, immediate steps should be taken to clean the watersides unless a condition of low water is

known to have been the cause. Cracking of alloy tubes is believed to result from a combination of the following: high temperature; relative differences in temperature between the furnace side and the rear side of the tubes, with strains resulting from the unequal expansion between these sides; characteristics of the material; and material fabrication difficulties.

The conditions that cause pitting and scaling involve diverse factors beyond the scope of this text, but firesides attentively cleaned and kept free from moisture are less likely to show defects of this nature. Reduction of metal surfaces by oxidation is found mainly on exposed studs of stud tubes and less frequently on other tubes. The visible tube seats, especially the superheater tube seats, should be inspected for evidence of leakage. Unless such leaks are quickly detected, they will result in cut tube seats and will cause wetting of the firesides on securing, with consequent corrosion of tubes or drums.

Heat-resisting steel baffle, seal, superheater support, and protection plates should be inspected regularly. Their service life is greatly shortened if hot gases of combustion are forced over them in excessive amounts because of a fouled tube bank.

Furnace brick pans should be inspected from beneath for evidence of overheating, buckling, and warping. These conditions are usually the result of the insulating qualities of the brick floor having been destroyed by slag. Visible parts of the casing should be inspected carefully for open seams, sheared bolts, buckling, or sheared welds.

It is important to remove soot and foreign matter from expansion joints, to permit the brick to expand when hot. After boilers are cleaned, fireside scale and soot must be cleaned off the brickwork.

Corrosion of Tubes Caused by Sulfur Content in Fuels

It is an established fact that if soot is allowed to remain in a boiler for any length of time, sulfur products present in the soot will corrode the metal. The extent of such cor-

rosion depends upon the amount of moisture present, which in turn depends upon the humidity of the atmosphere, as well as upon rain, fog, condensing steam, and spray. Loss of new tubes from this cause has occurred in the express type of boiler; pinholes have developed where the tubes enter the water drums and at other points where tubes are not readily accessible for cleaning. It is, therefore, important to have periodic inspections and removal of these deposits.

One of the principal causes of tube renewals in boilers having horizontal drums is external corrosion of tubes near the ends where they enter the lower drums. This deterioration of the metal is the result of those corrosive products present in the soot becoming chemically active with moisture (formation of sulfuric acid). The moisture may be absorbed from the atmosphere; it may be due to rain when smokestack covers are off; or it may come from leaky boiler, economizer, or superheater tubes, or from leaks passing through boiler-casing joints from machinery installed above the boiler. Figure 11-1 illustrates the results of neglect in cleaning the tubes near the water drum.

Corrosion can be kept to a minimum by preventing the entry of water or moisture, from whatever source, to the uptakes and firesides of idle boilers. After rain, fog, snow, or a period of excessive atmospheric humidity whereby the firesides of idle boilers may have become damp, the boilers should be promptly inspected and if necessary dried out. This inspection should be made whether or not smokepipe

i

L

Figure —Soot corrosion (sulfuric ocid corrosion).

covers were in place. The firesides of an idle boiler may be dried out by (1) lighting off a burner and raising the temperature of the water to about 212° F.; (2) circulating hot water from the feed system through the boiler; (3) placing electric heaters or lamp banks in the furnaces; or (4) lighting fires in pans in the furnaces. It is important that the firesides of idle boilers should not be permitted to remain damp.

METAL-CONDITIONING COMPOUND

When warranted by conditions, the cleaning of fireside surfaces may be assisted by the use of an approved metal-conditioning compound.

When compounds are used aboard ship, the operating schedule of the ship should be such that one or more of the boilers may be laid up for a sufficient period of time to make the application of the compound therein effective. In order that the compound may effectively penetrate the scale and rust on the metal surfaces, a conditioning period of at least 1 to 3 weeks is desirable; the longer period is preferable.

Before application of the metal-conditioning compound, the firesides must be thoroughly cleaned of all loose soot, dust, etc. (as covered in the preceding paragraphs), since such loose material will absorb compound and hinder its attack on the scale and corrosion. The boiler access doors and such side casing panels as are necessary to obtain ready access to the tube banks should be removed, thus facilitating both the cleaning and the application of the compound.

After cleaning, thoroughly spray the fireside surfaces of the tubes and drums with the metal-conditioning compound, using from 10 to 20 gallons per boiler, depending upon the size and condition of the boiler. The fireroom and furnaces should be well ventilated while the compound is being applied because the compound has a sharp, pungent odor when sprayed. The use of respirator masks may prove desirable for the comfort of the personnel working in the fireroom. After the compound has been applied, close the firesides.

After the compound has been allowed to take effect, the boilers may be lighted off; and when the boilers are hot, the tubes should be thoroughly blown. As a large amount of material will be removed from the tubes when they are blown, the ship should be clear of other vessels and the wind favorable. As soon as practicable after the tubes are blown, the firesides of the boiler should be opened up again and manually cleaned, and the accumulations not removed by the soot blowers should be brushed out. In most cases, any material found adhering to the tubes or drums will be fairly soft and easy to remove.

The metal-conditioning compound may burn under certain conditions or break down under heat into inflammable ingredients. Certain precautions must be rigidly complied with to prevent the possibility of a fire; among these precautions are the following:

1. Smoking or the use of naked lights should not be permitted in the fireroom while the compound is being applied. Take particular care to ensure that all portable lights are properly insulated, protected with wire or rubber guards, and fitted with steam-tight globes.

2. Do not apply compound to a boiler if another boiler in the same fireroom is steaming.

3. Compound must not be applied to a boiler connected to the same smokestack with a boiler which is steaming, even though the steaming boiler be in another fireroom, unless division

plates are installed to the top of the stack, preventing the mixture of the gases from the two boilers.

4. Particular care should be exercised when applying the compound in a boiler to avoid the use of excessive amounts which might collect in pockets subject to high temperatures when steaming.

5. After application of the compound in a boiler, particular care must also be taken when the boilers are first steamed, to detect any fires which might occur as a result of the compound soaking through casing joints into the drum insulation.

6. After compound has been applied to the boiler, work inside the boiler should be limited to emergency work until the compound has been removed by firing the boiler.

WATER-WASHING OF FIRESIDES

When gas passages have been clogged to such an extent that mechanical cleaning is ineffective, the firesides should be cleaned by hot-water washing. This procedure should not be used indiscriminately and is ordinarily necessary only for inaccessible superheaters and economizers, or in cases of extremely hard slag. The following should be kept in mind when resorting to water-washing: (1) possibility of damage to brickwork, (2) possibility of acid corrosion of tubes and drums, (3) availability of sufficient fresh water, and (4) protection of electrical equipment around or under the boiler. Boilers must be dried OUT carefully immediately after water-washing.

Methods of water-washing. —In general there are two approved procedures for water-washing: (1) use of a water lance for locally applying the water, and (2) use of soot blowers for distributing the water in the tube banks. The first method requires less water and results in less wetting of the boiler and refractories, but it has the disadvantage that all parts of the tube banks cannot be reached. The second method requires more water and results in considerable wetting of the boiler. In removing slag by means of water-washing, the soaking time to permit softening of the slag is more important than the application of large quantities of water. When sufficient water is available, best results can be attained by employing a combination of the two methods. The procedure described below involves the use of both methods. When necessary to conserve water, use the lance method. The latter will ordinarily be found sufficient for economizers and (when necessary to wash) the generating, water wall, and water screen tube banks. Figure 11-2 shows details of a water lance.

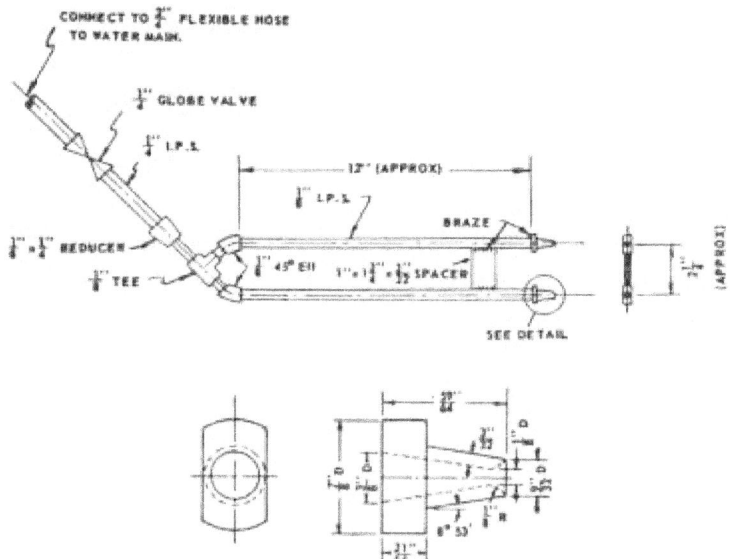

Figur* 11-2.—Water lanes.

Preparations for Water-Washing

1. Open, remove, or loosen access doors and panels to provide access to the firesides and drainage from the furnace, and around drums and headers. It may be found desirable only to loosen certain panels, to reduce splashing on adjacent machinery or personnel. Line up a fire and bilge pump for pumping the waste water overboard.

2. Install canvas shields or gutters where practicable to reduce wetting of the refractories. When an

economizer must be washed, the boiler can be protected by installing canvas under the economizer and directing the water through the side casings. Refractory corbels at the water drums should be protected as much as possible to prevent water from flowing between the corbel and the drum. It will be found that a tight joint at this location is difficult to make. A rubber strip gasket with proper bracing should be used when practicable.

3. Soot blower elements not to be used for spraying water should be blanked off so that leaky valves on such elements will not result in wetting of undesired places and in a waste of water. Close the soot blower root valve. Make arrangements for attaching the hose to a convenient place in the soot blower piping.

4. Prepare the spray nozzle for attachment of the water hose.

5. Line up a source of hot water (150° to 200° F.) to a reciprocating feed pump. Attach a hose to the discharge side of the pump.

6. Provide gloves, goggles, slickers, and such other equipment as advisable and give personnel detailed instructions on the safety measures they should take to avoid being scalded with hot water.

7. Provide litmus paper for testing the waste water for acidity. (The acid from scale or soot will turn the litmus paper red.)

8. Rig canvas over electrical equipment and machinery likely to be splashed during water-washing.

9. Provide a compressed-air lance for use in knocking off loosened scale following water-washing.

Washing economizers. —When economizers are to be washed, apply the water spray at 150 to 200 psi pressure over the top of the tubes until the economizer surfaces are clean. If soot blowers are to be used, attach the hose to the soot blower piping and rotate the elements several times with 200 psi pressure on the line. Repeat the lanc-

ing until satisfactory cleanliness has been obtained throughout the economizer.

The smokestack and uptakes can be conveniently water-washed at the same time as the economizers. To clean the insides of the smokestacks and uptakes, connect a %-incli water hose to the auxiliary feed pump discharge, with the hose long enough to reach to the top of the smokestack. Using the proper lance, start water-washing the inside of the smokestack and uptakes, working from the top down and using about 100 psi water pressure. When the economizer is reached, use the same lance to direct a hig-h-velocity stream of water down through the tubes. The water pressure for this latter operation should be about 200 psi. After water-washing, it is desirable to complete the job by cleaning the bottom tubes with a steam lance, working from the under side.

Washing superheaters. —Inaccessible superheaters will usually require the use of soot blowers. Attach the hose to the soot blower piping and, starting with the upper element, rotate the blowers 5 to 10 times with 200 psi pressure on the line. When the water runs clear from the superheater, secure the soot blowers. Apply the water lance to areas where deposits still remain, until satisfactory cleanliness has been obtained throughout the superheaters.

Washing generating tubes. —Apply the lance to the various parts of the bank. Access may be had from inside the furnace, from the end casings where doors or panels are removed, and from the outer side of the banks. Soot blowers should not be used unless the lance method is ineffective and plenty of spare fresh water is on hand.

Drying Out After Water-Washing

Immediately on completion of water-washing, carefully drain all excess water from the boiler and remove scale deposits on drums, in casing corners, and on refractories. Restore the soot blower piping to its service condition. Close up the boiler and light off one burner, using the

smallest plate available and 200 psi oil pressure to ensure good atomization. Every 15 minutes secure the burner for 15 minutes to allow the moisture in the refractories to evaporate slowly. Rotate the burners occasionally and continue the alternate firing and idle periods for about 5 hours. Allow the boiler to stand idle for an hour for air drying and then light off and steam it to an auxiliary load. If not possible to do this, repeat the alternate firing and idle periods for another 3 hours.

During the drying-out period superheaters should be protected in the usual manner—such as protection steam, bleeding to bilge and auxiliary exhaust line, or steaming the boiler to an idling generator.

After the drying-out period it is advisable to secure the boiler and open the furnace for inspection of the refractories. Particular attention should be given to the refractory corbels and for any signs of shrinkage or damage caused by too rapid evaporation of the water. Subsequently, when firing the boiler while under way, close watch should be kept until it is plain that no damage has been done to the refractories.

If, after drying out, the boiler is to be idle for some time, it is good practice to spray it with metal-conditioning compound.

WET STEAM METHOD OF CLEANING FIRESIDES

A method of cleaning firesides of boilers by lancing with wet steam as the cleaning medium has been used with reportedly good results.

The method is based on the solubility of the binder in the slag and removal of the loosened slag by the scouring action of the steam jets. As the dissolving medium is the moisture in the steam from the lance, it is important that the boiler be cold during the cleaning in order that the wet steam will not dry up immediately upon contact with hot tubes.

The following instructions should be followed when occasion arises for using this method.

Provide a 1-inch branch connection on a 70 to 150 psi auxiliary steam line in the fireroom. From this connection run a 1-inch temporary pipeline to the boiler front. In order to obtain a steam of the desired high moisture content, this pipeline should be provided with a loop or coil inserted in a 30-gallon drum or other suitable container full of water. A 40-foot length of 1-inch pipeline, uninsulated and with a loop inserted in a 30-gallon drum full of water, should produce the desired moisture content in the steam. Provide two %-inch hose connections at the boiler end of the 1-inch steam line for attaching %-inch steam hose and lances.

«M J /- 4 HOLES, 3 ON EACH SIDE SPACED ■ ^ DIAMETERICALLY OPPOSITE.
#6 DRILL AND^3 REAMER.
SEAL WELD END
DETAIL OP STEAM LANCE END

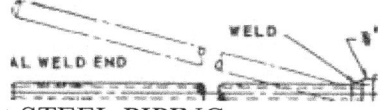

t STEEL PIPINC
^EAL WELD END iC^-^fj^j^^l t V," STEEL PIPING
LACCINC
THREADN. FOU VALVE^
Figurt 11-3.—SUam lane* for generating tub**.

For maximum cleaning efficiency, it is advisable to use several different types of steam lances, as illustrated in figures 11-3, 11-4, and 11-5. The lance shown in figure 11-3 is recommended for the generating tubes, the lance shown in figure 11-4 for the superheaters, and the lance shown in figure 11-5 for bare tube furnace walls, economizers, smoke pipes, and uptakes. These steam lances may be either straight or bent, as indicated by the dot and dash lines in figure 11-3, to best suit the boiler to be cleaned.

Using the proper lance, start cleaning the underside of the steam drum and work all the way back through the tubes. Clean small sections at a time to ensure thorough cleaning. If practicable, it is better to start with the tubes farthest away from the operator, because deposits

REAM WITH#S REAMER TO DEPTH OF J".
STEEL PIPE

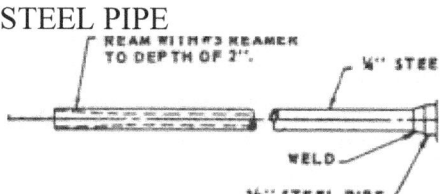

LACCINC-^ THREAD FOR VALVI Figure 11-4.—Sfeom lone* for tuporh«at«r tubot.
WELD. 9Jk" STEEL FIFE.

' ^ ' ' w^'^^' ^""'"^^ STAINLESS STEEL.
1 DRILL 'V HOLE THROUGH ENTIRE LENGTH. TAPER WITH *0 REAMER
IDEPTH OF TAPER

DETAIL OF NOZZLE AT ONE END AND T DEPTH
AT OTHER END.
- i" , i"\
■2t
WELD-

I
21

'4 STEEL PIPE-^ WELD
V STEEL PIPE^ LAGGING-^ THREAD FOR VALVE

Figure 11-5.—Steam lance for economizer tubes.

on the tubes nearest the operator will be softened up for cleaning as the lance is pushed in and out.

Clean the economizer, starting from top and working down. The design of the economizer in some boilers makes it very difficult to clean by using a steam lance. Inasmuch as economizers are almost separate parts of the boiler, and since the deposits on the economizer are usually soot deposits rather than hard slag, water instead of steam may be used as a cleaning medium.

AS soon as possible after the boiler has been cleaned, it should be lit off in order to dry out the damp insulation and prevent possible corrosion.

CLEANING OF SLAGGED SUPERHEATERS BY THE "SWEATING" METHOD

A method of cleaning fireside slag from convection type superheaters by forming a "sweat" on the outside of the superheater tubes (induced by circulating cooled water through them and thereby condensing moisture from the air) has been found to be successful and is less hazardous to brickwork and insulation than hot-water-washing.

This "sweat" is sufficient to convert the hard slag into mud, as the predominance of sulfates and chlorides in superheater tube slag renders ordinary deposits water-soluble. Most of this mud will then drip off and the remainder can be blown off by means of an air or steam lance.

To accomplish this cleaning process, use a portable open water tank or drum placed either on the ship's deck or fireroom floor plates. Fill this container with water and ice. Pump water from this tank with a portable submersible or other suitable portable pump to the superheater drain connection, using available hose as required. Let the water circulate through the superheater, out through the superheater air vent, and back to the tank through another section of hose.

The superheater inlet and outlet flange connection should preferably be blanked off to isolate the superheater. However, if this is not practicable, the whole boiler and superheater may be filled and maintained full of water during the circulation process (circulate the superheater water only), as this will not materially affect the chilling of the superheater tubes.

If the humidity of the air is high and the water is cold, condensate from the air will soon moisten the slag. It will usually be found advisable to speed up the formation of "sweat" by a very mild use of steam from a steam lance

introduced into the tube bank during the circulation period. Circulation of air through the tube bank and through the uptake should be held to a minimum, and ice in the tank replenished as necessary.

When the mud starts to drip from the tubes, start blowing it off with an air or steam lance. Continue circulation of water and blowing of tubes until the tubes are clean. It is important that once the above cleaning is started it be kept up without interruption, as an interruption may result in the slag returning to lis original hardened condition.

BOILING OUT BOILERS

It is seldom necessary to boil out a boiler; it should not be done except for a newly erected boiler, or after a major tube renewal, or in cases where careful examination has disclosed scale conditions so serious that boiling out is necessary. The boiling out of new boilers must be done thoroughly in order to remove all trace of grease or preservative compounds on the interior surface. During the first 6 months of steaming, new boilers should be washed out frequently in order to remove all scale and foreign material from the feed system which may have collected in the boilers. The following procedure should be used when boiling out:

1. Provide steam connections to the boiler bottom blow fittings and to the lowest opening in the superheater and in the economizer. Blank off the piping connection between the economizer and the steam drum and provide independent lines from the economizer and the superheater vents. The piping arrangement for boiling out a boiler is shown in figure 11-6.

2. Make up the boiling-out mixtures separately for the boiler, the economizer, and the superheater. When boilinfir out to remove scale, use 40 pounds of Navy boiler compound for each 1,000 gallons of water capacity. When boiling out to remove preservatives, use 40 pounds of Navy boiler compound and 10

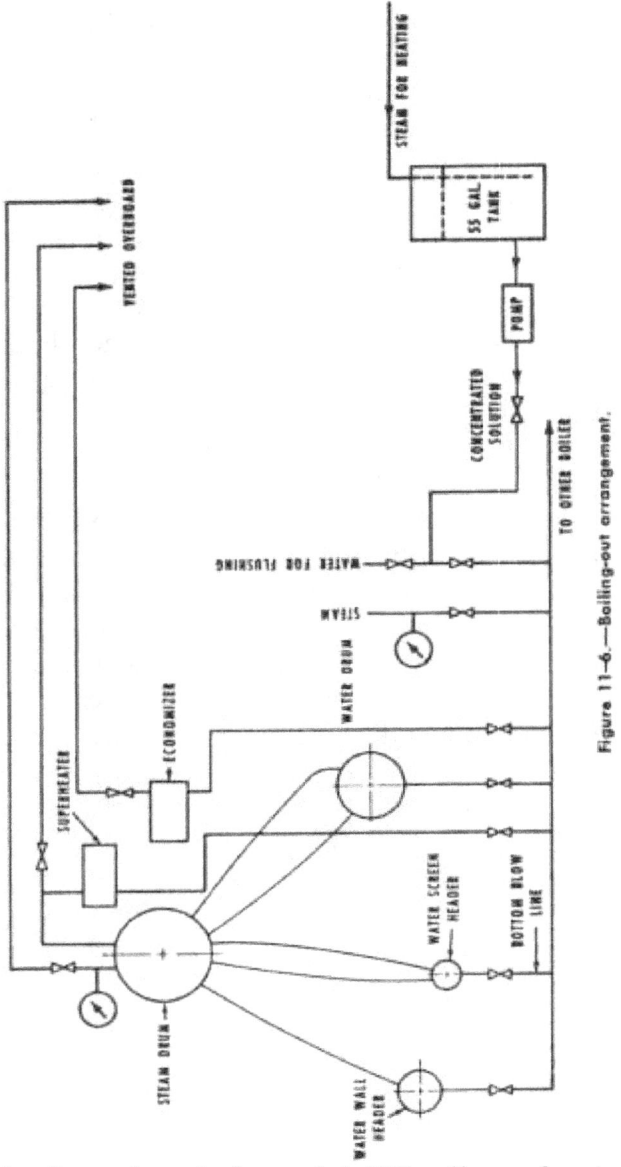

Figure 11-6.—Boiling-out arrangement.

pounds of caustic soda for each 1,000 gallons of water capacity. Use just enough hot water to dissolve the compound (and the caustic soda, if used).

3. Pump the boiling-out mixtures into the generating, superheater, and economizer sections of the boiler. The mixture for the generating section should be divided between the water drums and the headers. Add only enough fresh water to the generating section to bring the water level in sight at the lowest visible point in the water gage glass. Do not add water to the mixtures for the superheater and the economizer.

4. Supply steam through the connections at a pressure sufficient to maintain about 50 psi on the boiler.

5. When the water level in the steam drum reaches 2 or 3 inches above normal level, give a surface blow to 1 or 2 inches below normal level. Allow the economizer and the superheater to vent continuously, but restrict the flow. Care must be taken to ensure circulation throughout the superheater; for example, if the boiling-out connection is made at the drain of the last pass, the vent valve in this pass and all drains or vents in any intermediate passes should be closed, thus

allowing only the vent in the most remote pass to act as such. Throttling of the latter will avoid waste of compound by effecting circulation back to the steam drum. Continue boiling, blowing down, and venting for 48 to 72 hours. One or two bottom blows may be given, but too many will cause excessive loss of compound. Samples from the bottom blow line should be taken periodically in glass bottles and kept during the boiling-out process. A visual examination of the sludge in the bottles will indicate the results being obtained and whether or not it is profitable to continue the boiling. When using either the bottom blow or surface blow valves to give a short blow-down to the boiler, great care must be taken to maintain a pressure within the

boiler of about 25 psi above the pressure of the sea water on the overboard discharge valve. If this is not done, back flow of sea water into the boiler may occur.

6. At the end of the boiling-out period, shut off the steam supply and vent the boiler to release the steam pressure. Empty the boiler through the hose connection in the bottom blow line. If a fire and bilge pump cannot be connected to the boiler, drain the water to the fireroom. Drain the economizers and superheaters. Wash out the steam drum, tube banks, and water drums with fresh water. Economizers may be flushed either through the temporary boiling-out piping or through the feed piping; superheaters may be flushed either through the temjx)-rary piping or by filling the boiler and flooding the superheater.

7. Inspect thoroughly for cleanliness. It is advisable to run a rag on a rod down various tubes to determine the condition. A black smudge-like coating will frequently be found at tube ends of new boilers, although the remainder of the tube may be clean. Slight traces of this black deposit are not harmful and may be wiped out with kerosene. If any grease is found, or if the black smudge at the tube ends is greasy, it will be necessary to boil out again. Unless the boiler is scrupulously clean, a second boiling out is advisable.

8. As soon as men can work in the boilers, wire brush the drums and the ends of all tubes. .The interior of all tubes should be cleaned with an approved type of power-driven tube cleaning outfit.

9. Follow up the cleaning of all tubes by blowing them out thoroughly with a strong air jet; if the boiler is to be closed and filled immediately after cleaning, the drums and tubes may be washed out with fresh water.

If the inspection of the boiler reveals the necessity for a second boiling out, follow the same procedure in the second boiling out as in the first. If the boiler is ready to have the fires lighted, an alternate method is to boil, using an outside steam supply, for 24 hours; then to light a burner with a very small sprayer plate, and steam for 24 hours. Before lighting off, be sure the piping connection between the economizer and the steam drum is not blanked off. Every half hour, with the water level at three inches above normal water level, blow down with surface blow to 1 or 2 inches below half glass and feed back to 3 inches above normal level. Vent the superheater continuously to hold the steam pressure at 50 psi. When this second boiling-out is completed, open and examine the boiler for cleanliness as before.

MANHOLES AND HANDHOLES

Whenever manhole and handhole plates are removed so that a boiler may be inspected and cleaned, the manhole and handhole plates, gaskets, and gasket seating surfaces must be cleaned and tested for accuracy of fit.

The clearance between the shoulder of the manhole plate and the drum head must not exceed Yiq inch, when the manhole plate is centered accurately. If the clearance is found to be greater than i/ig inch, the plate should be built up by electric welding at the inner edge of the shoulder. Except in an emergency, this welding should be performed by a naval shipyard, so that the plate may be stress-relieved after it is welded, and so that the welded surface may be refaced.

If a manhole or handhole plate is so warped that the gasket flange cannot be satisfactorily trued up, the plate should be discarded and a new one installed.

Whenever manholes or handholes are opened, new gaskets should be fitted. Always be sure to use the correct size and type of gasket. Never use graphite, oil-and-graphite mixtures, or any other make-up compound when

installing gaskets on manholes or handholes on modern, high-pressure boilers.

Before new gaskets are installed, the two gasket seating surfaces (one on the plate and one on the drum head) must be cleaned and examined for evidence of pitting or corrosion. Any pieces of old gasket material adhering to the seating surfaces must be removed by scraping, before the new gaskets are installed.

To ensure proper positioning of a manhole gasket, contract it on the long axis until the inner edge of the gasket fits the shoulder snugly at the ends of the long axis of the manhole plate. The clearance between the gasket and the shoulder should be equalized at the top and bottom of the short axis. Do NOT allow the outer edge of the gasket to protrude at any point beyond the gasket seating surface in the drum head. If an edge protrudes, the gasket may unravel when it is compressed by the tightening of the manhole cover. Any gasket which protrudes beyond the edge of the gasket seating surface should be discarded.

When the boiler is given a hydrostatic test, the pressure of the water usually forces the manhole and handhole gaskets into place and thus ensures proper seating. The plates are first set up lightly. When the boiler is ready for testing, the pressure should be pumped up to within 50 psi of the hydrostatic test pressure, regardless of any leakage from the manhole or handhole plates. Leakage is likely to be general at first, but it will decrease as the pressure is increased. When the pressure is within 50 psi of the test pressure, most of the leakage will stop, although the nuts will still be loose.

If some plates are leaking very badly, the trouble is probably due to improper seating of the gaskets. As a rule, you will find that the gasket is caught on the outer edge, between the edge of the plate and the edge of the counterbore for the seat. A light blow with a hand hammer on the outside of the plate will usually relieve the tension on the gasket and allow it to seat properly.

After leaky gaskets have been adjusted, and while full

test pressure is on the boiler, all plates should be set up firmly. Use only the wrenches specified for this purpose.

Some economizer headers, and occasionally some superheater headers, are fitted with handhole plugs instead of handhole plates. In addition, some economizers have bayonet-type clean-out plugs on the front ends of the tube loops to allow access to the tubes at the return-bend end. Detailed instructions for installing and removing the plug-type handhole fittings and the return-bend clean-out fittings are given in appropriate manufacturers' instruction books.

HYDROSTATIC TESTS

Hydrostatic tests are conducted on a boiler either to prove the tightness of all parts or to prove the strength of the boiler and its parts. A boiler should not be tested to a higher pressure than that required to prove the object of the test.

Test for Tightness

To prove the tightness of all valves, gaskets, and fittings of boilers under operating pressures, the following hydrostatic tests should be made upon the completion of each general overhauling or repair affecting such parts. The test should also be made at any time that it is desired to inspect for leaks.

Examine, wash out, and close the boiler. Close all connections except air cocks and vents, and use the emergency feed pump to fill the boiler completely. When all air has been expelled, close the air cocks and the vents, and apply a water pressure equal to the boiler design pressure. After all gasket and plate leaks have been remedied, pump again to this pressure and close the valve in the pressure line, and, unless the boiler is required for immediate use, note the drop in pressure during a considerable number of hours.

To avoid complications arising from a change in pressure caused by changes in temperature of the boiler, of

the water, or both, the water used should be about the temperature of the boiler and the fireroom, and the latter should remain at approximately the same temperature throughout the test. Under these conditions the drop in pressure, if the boiler and fittings are tight, should not exceed 1.5 percent over a 4-hour period. Where loss through leakage is demonstrated to be excessive, and such leakage exists at tube joints, the leaky tubes must be rerolled. A high loss of pressure, when very slight leakage exists through tube joints, drum seats, and gaskets, is almost certainly caused by leaky valves or fittings and these should be overhauled as necessary. A tube seat should not be considered tight unless it remains entirely dry under hydrostatic test. Any tube that cannot be made tight should either be renewed or plugged and then renewed at the earliest opportunity.

Test After Renewal of Any Pressure Part

A hydrostatic test at times the boiler design pressure should be made after each renewal of any pressure parts and at any other time when considered necessary by the commanding officer.

Hydrostatic Tests at 5-Year Periods

A hydrostatic test at $1^\wedge$ times the boiler design pressure should be made for strength of parts 5 years after the boiler has been placed in service and at 5-year periods thereafter. If practicable, this test should be made while the ship is at a naval shipyard. While under pressure, the boiler should be subjected to a careful visual examination accompanied by a moderate hammer test.

IDLE BOILERS TO BE KEPT FULL

Except in emergencies which require the boiler water to be maintained at steaming level, idle boilers and boilers not open for inspection should be kept full of fresh water having alkalinity between 2.5 and 3.5 epm.

As soon as practicable after securing, the boiler should be filled with hot water to which sufficient compound should be added to ensure the prescribed alkalinity. Filling should be commenced, if practicable, by the time all pressure has disappeared from the boiler. Even if the boiler is to be examined within a few days after completion of steaming, the water should not be allowed to remain at steaming level; the boiler should be pumped full under about 50 psi hydrostatic presure. It is essential that the alkalinity in all parts of the boiler, including the economizer and superheater, be as specified.

When the boiler (except for the superheater) is full to the top of the steam drum, commence circulating the water. Circulation is accomplished by pumping water from the bottom blow line through the emergency feed pump and back into the boiler by way of the feed line; a hose connects the bottom blow line and the emergency feed pump.

Take samples from the feed pump discharge, and test them for alkalinity. Recirculation should be continued until at least three successive samples taken near the end of the recirculation period show approximately the proper alkalinity. Recirculation should be carried on at a high pumping rate so as to ensure thorough mixing of the chemicals in the boiler water.

The superheater should be filled after recirculation in the rest of the boiler has assured proper mixture of the boiler compound and the water. Pumping during this stage should be kept at a low rate in order to ensure that only water of the proper alkalinity enters the superheater. The water in the superheater should be tested for alkalinity, and necessary steps taken to correct any deficiency. In boilers that have provision for recirculation both from the superheaters and from the bottom blows, the superheaters may be filled in the beginning, and the recirculation proceeds as described above.

Care should be taken to ensure that all air is excluded
from superheaters and economizers, and from connecting lines between the steam drum.

Idle boilers should not be used for trimming ship nor used as reservoirs for storing any other water than that intended for steaming purposes. Run down or empty the boilers only when necessary for lighting off, examination, cleaning, or overhauling, and change the water only when it has become unfit for further use.

Water may be used from boilers for purposes of makeup feed when no other fresh water is available and none can be supplied by the evaporators. However, when this is done, the water in the boilers from which such water is taken must not be left at a level among the tubes, and the boilers must be pumped full again as soon as a sufficient supply of fresh water can be obtained.

MAINTENANCE OF BOILER CASINGS

Overhauling boiler casings and uptakes to make them tight is work that is tedious and difficult. Unless this work is carefully followed through, the returns may appear poor for the labor involved. Where this work has been conscientiously and thoroughly done, a marked reduction in fuel consumption will be the result. When boiler casings are inspected, they must be examined very carefully. No one item is too small to overlook. Special care must be taken to see that the boiler-casing doors are tight; the gaskets must be kept in good condition, and the dogs must be properly secured.

Waste in fuel due to air leaks in boiler casings may amount to 10 percent or more.

Chilling Effect of Air Leaks

Air leaking through boiler casings into the furnace does not become intimately mixed with the fuel and, therefore, does not aid in combustion. On the contrary, it has a decided chilling effect on both the gases of combustion and the heating surfaces. No air should enter the boiler at any place except through the air registers.

Testing for Tightness

During each naval shipyard overhaul period, and at such other times as are found to be required with any particular installation, boiler casings should be gone over completely and tested for tightness. Casing tightness tests should be conducted by competent personnel.

The OUTER casing of an air-encased boiler can be tested for tightness by spreading soap and water along the casing joints, access doors, and fitting entrances, and putting the casing under air pressure. Leakage will be shown by bubbles along the joints, doors, and fitting entrances that are not tight. The tightness of the outer casings is not of vital importance, from the standpoint of boiler efficiency. However, the outer casings must be tight in order to provide maximum protection to fireroom personnel, in the event of atomic, biological, or chemical attack.

The INNER casing of air-encased boiler installations should be gone over completely and tested for tightness at intervals of not more than 5 years, or whenever serious inner casing leakage is suspected.

Casing air-relief valves are installed on some air-encased boilers, to prevent possible damage to the air casings in the event that the burners are secured and registers closed without a simultaneous shutting down of the blowers. Such valves should be checked at each boiler cleaning period.

Corrosion Due to Water

Air leaks in boiler casings are usually due to the fact that the casings were not properly sealed when overhauled. Another source of air leaks is corrosion where water has dripped onto the casing. To avoid this, install drip pipes from the air cocks to carry drips clear from the top of the boiler casings and to prevent drips from falling on the top of the boiler casing and running under it at breaks. This fitting prevents much external boiler

corrosion caused by leaky cocks or by overflow when a boiler is filled.

Inspect around the base of the stack for any leaks that would allow water to get inside or on the boiler casings.

INSPECTION OF DRUM EXTERIORS

The exteriors of boiler diums must be watched for indications of corrosion under the coverings. Any signs of rusty streaks from the covering, or of corrosion around the edges of the covering or around the tubes or pads on the drums, should be investigated immediately. Where machinery or joints in piping are located above the boileis, there is always danger that water will drop onto the boiler drums and find its way under the boiler covering, thus accelerating corrosion of the drum exteriors.

Whenever the exteriors of the drums are exposed through the removal of the insulation or the lifting of boiler casings, the condition should be recorded in the boiler record sheet. Where the indications are that corrosion is progressing and corrective measures beyond the capacity of the ship's force are imperative, work request for the necessary assistance should be submitted without delay.

SMOKE-PIPE COVERS AND RAIN GUTTERS

Smokestack covers should be habitually kept on the stacks of idle boilers to prevent the access of moisture into the uptakes and boiler firesides. When unusual circumstances prevent the use of covers on the stacks of idle boilers, the firesides should be inspected for the presence of moisture. If moisture is present due to the omission of the covers, necessary action should be taken to dry the firesides.

Smokestack rain gutters, where installed, should be frequently inspected and, if necessary, cleaned to remove soot accumulations and to ensure that the drains are not plugged.

EXAMINATION OF SLIDING FEET

At each boiler cleaning period, the sliding feet should be examined to ensure that they are free to move. They should be cleaned of dirt and rust insofar as is practicable and then oiled or greased. The sliding feet of boilers of recent construction are provided with grease fittings for this purpose.

QUIZ

1. Why should a boiler be completely dried out Insfore it is inspected?
2. In what records must the results of boiler inspections be entered?
3. How many times must the brushes be passed through each tube?
4. After a boiler has been cleaned, blown out with air, washed with water, and dried, what should be done to check the adequacy of the cleaning job?
5. Why is hard scale sometimes difficult to detect on boiler tubes?
6. Why should pits NOT be filled in with cement or "smooth-on" paste?

7. How frequently should boiler watersides be mechanically cleaned?

8. How frequently should the interiors of superheaters be mechanically cleaned?

9. In what way does the presence of soot represent a corrosion hazard?

10. What are the two most likely causes of tube bulging?

11. How long a period of time is required for the effective use of metal-conditioning compound?

12. What two kinds of damage are most likely to be caused by hot-water washing of the firesides?

13. What are the two generally approved methods of water-washing?

14. Which method of water-washing is most economical of fresh water?

15. Under what conditions should the soot blower method of water-washing be used?

16. What drying-out procedure should be used after water-washing?

17. Why must the boiler be cold during a wet-steam cleaning of the firesides?

18. Which method of cleaning the outside of superheater tubes is likely to cause least damage to the brickwork?

19. How is the sweat formed on the outside of superheater tubes, when the fireside slag is being cleaned by the sweat method?

20. Why is it important that the sweat method of cleaning be continued without interruption, once it has been started?

21. Under what conditions is it necessary to boil out a boiler?

22. What are the two reasons for conducting hydrostatic tests on boilers?

23. When should a boiler be given a hydrostatic test for tightness?

24. How often should a boiler be given a hydrostatic test for strength of parts?

25. How much water should be kept in idle boilers?

26. What alkalinity is specified for the fresh water which is kept in idle boilers?

CHAPTER

FIREROOM MAINTENANCE

A WORD TO THE WISE

A clean and well-kept fireroom is one indication of a well-trained efficient fireroom force. The upkeep of machinery, the prevention of rust and corrosion, and the proper care and stowage of tools and equipment, all reflect the interest and ability of the fireroom personnel.

CLEANING GEAR

Wire brushes, scrapers, fox-tail brushes, brooms, soap, and rags are some of the necessary equipment for keeping the fireroom clean. You should teach your men how and where to use this gear. They should have a special locker for stowing gear which is not in use, and they should be instructed to return gear to its proper stowage place upon completion of their work. Gear left adrift may cause serious injury to personnel. Oily rags create a fire hazard for the whole ship, and should be cleaned or destroyed. If they are to be destroyed at a later time, they should be stowed in a fireproof metal bin.

Economy in the use of cleaning gear is important, since the Engineering Department must operate on a quarterly allowance. If you use a 3-months* supply in 2 months, it will be necessary either to draw on the following quarter's allowance or do without gear required for good maintenance.

RUST PREVENTION

Corrosion can be almost entirely prevented by keeping the fireroom and equipment dry. Sources of moisture must be found and eliminated if possible, and equipment must be protected from the effects of moisture by the use of rust-preventive compounds, special protective finishes, or paint.

Keeping Fireroom and Equipment Dry

All joints, valves, and cocks of the various lines of salt or fresh-water piping must be kept tight. They should be examined frequently for leaks: a leak that is allowed to continue becomes a source of corrosion. Leakage at a joint where a branch line joins another line usually occurs because of improper allowance for expansion in one or the other of the lines, or because of excessive vibration. You can usually eliminate the trouble by making a slight alteration in the anchorages, connections, hangers, or lead of the piping to allow for the required expansion and to prevent strain. You may have to fit supports to prevent vibration. Fittings with small gasket leaks should be taken up on immediately and the gasket renewed if necessary. Leaky packing glands on valves should be taken up on or else the valve should be repacked.

Keep fireroom bilges dry and make frequent inspections of the bilges for any corrosion. Metal surfaces that have become corroded should be scraped thoroughly and brushed to bare metal with a wire brush. Paint or a rust preventive compound should then be applied.

Rust-Preventive Compounds

The purpose of rust-preventi ve compounds is to provide a PROTECTIVE FILM over corrodible metal surfaces not protected by paint, galvanizing, plating, and the like. To a large measure, such a protective film will prevent corrosion due to moisture, salt water, salt-laden air, dust, or corrosive vapors. Rust-preventive compounds are used

for the preservation and protection of spare parts of machinery, such as boiler tubes, and tools which are not required for use for a long period of time.

PAINTING

Since every petty officer is faced at one time or another with questions in regard to painting, a brief discussion of how^, when, and where to paint should be of value to you.

In general, never repaint bulkheads, decks, and partitions, except where necessary for the preservation of the metal surface (that is, where the paint is flaking or peeling, or where there is evidence of corrosion underneath the paint). Normally paintwork should be brought up to standard by thorough cleansing with Navy standard cleansing agents, such as solvents and soap and water. The practice of painting instead of cleaning the paintwork is expensive in labor and in the cost of paint, and is not required for preservation. Where paint has been knocked off, however, you must touch up the damaged spots in order to preserve the surface.

For further information pertinent to the different phases of painting, handling of paints, paint materials, use and care of painting equipment, and safety precautions, consult the latest edition of appendix 6 to the General Specifications for Building Vessels of the United States Navy, and chapter 19 of BuShips Manual. Since these publications are always subject to change and revision, you'll have to keep up with such changes and revisions by periodically making a check of the latest issues on file in the engineering department or in the log room.

Materials Used in Painting

All materials used in painting should conform to the requirements in the applicable Federal or Navy Department specifications. These materials are made to give good hiding and durability in thin films and should be thinly applied in order to minimize the fire hazard, to prevent cracking and peeling, and to conserve weight.

Preparing Surfoces

Never apply paint over damp, oily, or greasy surfaces, nor over rust or scale. Remove grease or oil with a nonflammable solvent, and do not apply paint until the surface is dry. Before painting ferrous metals, thoroughly scrape the surfaces and brush to bare metal; then clean pits and holes so carefully that no trace of rust or scale remains. The tools used for removing scale should have rounded edges (about %4-in. radius) to lessen the danger of damaging the surface upon which they are being used. Loose dirt and dust should be swept out of the way, so that the workman will know when all rust and scale have been scraped from the surface. Care should be taken to remove all dust from the cleaned metal just before application of the primer.

The bulkheads, shell, and other plating on destroyers are extremely thin, and are built partly of galvanized material which is easily damaged by the use of scaling hammers. Therefore you should always remove old paint and rust on structural work or fittings of such vessels with WIRE BRUSHES and SCRAPERS. Use scaling hammers only where you find scale that cannot be removed in any other way. Take particular care to distinguish between old paint and rust or scale.

To prepare the painted surface of metal for repainting, first scrub with soap and water and then rinse thoroughly. (Be careful not to allow the water containing soap or soap powder to run down upon or over other painted surfaces, as this causes unsightly spotting.) Where the old paint is lumpy or thick, use fine sand with the soap, but under no circumstance should you use steel brushes or scrapers unless the complete surface is to be cleaned. Be sure that the surfaces are thoroughly dry before applying the paint.

Never use paint remover for extensive removal of paint; use it only where chipping and scraping cannot do the job satisfactorily. When paint remover is used, allow it to remain on the surface as long as is necessary to blister and lift the paint; the paint can then be removed

by scraping. Always take precautions against careless use of the paint remover, which frequently contains highly flammable solvents. There are two types of paint remover available for issue—the flammable solvent type and the nonflammable type. The use of the first type in the interior of ships is a potential fire hazard. On the other hand, nonflammable paint remover contains chlorinated hydrocarbons, which possess toxic properties. Both types, therefore, should be used with caution.

As a general rule, galvanized surfaces or other surfaces with special protective finishes should not be painted. But in the event that you are going to paint a galvanized surface, first apply one of the approved cleaners with a large brush. These cleaners contain phosphates, phosphoric acid, and suitable solvents and wetting agents to coat the surface with a thin phosphate coating. The coating deposits a thin, crystalline film on the surface of the metal, and this film acts as a bonding surface to furnish a good adhesion surface for paint. Phosphates are used because they retard corrosion, especially under-film corrosion.

After applying the cleaning solution, rinse the surface with fresh water (preferably hot) and apply the specified primer. Do not apply the primer until the surface has dried thoroughly. You should also take all possible measures to minimize the handling of the surface before applying the primer, because moisture deposited from hands will cause corrosion.

Applying Paint

Painting aboard ship should be done only under the supervision of experienced men. The supervisor should be responsible for proper care, adjustment, and use of the equipment, and for the quality of the work done. He should see that all safety precautions are observed.

Painting should never be attempted at temperatures below 32° F., except in urgent cases, and then only when proper authorization has been obtained. Painting should

not be done under conditions of high humidity, if water is condensing on the surface. Drying the air in the space, or raising the temperature of the surface being painted, will aid in preventing condensation.

Never apply paint until it has been thoroughly mixed in the container. Paint is composed of a liquid portion (vehicle) and a solid portion (pigment). On standing, the pigment tends to settle to the bottom, often leaving a clear liquid on the top. When the container is full, it is practically impossible to thoroughly mix the paint with a paddle. The best way is to pour off approximately Ys of the contents into a clean, empty container of the same size. With the paddle, mix the remainder until it is smooth (all lumps broken up), then gradually add the portion poured off, stirring after each addition. Finally, pour the contents from one container to the other several times, until thoroughly mixed.

The proper method for applying paint with a brush is as follows: Dip the brush into the paint not over halfway and then remove the excess paint by wiping the brush on the side of the container. In painting, the brush should be held at right angles to the surface. The paint should be applied horizontally and in one direction only; lift the brush at the end of each stroke and start the next stroke parallel to the previous one. This is termed LAYING ON. After laying on a small area, the work should then be crossed at right angles to eliminate streaks and brush marks. This is called laying off. The surface is finally smoothed to a thin coat by lightly brushing out.

Solvent-type paint, enamels, and varnishes may also be applied with a SPRAY GUN. In using this method, it is extremely important that you mix the paint thoroughly and strain it through a wire screen in order to remove foreign particles that might clog the nozzle of the gun and cause uneven application of the paint. An air filter must be connected to the main air supply line to prevent mixing of moisture or oil particles with the paint discharged from the gun.

Keep the air pressure as low as possible. The proper air pressure depends on the consistency of the paint, the length of the hose (or the height of the gun above the paint reservoir), and the speed of application. Experience will show you that most paints, varnishes, or enamels should be sprayed at hand-brushing consistency or slightly thinner. The use of too high an air pressure and an excess amount of thinner causes premature loss of the volatile material, which in some paints may prevent flowing and result in an "orange peel" effect.

Set the nozzle of the gun to obtain a width of spray that will give an even, cone-shaped discharge of paint on the surface. Hold the gun about 6 or 8 inches from the surface, and at right angles to the surface. Point the gun directly at the work, and pass it back and forth horizontally at a steady speed. A slight pull on the gun trigger emits air only; hold the trigger all the way back when painting.

When painting machinery, take special precautions to see that no paint is applied to the working surfaces, screw threads, oil holes, grease cups, or surfaces in contact with lubricating oil. Do not paint glands, stems, yokes, toggle gear, or any machined part of valves. Do not apply paint to fittings in such a way as to cause the bearings to become frozen. Rubber gaskets on light fixtures, manholes, doors, hatches, or any other fitting should not be painted, because paint causes the rubber to deteriorate. Paint that has been accidentally applied to gaskets should be removed immediately.

Care of Painting Equipment

Proper care of paint brushes is essential for their long life and good service. Before a paint brush is used, it should be rinsed with paint thinner to remove any dust or foreign matter which may be on the bristles. If you are planning to re-use a brush on the following day, remove excess paint and suspend the brushes by the

handles, with the bristles immersed to just below the bottom of the ferrule in paint thinner or linseed oil. The weight of the brush must never rest on the bristles. Brushes that are not to be re-used immediately should be carefully cleaned with thinner and then thoroughly washed with soap and water, and rinsed. The brushes should then be suspended from the handle racks or laid flat.

After using the spray gun, paint container, and hose, clean them thoroughly in the following manner: Disconnect the paint supply line, connect a container of paint thinner to the paint inlet, remove the nozzle of the gun, and pull the trigger. This will force the fluid through the unit and clean it thoroughly. Never soak the assembled gun in thinner, since this has a detrimental effect on the packing around the fluid valve, as well as on the grease and oil trigger action. The nozzle, however, may be soaked in thinner to remove paint from its jets. The adjustments at the back of the gun and the trigger action should be lubricated occasionally.

Manufacturers of spray guns usually provide permanent instruction plates regarding care and operation.

CARE AND STOWAGE OF TOOLS

Tools are indispensable in the repair and maintenance of Navy equipment; even the simplest routine jobs involve the use of tools. In order to keep these tools in good condition, see that they are handled carefully. Instruct your men to use the proper tool for each job. For example, the screw driver is designed to loosen and tighten screws. Don't allow it to be used for a cold chisel or pry bar, since this may ruin the handle, chip the point, or bend or break the blade.

Tools not in actual use should be kept in a safe place; if they are left lying adrift, they may cause injury to fire-room personnel, or they may be broken or lost. To save time and to avoid needless waste, assign each tool to a place in the tool locker or tool box and make sure that each tool is kept in its proper place. After working in a

boiler, make sure that no tools are left inside the steam drum, and as a double precaution, check to see if any tools are missing from their assigned spaces in the tool locker. Before your tools are put away, inspect them for dirt and rust. If there is evidence of either, have them cleaned and coated with a light oil. Dirt and rust are vicious enemies of all tools.

While using tube expanders, clean them frequently to remove any grit or other foreign matter which will cause the rollers and mandrels to flake. Flaked mandrels and rollers should be replaced, as they cause excessive wear of other parts. When expanding tube operations have been completed, clean the expanders thoroughly and replace all worn parts. Then immerse the entire tool in an oil bath for protection against corrosion. In this way, you ensure that the tool will be in operating condition for future use.

When using portable power tools, lubricate them periodically. Most portable power tool motors run in self-lubricating bearings and do not require much attention, but check the manufacturer's instructions in this respect. Metal stowage cases are usually furnished with portable power tools; when the tools are not in use, keep them in this case. Make your men disconnect the power supply to a tool before they put it down on the work bench. You should read, and insist that every man in the B division read and put to use, every word of the manufacturer's maintenance instructions pertaining to each type of power tool used.

All precision tools and instruments must be handled with great care and must be cased in special boxes or containers when not in use. Precision tools cannot retain their accuracy if allowed to become rusty, bent, or dented. To prevent rust and other damage, coat these tools with a thin film of clean oil, and adequately case them. Your men must be taught that they cannot be too careful when using precision tools, gages, and indicators. It is inexcusable to drop them at any time. Emery cloth, sand-

paper, steel wool, or any other abrasive must never be used to clean the moving parts of any precision tool or instrument. The moving parts are machined and ground to exceptionally close tolerances; and if some of the metal should be worn away, the fit of the parts would be ruined and the accuracy of the instrument impaired.

Whenever tools become broken or worn out, replace them as soon as possible. This will save much time and labor when an emergency job arises.

CARE AND MAINTENANCE OF PIPING

Care must be given the various piping assemblies, as well as the machinery connected by the piping. All joints, valves, and cocks in the lines must be examined frequently and kept tight. On active ships, the main and auxiliary feed systems and all salt-water piping must be tested quarterly under full operating pressure for a sufficient length of time to detect any leaks or other defects. On inactive ships, the piping must be kept drained, and no pressure tests should be run. Never subject piping to strain by using it for hand or foot holds, by securing chain falls to it, or by making use of it to support weights.

Piping Materials

Materials generally used in the manufacture of pipes and tubing for naval use are steel, low-alloy steel, copper, brass, and various copper-nickel alloys. Nonferrous piping is used for transferring such fluids as steam condensate, fresh and salt water, lube oil, compressed air, and refrigerant liquid and vapor. Steel piping is used for transferring feed water, fuel oil, and steam, for high-pressure, high-temperature steam, carbon molybdenum alloy steel piping is used.

Piping materials to be used must not be picked according to appearance; they must be carefully identified from specified standards (BuShips schedule for piping systems). For example, the 7-inch steel tubing employed for transferring 400 psi of 650° F. steam may look the same to you as the 7-inch steel tubing employed for transferring 600 psi of 850° F. steam. The material of the first, however, may be of carbon steel, whereas that of the second must be molybdenum alloy steel, capable of resisting high temperature "creep." (Creep is a slow plastic deformation at high temperatures under constant loads which are considerably lower than the designed tensile strength of the metal.)

Bolted Flange Joints

Bolted flange joints are suitable for all pressures now in use, and are specified for piping systems which ordinarily operate under a vacuum. The flanges are attached to the piping by welding, brazing, screw threads (for some low-pressure piping), or rolling and beading into recesses. The Van Stone joint is used extensively where piping is subjected to high pressures and heavy expansion strains. The gasket materials employed in these flange joints are:

1. Asbestos compound fibrous gaskets;

2. Metal gaskets of either monel or soft iron;

3. Metallic-asbestos spiral-wound gaskets;

4. A plastic gasket sealing compound which spreads like a paste and then hardens into an effective seal.

In RENEWING gaskets in flanged joints, special care must be exercised when breaking a joint. This particularly applies with steam or hot-water lines, or in saltwater lines which have a possibility of direct connection with the sea. Always observe the following precautions:

1. Be sure that there is no pressure on the line.

2. Secure, wire closed, and tag the line pressure valves, including the bypassing valves.

3. Drain the line completely.

4. Leave at least two diametrically opposite flange-securing bolts and nuts remaining in place until the others are removed. These can then be slackened to allow breaking the joint, and

can be removed after the line is clear.

5. Take all possible precautions to prevent explosions or fire, when breaking joints of flammable liquid lines.

6. Ensure proper ventilation before breaking joints in closed compartments.

The observance of these precautions may prevent serious explosions, personnel scaldings, or compartment floodings.

In MAKING UP a flanged joint, first make sure that all parts are thoroughly clean. The gasket seats should then be checked with a surface plate, and necessary steps taken to afford uniform contact.

With the pipe in place and properly supported, the pipe flanges should be checked for accurate alignment, both horizontally and vertically along the faces. After the flanges are placed in position, bolts are inserted into the

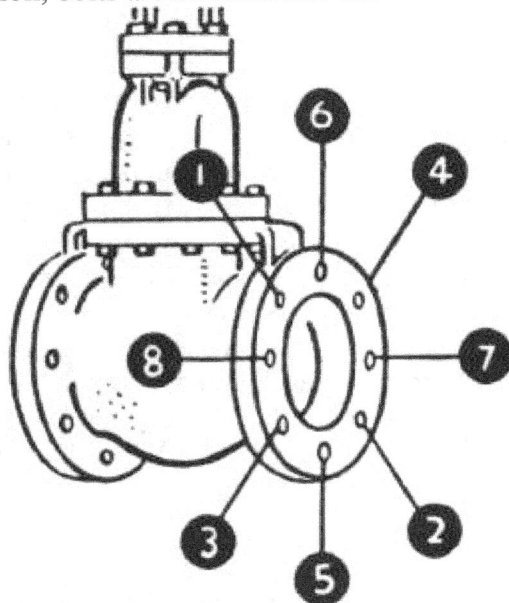

Figure 12-1.—BolMightening sequence.

lower half of the bolt holes to hold the gasket in place. In cases of high-pressure steam lines, new bolts of proper specification should be installed. Slip in the remainder of the bolts, apply thread lubricant (Antiseize compound) to the bolts, and turn the nuts by hand as far as they will go. The nuts should then be pulled up with a wrench, not in rotation, but by the crossover method shown in figure 12-1. This method loads the bolts evenly and eliminates any concentration of stresses on the flange. When making up flange joints for high-pressure steam lines, such as the main steam line, you should be familiar with the instructions given in chapter 48 of BuShips Manual.

INSTALLATION AND MAINTENANCE OF VALVES

Before installing any valve, you should know the range of materials from which valves are made. Each material has its limitations of pressure and temperature, and must not be used for services beyond the recommended maximum.

When and where to use each type of material is best determined by following the markings on the equipment. However, the following rules should be applied without exception: Don't use brass for temperatures exceeding 550" F.; don't use iron for temperatures in excess of 500" F.; use steel for all services above 550° F. Steel should also be used in cases where either internal or external conditions, such as high pressure, vibration, or shock, may be too severe for brass or iron to withstand.

Valve Identification

In addition to the maker's brand and size marks, there appears on most valves a basic service rating. Pressure and temperature are always expressed as steam ratings unless otherwise indicated. Steam ratings are used as a basis, since temperature is the factor which determines the suitability of a material for a given application. For a material at room temperature, the safety factor in terms of pressure is much greater than would be indicated by the steam rating.

The valve shown in figure 12-2 is rated at 125 pounds saturated steam pressure at its equivalent temperature of 335° F. For services at about room temperature, this valve is suitable for pressures up to 300 psi.

Figure 12-2.—Typical brats globe valve, showing basic service pressure.

Valves for general purposes may show two service ratings. For example, the valve shown in figure 12-3 is marked 250 S. This indicates that the valve is to be used at 250 pounds steam operating pressure. Its cold service rating is designated by the mark 500 WOG, which means 500 pounds cold water, oil, or gas, nonshock pressure. Cast and forged steel valves bear a mark such as 150, 300, 600, etc. These figures denote the maximum pressure at a given maximum temperature for which the valve is suitable.

Figure 12>3.—Standard iron gate valve, shov^ing dual service ratingt.

General Cautions on Installing Valves

While most valves can be installed with the stem at any angle, the best position is with the stem straight up. Should the valve be installed with the stem downward, the bonnet acts as a pocket for scale and other foreign matter. Any collection of scale will interfere with the valve operation and may eventually destroy the inside stem threads.

Special care should be taken to install check valves so that the disks open with the flow of fluid and close by gravity when back flow occurs.

Globe valves may be installed so that the higher pressure is above the disk or below the disk. The method of installation should be governed, in each case, by consideration of the service conditions.

Pressure from below the disk is desirable in cases where the flow must be continuous. For example, a globe valve installed in a boiler feed line should be installed with pressure below the disk, since pressure from above might cause a detached disk to seat and thereby shut off the flow

The temperature to which the valve will be exposed must also be considered in installing globe valves. If a globe valve for high-temperature service is installed with pressure under the disk, the upper part of the valve is likely to cool when the flow is shut off. Cooling of the stem might cause sufficient contraction to unseat the valve just enough to cause leakage. The resulting extremely high-velocity flow might cause severe erosion of the disk and seat. In this case, therefore, it would obviously be better to install the valve with pressure from above the disk. Remember that any valve in high-temperature service operates more efllciently if its mechanism is exposed to a constant temperature, and that this condition can be met by installing the valve so that the pressure will be above the disk.

In summary, the general rule for installing globe valvea may be stated as follows: always install the valve with pressure above the disk, unless there is some special reason for installing it with pressure below the disk.

Valve Leakage Causes and Remedies

Valve leakage is generally a result of the disk and seat failing to make a tight joint, and this failure is usually due to one of the following causes:

1. Foreign substances, such as scale, dirt, waste, or heavy grease, are lodged on the seat in such a way that the disk cannot be seated. If the obstructing material cannot be blown through, the valve will have to be disassembled and cleaned out.

2. Scoring of the seat or disk has been caused by attempts to close the valve on scale or dirt, or by corrosion. If the damage is slight, the valve may be made tight by grinding; if the damage is extensive, the valve will have to be reseated and then ground.

3. The disk is cockbilled because the feather guides fit too tightly, or because of a bent spindle guide or a bent valve stem.

4. The valve body or disk may be too weak for the purpose for which it is used, permitting distortion of the valve seat or disk under pressure.

5. In valves fitted with seat rings, leakage through the valve may occur as the result of leakage around the threads of the seat rings. To correct this defect, remove the seat ring, clean the threads, and remake the joint. It may be necessary to recut the threads in the valve and to renew the seat ring to secure tightness.

REDUCING VALVES

Reducing valves are used to provide a steady discharge pressure lower than the supply pressure. They are used on gland seal lines, galley steam lines, heating system lines, and on many other reduced-pressure lines. A reducing valve can be set for any desired discharge pressure, within the limits of the design of the valve; after the valve is set, the reduced pressure will be maintained regardless of changes in the supply pressure, as long as the supply pressure is at least as high as the desired delivery pressure.

Two general types of reducing valves are in common use: the spring-loaded reducing valve, shown in figure 12-4; and the pneumatic pressure-controlled reducing valve, shown in figure 12-5.

All reducing valves should be inspected, cleaned, and repaired semiannually, or whenever they do not operate properly. Kerosene may be used for cleaning the valves.

If a spring-loaded reducing valve fails to operate properly, the trouble may be due to one or more of the following causes:

1. The adjusting spring may have taken a permanent set. Readjust it or install a new

spring.

ADJUSTWG SCREW

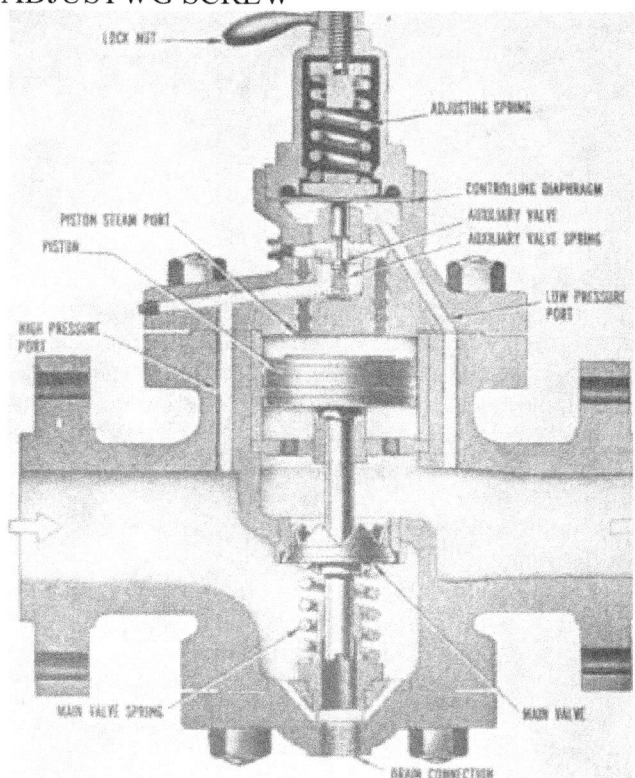

Figure 12-4.—Spring-loaded reducing valve.
VALVE SPRING
STEAM INLET
WATER SEAL FILLING PLUG

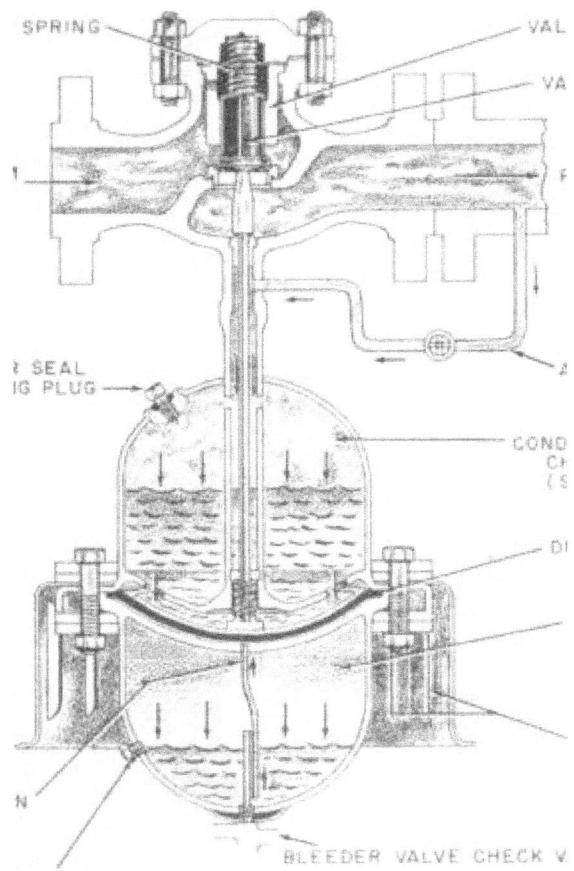

VALVE GUIDE
VALVE DISK
REDUCED PRESSURE STEAM
ACTUATING LINE
CONDENSATION CHAMBER (STEAM)
lAPHRAGM
LOADING CHAMBER (AIR)
SIPHO TUBE
GLYCERINE SEAL FILLING PLUG
COOLING FINS
VALVE
AIR GAUGE AND PUMP CONNECTION HERE

Figurt 12-5.—Pneumatic-pr«tsur»*controll«d reducing valv*.

2. The diaphragm may be damaged or excessively deformed. Install a new diaphragm or—in the case of slight deformation—make a proper adjustment of the adjusting spring.

3. Leakage may be caused by failure of the main valve or the auxiliary valve to seat properly. Check the valves for wear, and for the presence of dirt or scale. Correct the trouble by cleaning and grinding in the main valve and the auxiliary valve. After grinding in the auxiliary valve, the auxiliary valve stem may be too long. In this case, face off the end of the auxiliary valve stem until the proper clearance is obtained between the diaphragm and the end of the valve stem.

The pneumatic pressure-controlled reducing valve has a water seal in the upper half of the dome, and a glycerine seal in the lower half of the dome. The glycerine seal is put in at the factory; the water seal is put in when the valve is installed. The condensation of steam is sufficient to maintain the water seal at the proper level, after the valve has been placed in service; when the valve is being repaired, however the water seal will probably be lost. Be SURE to replace the water seal before putting the valve back in service, since steam must not be allowed to come in contact with the diaphragm. The glycerine seal does not, as a rule, require replacement in service. However, if it is necessary to replace or replenish the glycerine seal, place the dome in its normal vertical position and fill it with glycerine to the level of the filling plug. Screw the plug in and tighten it. In an emergency, water may be used temporarily instead of glycerine for the lower seal.

If a pneumatic pressure-controlled reducing valve fails to operate properly, check the following points:

1. If the pressure in the lower dome becomes excessively high soon after the valve has been put into service, the extra pressure may be caused by expansion of the air due to temperature changes. Bleed enough air from the dome so as to maintain the proper pressure in the dome, at the operating temperature.

2. If there is a gradual loss of pressure in the lower dome, check the bleeder valve, the air-loading connection, the pressure-gage connection, and the filling plug for air leakage.

3. If the reduced pressure builds up beyond the set pressure, steam may be leaking past the valve. Check the valve for wear and for the presence of dirt or scale; also, check to be sure that the valve stem is not binding and so holding the valve open.

4. If the reducing valve closes and fails to deliver steam, check the dome pressure gage; if it reads the same as the outlet pressure gage, the diaphragm has probably failed.

QUIZ

1. What are the disadvantages of the two types of paint remover used on board ship?

2. How often must the piping for the main and auxiliary feed systems and all salt-water piping be tested under full operating pressure?

3. When making up a flanged joint, why is it important to tighten the bolts in a *'cross-over*' sequence?

4. For what temperatures may brass valves be used?

5. What material is used for valves in services above 550° F?

6. In the basic service rating that appears on most valves, how are pressure and temperature expressed?

7. As a general rule, valves should be installed with the valve stem in what position?

8. Is it better to install a globe valve with the higher pressure above the disk or below the disk?

9. How often must reducing valves be inspected and cleaned?

10. In a spring-loaded reducing valve, what can be done to compensate for a slight deformation of the diaphragm?

11. Why is it essential to keep a water seal in the upper dome of a pneumatic pressure-controlled reducing valve?

12. In a pneumatic pressure-controlled reducing valve, how is diaphragm failure indicated?

CHAPTER
NAVY REPAIR PROCEDURES

Ships can operate only a certain length of time without repairs. To keep them in prime condition, definite intervals of time must be allotted for their overhaul and repair; and material upkeep should be given constant attention.

Yet in spite of regular maintenance procedures, accidents and derangements will occur, necessitating emergency repair work. Defects and deficiencies within the capacity of the ship's force to correct should be repaired as soon as possible after discovery. Repairs beyond the capacity of the ship's force to accomplish, and ship's force items that cannot be undertaken immediately, should be recorded in the current ship's maintenance project, for early accomplishment.

REPAIRS AND ALTERATIONS

In general, maintenance work is divided into three categories: (1) repairs, (2) alterations, and (3) alterations equivalent to repairs.

A REPAIR is defined as work necessary to restore a ship or an article to serviceable condition without change in design, materials, or the number, location, or relationship of component parts. Repairs may be accomplished by ship's force, by repair ship or tender, or by a naval shipyard.

An ALTERATION is defined as any change in hull, machinery, fittings, or equipment which involves changes in design, materials, or the number, location, or relationship of component parts.

An ALTERATION EQUIVALENT TO A REPAIR is an alteration which meets one or more of the following conditions:

1. The substitution, without other change in design, of different materials which have previously been approved by the Bureau for similar use and which are available from standard stock;

2. The replacement of worn out or damaged parts requiring renewal by those of a later and more efficient design previously approved by the Bureau;

3. The strengthening of parts which require repair or replacement in order to improve reliability, provided no other change in design is involved;

4. Minor modifications involving no significant changes in design or functioning of equipment, but considered essential to prevent recurrence of unsatisfactory conditions.

Alterations that affect the military characteristics of a vessel are know^n as NAVALTS. Alterations under the technical cognizance of the Bureau of Ships are known as SHIPALTS, regardless of whether or not the alterations affect the military characteristics of the vessel. Thus, an alteration may be both a NAVALT and a SHIPALT; or it may be only a SHIPALT.

CSMP Alteration Record Cards

The Alteration Record Card (NavShips 530), shown in figure 13-1, is used to record all approved alterations. The information and data are obtained from the SHIP-ALT (NavShips 99) and from various BuShips letters that may be received regarding the alteration. The use of these cards makes it easier to list the required information for alterations in the Current Ship's Maintenance Project (CSMP).

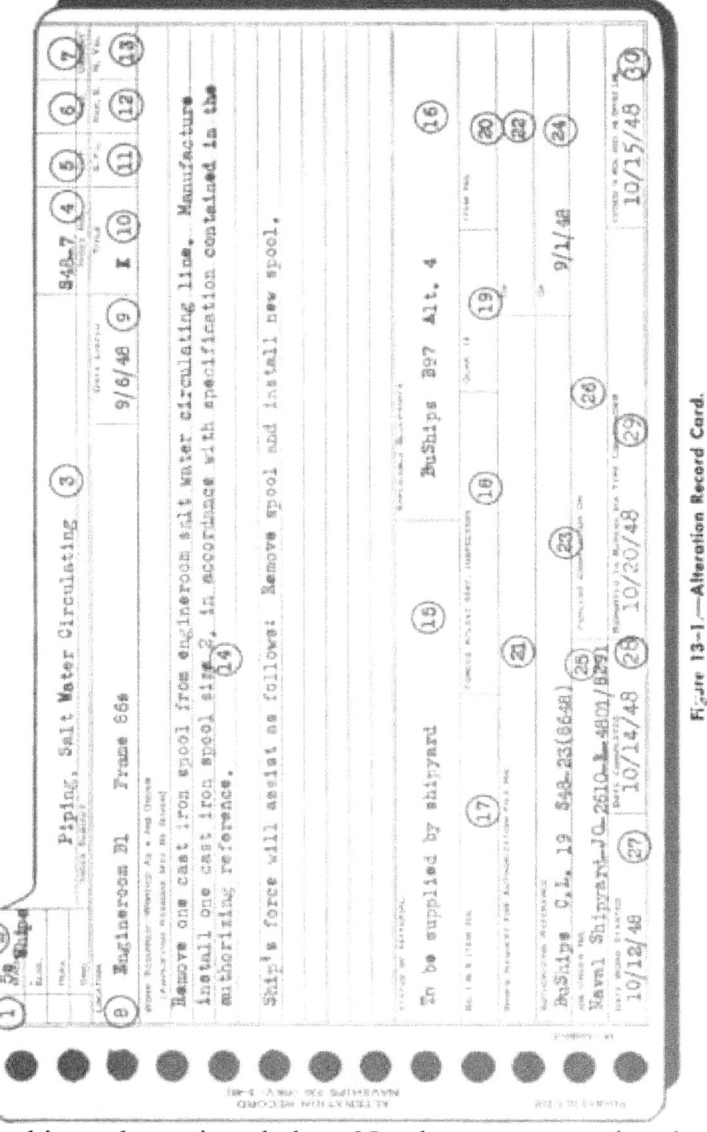

Figure 13-1.—Alteration Record Card.

Entries on this card are given below. Numbers correspond to those encircled in figure 13-1.

1. Priority number assigned by the ship. This number is not of a permanent nature; it may be changed in making out work requests for the different availabilities that the ship may have until the alteration under consideration has been completed.

2. The name of the bureau having cognizance of the alteration. This is always listed as "Ships" for the engineering department CSMP, as alterations will be either mechanical, electrical, or hull.

3. The number and name of the unit on which the alteration is to be done.

4. The index number of the unit. This number, as well as the index subject, will be assigned in accordance with the Navy Filing Manual instructions. This number will give the location of the CSMP alteration card in the Material (Machinery) History. The number and name

are the same as the ones used on the corresponding Machinery History Card.

5. 6, and 7. An "X" entered on one of these spaces indicates the priority of the alteration—whether deferred, routine, or urgent. This priority is assigned by the ship except in a few cases where a certain priority may be recommended by a higher authority, such as the type commander.

8. The name and number of the compartment in which the unit is located. Where necessary, frame numbers are added.

9. The date of the alteration. This date is copied from the SHIP ALT (special form, NavShips 99) or authorizing letter.

10. The title to which the alteration is charged. This information is obtained from the SHIP ALT or authorizing letter.

11, 12, and 13. An "X" indicates who is to do the work—whether ship's force, forces afloat (repair ships or tenders), or a naval shipyard. This information, also, is obtained from the SHIPALT or authorizing letter.

14. A brief description of the work to be done.

15. The status of the material to be used for the alteration. The SHIPALT or authorizing letter will indicate who is to do the work (see numbers 11, 12, and 13). If the Naval Shipyard is designated to do the work, it will also supply the material. For alterations that are to be accomplished by ship's force, the material will be requisitioned by the ship.

16. This will be the BuShips plan number and the identification number of the applicable blueprint for the alteration.

17. The Board of Inspection and Survey list of authorized alterations item number, if such an inspection has occurred.

18. The title of the type commander convening a forces afloat material inspection, and any recommended priority made by such inspection.

19. The number of the quarter and the year the forces afloat material inspection was held.

20. The item number from the list of recommended alterations by the forces afloat material inspection, if listed.

21. The file number of the letter from the ship requesting authorization for a repair activity to accomplish the approved alteration.

22. The date of any such letter as indicated in number 21.

23. The number of the SHIPALT or letter authorizing the alteration to be done.

24. The date of the SHIPALT or letter indicated in number 23.

25. The shipyard or repair ship job order number, when assigned by the activity undertaking the work.

26. In cases where the work has not been completed, the date and percentage of completion should be entered, in this space, for reference for the next shipyard or repair ship (or tender) availability.

27. 28, 29 and 30. These items are self-explanatory. The back side of the card is used for any additional
information that may be necessary. If the detailed description of the work is long (see number 14), it may be continued on the back side.

Alteration and Improvement Program

The type commander maintains a master record of all outstanding SHIPALTS and alterations for the ships of his force. Sometimes this list of alterations is divided into 3 sections— SHIPALTS (including NAVALTS), ORDALTS (ordnance alterations), and alterations equivalent to repairs.

In order to keep the master record of the alteration program the type commander requires the ships to report the completion of any alterations, either by letter or by the pink copy (NavShips 99) of the completed SHIPALT. Also, each quarter, copies of the master list (or record) are sent to each ship. The ship will inspect this list of alterations to see that it is correct and up to date. Corrections, if any, are made and one copy of the list of alterations is returned to the type commander. New alterations are added to the list when they are approved. When an alteration has been completed by all ships concerned it is dropped from the list. Copies of this alteration list, commonly called the "A & I program" are kept on file in the log room.

Requests for Alterations

A definite policy has been set up to prevent addition of weight to naval ships by unnecessary or unimportant alterations. The type commanders i.ssue instructions concerning the procedures for requesting approval of new alterations. In general, the procedure is as follows:

1. Requests shall be submitted as soon as the need for the change becomes known.

2. Each alteration request will contain specific details, including any defects of the present installation that may be encountered, advantages of the proposed alteration, reference to applicable blueprints, sketches of proposed changes, etc.

3. The request must specify the estimated change in weight and vertical moment, and recommend an adequate weight compensation for the proposed alteration.

4. If the proposed alteration should result in a reduction of space for accommodation of the crew, details of the reduction of space and the reasons for accepting this loss shall be included.

5. Requests will be forwarded in the following manner:

a. The commanding officer will submit requests via their unit commander with sufficient copies for the required distribution.

b. The squadron commander (when applicable) will forward the original and sufficient copies of the request to the type commander, including appropriate comments in his endorsement. Copies of the request shall be furnished to all like commands.

c. Recipients of these copies shall investigate the proposed alteration and submit their recommendations to the type commander.

d. The type commander will forward the request to the appropriate bureau, together with his recommendations and a statement as to the applicability of the proposed alteration to other ships of his force.

PREPARATION OF WORK REQUESTS

In wartime a work request was simply a list of items with the information necessary to define each job. Often, in the case of urgent work, the work request was sent out in the form of a dispatch. In peacetime, work requests are made on printed forms. Figure 13-2 illustrates a common type of work reqUest form used by a type command.

Work requst forms are filled out by ship's personnel and sufliicient copies are prepared for distribution as follows: original and one copy for the repair ship (some repair ships require two copies), one copy for return to originating ship after approval by the type command's representative, and one copy for the type commander's maintenance files. The entries made on the form shown n figure 13-2 may be described as follows.

Bureau: The title of the bureau having cognizance over the item to be repaired is filled in here.

Title: This space is no longer applicable.

WORK REQUEST

Brief of Krpair - Include any fK-cosAAry data. coinm*rrl on m«tcn»l.«r»J list hlue prinU where eiuential (attentinn invite«l to Chapter V. S«Tvice Force Aclivitie« Rullelin). Ujir if r»efeM*rv.

Manufacture one (1) liquid piston, rod, stuffing box gland for No. 1 Fire and Bilge pump.

Punp Datat Mfg. - Wortnington, Site - 7x9x12, type - vertical sii^ plex, serial No. - 1099523> stea-ti press - hOO f*, capacity - 200 CPKe

The stuffing box f^land is bent and broken.

Manufacture one (1) f^land which is snovn as piece No. D£ 51 -SUlOO - 10 Alt. 2. Ships force will deliver old gland with the above blueprint and call for conpleted work*

Ship-to-Shop

Ships Personnell Lt* W.T. Johnson J. V. SBiith, BTC-

LEAVE BLANK

Figure 13-2.—Sample work request.

354

f

Group or index number: Each group of machinery, structure, or equipment aboard ship bears a file number such as S16, access openings; S35, laundry; S41, main propelling machinery; and S51, boilers.

Ship's serial number: Each ship numbers its work requests serially throughout a calendar year.

The ship's serial number is usually made up of a composite number—for example, DD785-EM25-52. This number is made up of 3 parts. The first part is the number of the ship, and indicates that it is a work request from the DD785. The first letter (E) in the second part indicates that it is a work request from the engineering department. The second letter (M) indicates that it is a mechanical job. The number "25" is the individual serial number of the mechanical job or work request. The last part (52) of the number, of course, means the calendar year of 1952. In other words, it is the 25th mechanical work request from the engineering department, in the year 1952.

USS: The name of ship is entered here.

Classification: Some items of equipment bear a security classification, which is indicated here.

Priority: This is the priority requested by the ship— urgent, routine, or deferred. (A different priority may be assigned at the time when the work request is approved.)

For Use of Type Commander: The type commander's maintenance officer, or his representative, uses this space to indicate the approval (or disapproval) of the request.

Ship's force, repair ship, shipyard: Check the activity which will accomplish the indicated repair.

Date: This line is self-explanatory.

J. 0. number: Job order number is filled in by the repairing activity.

Brief of Repair: A specific statement of the work desired is necessary; it is often desirable to include the symptoms of faulty operation. It is further desirable to list the applicable drawings and to indicate whether or not they can be furnished. Where applicable, a statement

is made as to which part of the work will be accomplished by the ship's force (for example, dismantling, reassembly, and delivery to and from the repair shop). Phraseology such as "do work as necessary," "check," or "open up, examine, and repair" is almost meaningless. It is frequently good practice to include the name of the ship's officer who should be contacted for further details, if necessary, and who will be responsible for inspection of the completed job.

Leave Blank: In this space the repair activity enters such data as (1) number of man-hours on the job, (2) stub requisition numbers of material drawn, and (3) signature of person signing for the job as being completed.

As a check-off for the ship being repaired, to aid in its record-keeping after completion of the repair, some notation is made in the machinery history.

CSMP Repair Record Cards

The Repair Record Card (NavShips 529), shown in figure 13-3, is used to record all repairs that are pending. The card is made out as soon as the necessity for a repair item or repair job becomes known. A separate card is made out for each item in need of repairs, except that similar small items such as valves are usually included on one card. The cards are kept in a current or active file until the indicated repairs have been completed.

The entries on this card are as indicated below. Numbers correspond to those encircled in figure 13-3.

1. For engineering department purposes, the "X" will indicate the type of repair work— mechanical, electrical, or hull.

2. Enter the name and number of the unit on which the repair is to be made.

3. Give the index number of the unit to be repaired, as listed on the subject machinery history card.

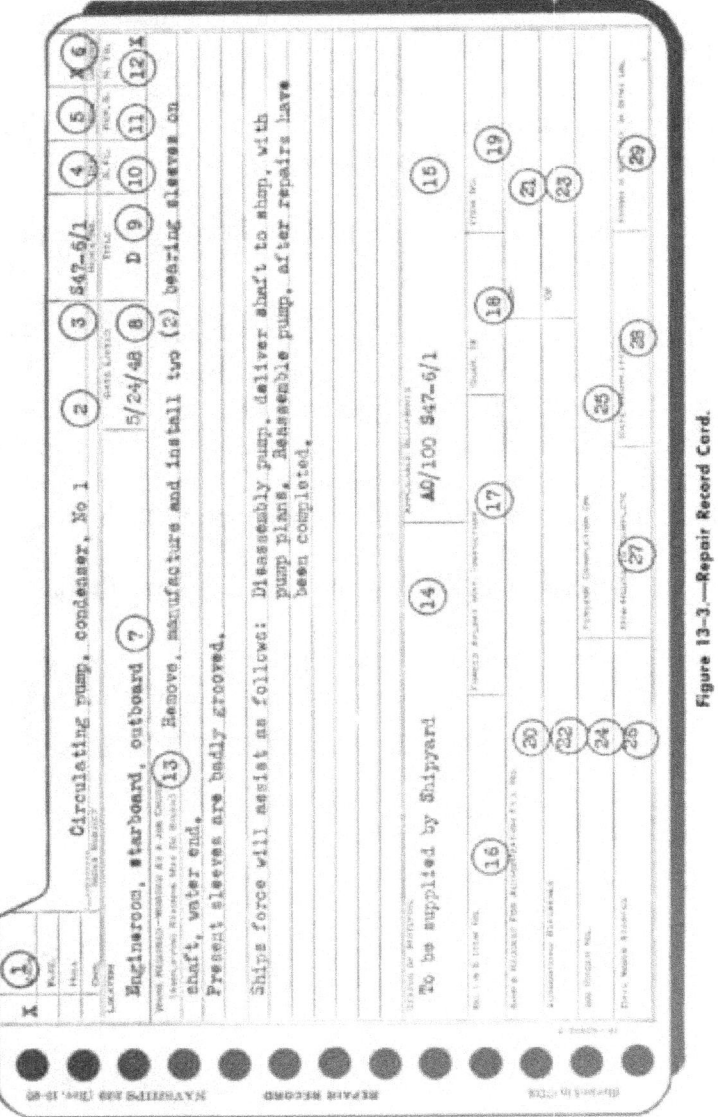

Figure 13-3.—Repair Record Card.

4, 5, and 6. An "X" in one of these spaces indicates the priority of the repair—whether deferred, routine, or urgent.

7. Give the location of the unit to be repaired. The compai tment name and number should be given. Give the location in the compartment when it is a large space. Frame numbers are added where necessary.

8. Give the date the card is started. This should be the date the repair is noted as being necessary.

9. Formerly, the title to which the repair was charged was entered here. This is no longer applicable, however, so this space should be left blank.

10, 11, and 12. An "X" indicates who is to do the work —whether ship's force, repair ship, or shipyard.

13. This must be written up so that it will give the following information:

a. The required repair work, including replacements of parts

b. A detailed description of the defective conditions or malfunctions

c. Where applicable, assistance that will be given by ship's force

d. When necessary, reference to subject matter and plans in available manufacturer's instruction books on board ship

e. The name of the person who can be contacted for additional information in regard to the job

If the space allowed for this entry (No. 13) is not sufficient, the information should be continued on the reverse side of the card.

14. Indicate the status of the repair parts or material to be used for the repair, such as "on board," "on order," or to be furnished by the repair activity.

15. Give the BuShips plan numbers of the applicable blueprints that may be required to accomplish the repair job.

16. Give the Board of Inspection and Survey item number, if such an inspection has been held. (There are special instructions for the assignment of this composite number which will not be given in this discussion.)

17. Give the title of the type or force commander convening a forces afloat material inspection, if such an inspection has been held.

18. Give the number of the quarter and the year in which the force afloat inspection was held.

19. Give the item number assigned to the repair job, under consideration, for the force afloat material inspection. (See comment for space number 16.)

20. Give the file number of any letter that may have been written in regard to the repair item.

21. Give the date of such a letter (No. 20).

22. Give the file number of the letter in reply to any letter that may have been written (No. 20).

23. Give the date of such a letter (No. 22).

24. Give the job order number assigned by the shipyard or repair ship where the repair work is to be accomplished.

25. In cases where the repair job has not been fully completed, give the date and percentage of completion for reference for the next shipyard or repair ship availability.

26. 27, 28, and 29. These items are self-explanatory.

Current Ship's Maintenance Project

The CSMP is maintained to ensure that no item of work will be overlooked, and to make possible the easy and orderly making up of work requests for an overhaul period. Also, it will ensure that the detailed information required for each repair job is readily available.

The CSMP is made up of the following three cards: Repair Record Card (NavShips 529) (blue); Alteration Record Card (NavShips 530) (pink); and Record of Field Changes (NavShips 537) (white). The repair card

and the alteration card have been explained. The record of field changes card is used in connection with authorized alterations of electronic equipment. Because you will not be concerned with this card it will not be explained in detail.

BuShips Manual requires that the CSMP be kept as part of the Material (Machinery) History, which is made up of two or more large binders. A binder or book is made up of various cards, of which the most important is the Machinery History Card. These cards are of the same size and shape as the alteration and repair cards. All cards follow a certain sequence which is in accordance with the index number. The CSMP cards (figs. 13-1 and 13-3) are the only cards that have a short extension (called a "tab") at the upper left-hand corner.

When an alteration or a repair card is made out, it is placed behind the appropriate machinery history card so that the tab of the alteration card (pink) or repair card (blue) is left showing. This indicates that the work item is pending. These exposed tabs will indicate the CSMP cards in the material history. During the making up of a list of repair work and alterations for a coming repair availability period, all the individual items can be readily checked. The various CSMP work items are then arranged in order of priority and the individual work request numbers are assigned.

Except for the Naval Shipyard routine overhaul period, the common procedure is to transfer the information from the CSMP cards to the work request forms (fig. 13-2). The appropriate copies of the work requests are typed out by the Engineer's Yeoman.

As a Boilerman 1 or C, you can see why it is so important to get all necessary repair items down in writing on the CSMP cards. Otherwise, small but important repair items will be overlooked or forgotten, and thus will not be accomplished during the overhaul period. The CSMP repair record cards should be made out as soon as a defective material condition is discovered. The details

are fresh in your mind and a complete description of the conditions and the required repairs can be written up. This will prevent delays, insufficient information, or a misunderstanding at a later date. A good CSMP must have all the detailed information as well as a complete list of all the required repair items. An experienced Boiler-man will write up all the necessary details on a repair item and give it to his chief or division officer so that the repair item can be entered in the CSMP. The repair item will then be down in black and white for the coming tender repair period or Naval Shipyard overhaul.

Blueprints

In order to be proficient in all phases of his work the Boilerman should have a good knowledge of blueprints and an understanding of the method of filing and keeping blueprints in the Navy. Individual ships, tenders, and repair ships are furnished with plans, and with such microfilm as is available and applicable to ships that may be repaired by the tender or repair ship concerned. The blueprint is a photographic print, blue, white, or any other color, used for copying drawings or plans.

All ships use the same system for filing blueprints, although the number and type of plans carried will be different on different ships. The number assigned to a blueprint for identification and filing purposes is composed of several groups. For example, in the Bureau of Ships plan number CA139-S5101-525802, the group CA139 is the class ship designator. It is the class of ships for which the plan applies. The next group is S5101, which is the "S" (subject-matter) group or material file number. In this example, the reference is to a blueprint on the general arrangement of a boiler front. The group 525802 is the individual plan number. Another example of a plan number is DE51-S4602-1, where DE51 is the ship class, S4602 is the subject-matter filing number (in this case condensers), and 1 is the individual plan number. In other words, it is the first plan under the S4602 group for DE51 class of ships.

In cases where an alteration or change has been made, the blueprint will have an alteration number also. For example, in blueprint number CA139-S5103-528155 Alt. 4, the "Alt. 4" is the alteration number. If all the alterations have been completed, the plan with "Alt. 4" is kept on file.

Care must be taken, when working from blueprints, to see that you have an up-to-date plan and not one that is obsolete. This can be done by checking on the "Alt." numbers. In case the last alteration has not been completed the ship concerned will keep two blueprints on file until such time as the alteration has been completed. In this example the two plans would be CA139-S5103-528155 Alt. 3 and CA139-S5103-528155 Alt. 4.

You must have an understanding of the S or material group classification and numbering system, because it is used in the filing of blueprints. The listing of filing numbers and their subject matter is given in the Navy Filing Manual. The same system is used for numbering the different chapters in BuShips Manual. For example, blueprints for pumps are in the S4700 group; main turbines would be S4100, condensers S4600, and boilers S5100.

The best method for finding a blueprint on board a ship is to refer to the Ship's Plan Index. Assume that repairs are to be made to a main feed pump and a blueprint is needed. The proper procedure would be to go to the engineering log room and obtain the Ship's Plan Index. If you know the filing number for pumps you can immediately turn to the S4700 group. Here you will see the numbers and titles of the various blueprints on pumps listed. From the listed titles you can usually spot the blueprint you want. The number opposite the title is noted down. The next step is to go to the filing cabinets where the blueprints are stowed.

The file drawers are labeled with the S-group numbers and the numerical plan number sequence of the blueprints in the drawer. Look for the file drawers marked S4700 and then check for the numerical sequence that the desired

blueprint would come under. For example, if the individual plan number of the desired blueprint was 527903 and the file drawer marked "S4700—527807 through 528014" was located, you would know that the desired plan was in this drawer. The individual blueprints inside the drawer are usually in large manila envelopes with the number and title in the upper left-hand corner, and are filed in numerical sequence.

By following this sequence, it will be a simple matter to locate plan number 527903. The blueprint is removed from the manila envelope, which stays in its proper place in the file, and checked over to make sure that it is the plan that gives the required information. In order to keep track of all plans removed from the files, the plans should be signed for, either on an index card or in a book kept for that purpose.

AVAILABILITIES

A vessel may not informally and on her own initiative come alongside a repair ship or tender or enter a naval shipyard for repairs. The control and disposition of a vessel is at all times a function of certain operating commands. Thus, when a vessel needs outside repair assistance, the vessel's type commander—and in certain cases the task force commander—assigns the vessel an "availability" at a repair activity.

The term availability indicates that the ship is available to a repair activity for repair, overhaul, and/or alteration. Navy Regulations defines the term this way: "The period of time assigned a ship by competent authority for the uninterrupted accomplishment of work at a repair activity."

The different conditions and purposes of availability are as follows:

A REGULAR OVERHAUL is an availability for the accomplishment of general repairs and alterations at a naval shipyard or other shore-based repair activity. Regular overhauls of ships are scheduled to occur in cycles, and the

period between overhauls (generally 18 months) is recommended by BuShips. With the periods for the type of ship established, each type commander prepares a regular overhaul schedule for each vessel under his command, by projecting from the date of completion of the last regular overhaul. The type commander then requires his ships to submit their respective work requests, transfers funds to the appropriate shipyards, and directs the ships to report at the assigned date to the shipyards. This type of availability concerns you only inasmuch as your ship will be scheduled for regular overhauls and you will assist in making out work requests.

A RESTRICTED AVAILABILITY is an availability for the accomplishment of specific items of work by a repair activity, with the ship present. For example, a restricted availability would be granted for the repair of a propeller blade of a ship. Many of the ships which come alongside a repair ship or tender will have been granted this type of availability.

A TECHNICAL AVAILABILITY is an availability for the accomplishment of specific items of work by a repair activity, with the ship not present. This type of availability is granted when a unit of auxiliary equipment, such as a pump, needs repairing—a unit that can be detached and left for repair while the ship continues on its mission. Since the ship will not be present during the availability, arrangements must be made for the ship to deliver the defective equipment and to call for it on completion of repairs, or to provide shipping instructions.

Voyage repairs is an availability for emergency work which is necessary to enable a ship to continue on its mission and which can be accomplished without requiring a change in the ship's operating schedule or in the general steaming notice in effect. This type of availability is very similar to restricted and technical availabilities, except that a change in operating schedule is not involved.

An upkeep period is a period of time assigned a ship, while moored or anchored, for the uninterrupted accom-
plishment of work by the ship's force or other forces afloat. Ships are assigned upkeep periods at more or less regular intervals, usually between cruises or periods of operations.

SHIP'S FORCE MAINTENANCE AND REPAIRS

Each ship should, in so far as practicable, be self-sustaining with regard to normal repairs. Each ship should be well equipped with material, repair parts, and repair tools and equipment, in order that much of its own repair work can be accomplished. Repairs should be undertaken under the supervision of the most competent and experienced personnel. In cases where personnel lack familiarity with the manner of making specific repairs and tests, they should be instructed to take advantage of shipyard or alongside repair ships overhauls, to observe how such work is undertaken. Budget limitations in peacetime, and military operations in wartime, require that a ship be as self-sufiicient as possible.

Ship's Allowance List

BuShips furnishes each ship with a set of allowance lists. The allowance list is the official source of information concerning the ship's allowance of machinery, equipment, and material.

The ship's allowance list is divided into "S" groups based on the Navy filing system, as previously described. On the top of each page, the major unit of machinery or equipment is listed; this is followed by the major component parts of the unit, and an itemized list af accessories, special tools, and replacement parts. Shore-based replacement parts are also indicated for some of the major units of machinery and equipment.

Administration of Maintenance Work

The necessity for complete indoctrination of personnel in the principles of self-maintenance cannot be overemphasized. The Current Ship's Maintenance Project, the Ma-
chinery History, the inspection methods (including work inspection and progress inspection during overhaul), and other methodical procedures will, when properly used, guarantee systematic inspection of material, provide adequate records of inspection and work, and make virtually certain that all material is given proper attention and care.

These, and other material administrative measures, are instituted to provide maximum efficiency within the funds and repair facilities available. They have the added effect of inculcating thoroughness and a sense of responsibility in naval personnel. The technical excellence and reliability of our Navy is due in large measure to our systematic material administration, carried on in conjunction with carefully planned and analyzed operation procedures.

Preventive Maintenance

The purpose of preventive maintenance is to see that all material conditions are satisfactory and that the equipment or machinery is in all respects ready for service. A regular schedule of cleaning, inspections, operations, and tests is required to ensure trouble-free operation and the detection of incipient faults before they develop sufficiently to be a major source of difficulty.

Check-off lists. —Ships are required to establish and maintain check-off lists for each station; these lists will indicate the daily, weekly, monthly, quarterly, annual, and other periodic tests and inspections required for each item of machinery or equipment at that station. The purpose of these check-off lists is to see that no phase of the ship's force maintenance work has been overlooked.

Routine inspection and tests. —The ship's force is expected to perform the great majority of routine inspections and tests. Most of them are quite simple in nature; others require planning so that they can be undertaken during upkeep or overhaul periods. Shipyard and repair ship assistance should not be requested unless the test or inspection is actually beyond the capacity of ship's force.

Routine Maintenance and Repairs

The material upkeep of the ship's engineering plant must be given constant attention at all times. Defects and deficiencies that are within the capacity of the ship's force to correct must be repaired as soon as possible after discovery. Repairs beyond the capacity of the ship's force to accomplish, and ship's force items that cannot be undertaken immediately, should be recorded in the Current Ship's Maintenance Project, for early accomplishment.

All the ship's routine maintenance work is accomplished by the ship's force. This type of work consists of cleaning boiler firesides and watersides, renewing packing in valves and pumps, replacing gaskets in piping joints, replacing insulation and lagging, general cleaning and painting, inventorying and requisitioning supplies and repair parts, and other jobs of a similar nature.

In addition to routine maintenance work, a certain amount of repair work may have to be done by the Boiler-man 1 or C. He may be called upon to make such minor repairs as patching furnace refractories, overhauling boiler fittings, replacing hand-hole plates, etc. And at times, emergency repairs will be necessary in order to keep the ship in full operating condition.

Maintenance work, or minor repair jobs, should never be unnecessarily postponed. If this sort of work is allowed to accumulate, it may be extremely difficult for the ship's force to get caught up again.

FUNCTIONS OF REPAIR SHIPS AND TENDERS

Ships are scheduled for routine overhaul periods (normally of 2 weeks* duration) alongside repair ships or tenders. The specified interval between such overhauls varies according to the type of ship: destroyers, for example, usually have a tender overhaul period every 6 months. These overhaul periods are planned to accord with the quarterly employment schedule of the ship concerned ; and under normal conditions, a ship has advance knowledge of when and where it will be alongside the repair ship or tender.

Since the Boilerman 1 or C will frequently have some responsibility in preparing records to be used in connection with scheduled overhaul periods, he should know something about the organization and the function of the repair ship or tender, and of the Naval Shipyard or other shore-based repair activity.

The basic difference between repair ship and tender is one of function. Repair ships are primarily concerned with maintenance, in support of all types of vessels and craft; on board a repair ship there will be general maintenance facilities for a number of types of vessels, and stocks of commonly used repair parts. Tenders render both repair and maintenance to the specific types of ships to which they are assigned; their facilities are specific to the type of ship tended, and the material items on board will be applicable to that type and to the particular class composing the squadron to which the ship is attached. The tender supplies the squadron not only with repair services, but also with general stores, medical facilities, ammunition, and often quarters.

Classes of repair ships. —There are several classes of repair ships: (1) the general-duty Repair Shii>—AR—designed to meet the maintenance requirements of capital ships, (2) the Battle Damage Repair Ship—ARB, (3) the Internal Combustion Engine Repair Ship—ARG, (4) the Heavy Hull Repair Ship—ARH, (5) the Landing Craft Repair Ship—ARL, (6) the Salvage Vessel—ARS, (7) the Salvage Lifting Vessel—ARS (D), and (8) the Salvage Craft Tender—ARS (T).

Classes of tenders. —Tenders are classified as follows: (1) the Destroyer Tender—AD—designed to meet the normal supporting requirements of 18 destroyers, (2) the Submarine Tender—^AS, and (3) the Motor Torpedo Boat Tender—ACP.

REPAIR PROCEDURES FOR REPAIR SHIPS AND TENDERS

When a ship receives its employment schedule, or is otherwise notified of its scheduled overhaul period, preparations for accomplishing the necessary paperwork can be

begun. An inspection is made to see that the Current Ship's Maintenance Project is up to date in all respects. The information and data from the appropriate CSMP cards is copied onto work requests for repair ship accomplishment. At the same time the ship's serial numbers are assigned to the work requests. By means of the ship's serial number, or otherwise, the work requests are typed in order of priority for their respective groups, such as mechanical, electrical, and hull. In cases of tender and repair ship overhauls, alterations which are marked for accomplishment by forces afloat are included in the requested work items and are processed in the same manner.

The work requests, with the required number of copies, are sent with a forwarding letter to the type commander or his representative. The staff officer handling material and maintenance screens the work requests. Most of the ship's overhaul work list items are approved or authorized. Some items are disapproved. Other work items are reworded so that ship's force will accomplish most of the work involved. Also, the ship may have to furnish more detailed information on some work requests. The amount of corrective action taken by the reviewing staff officer will depend upon how well the work requests are written and whether they follow established policies and procedures. Upon the completion of this screening, the ship's work requests are forwarded to the repair ship or tender.

Generally, the ship must prepare the work requests in sufficient time so that all major work items will reach the repair ship or tender not less than 30 days in advance of arrival alongside. Where necessary, supplementary work requests should reach the repair activity not less than 10 days in advance of arrival.

Arrival Repair Conference

When a division or a ship comes alongside for an assigned routine availability, an arrival repair conference is usually held immediately. Representatives of the ships, of the repair department, and, usually, of the type commander are present at the conference. The relative needs of the ships and the relative urgency of each job is settled. Jobs which were stated indefinitely in the work requests are specifically defined. The arrival repair conference serves to clarify all uncertainty for the repair department, which has received and studied the work requests in advance.

Services

For a routine overhaul period, the tender or repair ship provides the primary services of auxiliary steam, fresh water, and electric power. Other services such as compressed air or boiler feed water may, in certain cases, be available. Such services as electric power and fresh water are, in most cases, limited—auxiliary steam supply is usually insufficient to operate the ship's distilling plant. Therefore, the ships alongside must cut down as much as possible their use of these services.

The normal procedure is to secure the entire engineering plant, when alongside, so that repairs can be accomplished without delays or interruptions.

Work Requests and Job Orders

The terms "work requests" and "job orders" are sometimes used interchangeably, as though they meant the same thing. This is not technically correct because there is a difference in meaning of the two terms. Work requests are made up by the ship and forwarded through proper channels to a repair ship or Naval Shipyard. When the work request has been approved by the repair activity, and a job order number assigned, it becomes a job order. The term job order is used by repair personnel of a repair ship or Naval Shipyard because it is actually an order to do a certain repair job.

Submitted work requests which have been signed by the repair officer are used as job orders by repair ships anc tenders. On the other hand. Naval Shipyards will issue their own form of job orders when the work requests have been approved. In short, a job order is a work request which has been approved by the repair activity.

Getting the Work Started

As soon as the work requests ha /e been approved at the arrival conference, the jobs where parts or material have to be delivered to the tender shop should be started immediately. Getting these repair jobs started early is very important in getting all the necessary repair work completed. Also, a good start will help to prevent delays and headaches later on. The sooner you can get the repair jobs over to the tender shops the quicker they can get started with their work.

It is a good idea to plan as far in advance as possible for the work to be done during the tender repair period. Some of the equipment not needed for the operation of the ship may be disassembled in advance, so that the defective parts can be delivered to the tender as soon as the work requests have been approved.

All material delivered to the tender should be properly tagged and identified, preferably by means of metal tags secured with wire. The information on the tag should include: the number (and where practicable the name) of the ship; the department, the division or space; and the job order number (the tender's assigned number should be used when possible). Additional information may be added as considered necessary. Reference material such as blueprints and manufacturers' instruction books should bear the ship's name and number.

Ship-To-Shop Jobs

Many repair jobs will be designated by the ship or approved by the repair activity as "ship-to-shop" jobs. This means that the ship's force will do a large part of the repair work. Take, for example, a pump with a damaged shaft. Ship's force will disassemble the pump and remove the damaged shaft. The shaft is tagged, and brought over to the machine shop of the repair activity. When necessary, the pertinent blueprints are also delivered. The machine shop supervisor will check on the job and give an estimate of the time needed to complete it. When the shaft has been repaired or a new one made, ship's force will pick it up at the machine shop and bring it back to the ship. The new or repaired shaft is installed and the pump assembled by ship's force. Inspections and tests are made to see that all conditions are satisfactory.

Repair jobs for such items as gages, valves, and portable equipment are always written up as ship-to-shop jobs.

In some cases repair jobs are written up for a repair activity to assist ship's force in accomplishing certain repairs. Many of these jobs can be called ship-to-shop jobs.

Checking Progress of Tender Repair Jobs

The Boilerman 1 or C should know at all times the progress of repair work for his space or equipment. The progress of ship's force repair work should be carefully checked on and estimated. The tender repairs that are being accomplished on board your ship can easily be checked on by discussing them with the petty officer in charge of the repair detail.

The method of checking progress of work in the shops on the tender may require a little planning or coordination between your ship and the tender. It must be remembered that the personnel in the tender shops are busy with their repair work. Therefore, the method used in checking progress should be one that does not interfere with personnel working in the shops.

Aboard some tenders and repair ships a chief petty officer acts as a ship's superintendent. Generally, his duties are:

1. To act as liaison officer between the ships alongside and the tender in regard to repair department jobs;

2. To act as a coordinator of shop work for the assigned ships;

3. To report daily to a representative of the commanding officer of the ship, to ensure that the work is progressing satisfactorily as far as the ship is concerned ;

4. To maintain a running daily progress report or chart which will indicate (a) the percent completion of each job; (b) the availability of plans, manufacturer's instruction books, or samples; and (c) the availability of material required for each job;

5. To notify the ship to pick up completed material on the tender;

6. To notify ship's personnel to witness tests on machinery, compartments, and tanks occasioned by work performed;

7. To obtain signatures from officers concerned in case of cancellation of a job order;

8. To secure signatures from officers concerned on completion of job orders.

When a tender or repair ship provides a ship's superintendent, it is easy to check on the progress of work on the tender because your own ship will be furnished the necessary information by the ship's superintendent.

In cases where the services of a ship's superintendent are not provided by a tender or repair ship, it is advisable for the ship to appoint a chief or first class petty officer to perform similar duties for the division or for the engineering department.

A progress chart should be obtained and filled out for all the jobs that are going to be accomplished during the repair period. This chart should be kept up to date to indicate the status of each repair job.

The basic purpose of checking progress of repair work is to see that (1) jobs are not delayed unnecessarily, (2) no job is overlooked or forgotten, and (3) all jobs undertaken are satisfactorily completed at the end of the repair period.

SHORE-BASED REPAIR ACTIVITIES

Shore-Based Repair Activities are construed to include naval repair activities under the management control of

BuShips or CNO, and commercial ship repair yards under contiact to the Navy.

Organization

Shore-based repair activities vary in their organization, depending upon the size, the extent of the repair facilities, and local conditions at the repair base. Some shore-based repair activities are organized along the lines of a Naval Shipyard; others are organized more hke a repair ship or tender. The two systems of organization differ chiefly in the fact that in the naval shipyard type of organization, functions of planning and production are assigned to separate divisions, whereas in the repair ship or tender type of organization, they are assigned to the repair officer and his assistants. Generally, the larger shore-based repair activities tend to follow the naval shipyard pattern of organization, while the smaller activities are organized as repair ships or tenders.

Procedure for Obtaining Repairs

The peacetime repair procedure at a shore-based repair activity is similar to that of a repair ship or tender. The work requests proceed from the originating ship to the type commander's representative in the task force or group for screening and authorization of an availability, in accordance with operational commitments. Thence they go to the local service force representative, such as the service squadron or division commander, for assignment in line with the division of the work in the area. Finally, the requests go to the repair facility. Under wartime conditions the procedure is similar, but is likely to be very much simplified.

From time to time, between maneuvers, your ship will be anchored in places where shore-based repair facilities are available. With the approval of the type commander, who furnishes the funds for repairs, emergency and voyage repairs may be accomplished, depending upon the time required.

REPAIR PROCEDURES AT NAVAL SHIPYARDS

A Naval Shipyard is a component activity of a naval base. The primary mission of the Naval Shipyard is to render service to the fleet in the form of efficient and economical building, repairs, alterations, overhauling, docking, converting, or outfitting of ships and related special manufacturing and necessary replenishment of stores and supplies where required.

Naval Shipyards are designated as home yards and as planning yards. A home yard is the shipyard to which a particular vessel is usually assigned by CNO for accomplishment of repairs and alterations. A planning yard is a shipyard which has been designated by BuShips as the yard which undertakes the design work for the type ship allocated to it by the Bureau. Naval Shipyards perform many other functions—such as manufacturing, research, and design—which will not be considered in this book.

Administration of Repairs

CNO has delegated his authority to the fleet commanders to grant availabilities at Naval Shipyards for regular overhauls, voyage repairs, emergency repairs, and technical and restricted availabilities. The fleet commanders have further delegated to the type commanders this authority for voyage repairs and technical availabilities, and in some cases restricted availabilities. To describe the type commander's part in dealing with availabilities, only the regular overhaul will be discussed. It should be noted that, in general, repair work cannot be accomplished by any repair activity unless it is granted funds with which to do the job. Repair activities do not have money in their own right for this purpose. The MBS (Maintenance Bureau of Ships) repair allotment, administered by the type commander, is utilized to reimburse shore repair activities for labor charges and material costs, and to reimburse tenders for material costs incurred incident to repair to hull, machinery, and equipment under the cognizance of BuShips. Funds for improvements to vessels

(SHIPALTS) are administered by BuShips for all material under the cognizance of that Bureau.

The regular overhauls of ships are cyclical and the period between overhauls for each type is recommended by BuShips. This period is generally 18 to 21 months, depending upon the type of ship. With the periods for each type established, the next step in preparing regular overhaul schedule for each ship of his force rests with the type commander. This schedule is forwarded to the fleet commander. After appropriate action, the fleet commander forwards the combined regular overhaul schedule to CNO for approval. It is then the responsibility of the type commander to require his ships to submit their respective work requests, transfer funds to the appropriate shipyard, and direct the ships to report at the required date to the shipyard.

Procedure for Submitting Repair Lists

The rules for submission of shipyard work requests prior to a regular overhaul are laid down in general in Navy Regulations and in detail in fleet and type commander regulations. The following procedure is followed in the Atlantic Fleet. Commanding officers are required to submit their Naval Shipyard work lists to the type commander 60 days prior to a scheduled regular overhaul. The type commander will carefully inspect these lists. The various items are approved, disapproved, changed, or corrected in accordance with standard repair policies. The lists are then forwarded to the Naval Shipyard not less than 30 days prior to the start of the overhaul.

Ships having mimeograph machines are required to submit 30 copies of the work lists; others, an original and 6 copies. A separate (departmental) work list is made out for each of the following headings: hull, engineering (mechanical), engineering (electrical), electronics, and ordnance.

Items of work are listed in the relative order of priority for each work list of the groups listed above. After these work lists have been completed, a ship's priority index is

made up. The priority index is usually made up in a conference of all heads of departments and the executive officer. The various items are selected from the individual repair lists and assigned in an over-all order of priority for the ship. The ship's priority index usually consists of 2 columns of numbers; the first column is the order of priority, and the second column is the repair item number.

Certain procedures are usually followed in making out a departmental work request list. Some type commanders require that each work item should be submitted in the following form and contain the designated information:

1. Description of the item, including location, name plate data, and, where applicable, plan numbers;

2. Report of existing defects in the items to be repaired;

3. Complete and full description of repairs required to place the item in satisfactory

operating condition. (The repair parts which will be required should be clearly indicated.)

4. Reference to authorizing correspondence, where applicable. (If none is applicable, report "none.")

5. Specify whether "ship-to-shop" or any other assistance to be provided by ship's force;

6. Ship's inspecting officer or petty officer. (The person (s) specified should have detailed information on the repair item.)

The following repair request*item is given as a sample:

(a) Fuel Oil Heaters No. 1 & 2, location B-1-1 and B-3-1. BuShips plan No. DD692-S5503-12 Alt. 3.

(b) BuShips Manual requires that fuel oil heaters be tested every five (5) years.

(c) Test steam side of fuel oil heaters to hydrostatic pressure of ly, times designed operating pressure (900 psi). Test oil side to hydrostatic pressure of 114 times designed operating pressure.

(d) BuShips Mawwai, Article 55-103 (2) & (4).

(e) Ship's force will assist by preparing system for test.

(f) Ship inspector: Lt. Jones

Johnson, BTC

In order to be sure that all repair items are written up properly and that none are overlooked, it is imperative that the CSMP be well maintained and up to date at all times.

In the Pacific Fleet, the procedure for submitting routine Naval Shipyard repair requests is somewhat different in detail. Each item of work is submitted on a separate work request form, with sufficient copies. The sheaf of work requests is accompanied by a priority list.

Procedure for Accomplishing Alterations

The list of authorized alterations which are to be accomplished at a routine Naval Shipyard overhaul is prepared by BuShips. Approximately £0 days in advance of the ship's arrival, BuShips will forward to the shipyard, type commander, and the ship a list of approved alterations in the priority applicable to the individual vessel. Funds (in an amount based on shipyard estimates) are provided by BuShips.

Prior to the above procedure, BuShips usually provides the type commander with a prospective priority list of alterations to be accomplished on the ship due for a routine overhaul. The type commander may submit, to BuShips, recommended changes in the shipalt priority list. The type commander usually requests recommendations from the ship concerned in regard to the shipalts that should be completed during the shipyard overhaul period.

Shipalts marked "ship's force" or "forces afloat" are, as a rule, not undertaken by the Naval Shipyards.

Naval Shipyard Arrival Conference

When the ship arrives in the shipyard for a routine overhaul, an arrival conference is held. This conference is supervised by the Planning Oflftcer and attended by representatives from the ship, type commander, Naval Shipyard planning department, and other interested parties. The ship's repair request list, and individual item costs estimated by the shipyard planning depart-ment, are reviewed. When necessary, the details of the repair items are discussed and the work to be done is decided upon.

The limitations of the funds available determine to a great extent the amount of repairs that will be accomplished during a naval shipyard overhaul period. The estimated cost of each repair job, when approved at the conference, is added up to give the total cost. When the total cost approximately reaches the amount of funds appropriated, the shipyard will not, of course, accept any more repair requests. Under this condition, in case there are several important jobs that should be accomplished, the type commander may furnish more funds to cover these jobs.

Naval Shipyard Organization

Included in the Naval Shipyard organization under the control and authority of the Shipyard Commander are the Planning, Production, Public Works, Supply, Comptroller, Medical, Dental, and Administrative Departments. The staff of the Shipyard Commander also includes an Industrial Relations Division, and a Management Planning and Review Division.

Figure 13-4 shows a basic administrative chart of a Naval Shipyard. In considering engineering repairs, two shipyard departments will be of most interest to ship's personnel. These are the Planning and the Production Departments.

The Planning Department does all the planning with regard to the submitted work requests. A certain number of civilian planners, each one a specialist in his field, are assigned for the ship. After a certain amount of inspection and research work, job orders are written up for each approved work request. These job orders include such items as instructions and procedures, reference plans, information on material and repair parts, shops that are to do the work, and the estimated man-hours and cost. The Planning Officer, or his assistant, must approve

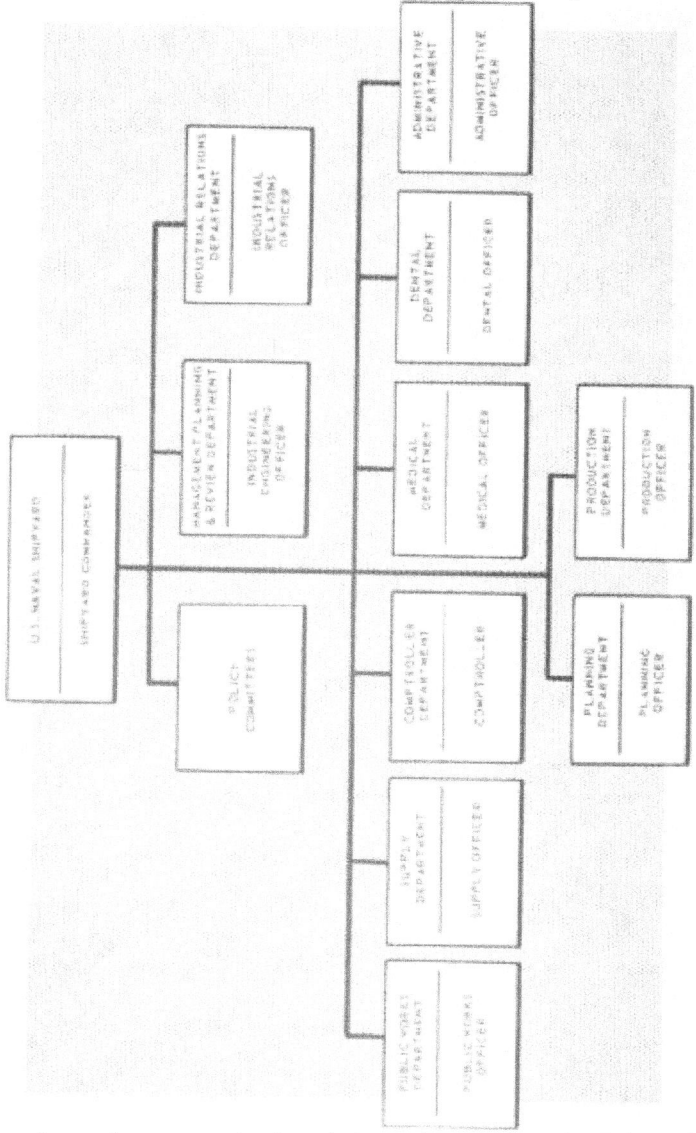

each work request before it is authorized and a job order made out and processed.

The Production Department includes the various shops and repair facilities of the Naval Shipyard. The actual work is done by the Production Department. The Production Officer is responsible for seeing that all work issued for accomplishment by the Production Department is accomplished within the time allowed and funds allocated, and in accordance with applicable instructions and sound engineering practice. The Production Department will not accomplish any work unless it is authorized to do so by a job order.

Naval Shipyard Shops

A shop in a Naval Shipyard is a separate unit assigned certain specific work, usually by trades, and measured with qualified men adept in the type of work assigned. A list of the various shops is shown in the following table:

Shop No.	Shop Name
01	Supply shop.
02	Transportation shop.
03	Power plant.
06	Central tool shop.
07	Public works shop.
11	Shipfitter shop.
17	Sheetmetal shop.
23	Forge shop.
25	Gas manufacturing plant.
26	Welding shop.
27	Galvanizing plant.
28	Plating shop.
31	Inside machine shop.
33	Director shop.
35	Optical shop.
36	Ordnance shop.
37	Electrical manufacturing shop.
38	Outside machine shop.
41	Boiler shop.
45	Salvage shop.

Shop No.	Shop Name
51	Electric shop.
55	Gyrocompass shop.
56	Pipe and copper shop.
61...	Shipwright shop.
63---	Joiner shop.
64	Woodworking shop.
68	Boat shop.
71	Paint shop.
72	Riggers and laborers shop.
74	Sail loft.
81	Foundry.
93	Print shop.
94	Pattern shop.
96	Paint manufacturing shop.

97 Ropewalk.

99 Temporary service shop.

All productive shops in the yard are under the supervision of the shop superintendent. Each shop is under the control of a civilian, usually a master mechanic. The different ratings of civilian supervisors in the shops and similar activities are as follows: Master Foreman

Chief Quarterman Quarterman Special Leadingman Leadingman Snapper

It is to your advantage to know the titles of the various Naval Shipyard personnel that you come in contact with on the various repair jobs.

Ship's Superintendent

The ship's superintendent is a Navy officer attached to the production department of the Naval Shipyard, and acting as a liaison officer between the ship and the yard. In most yards it is customary to assign one officer as ship's superintendent for each ship. However, conditions may vary according to the size of the ship and the number of ships present. A good ship's superintendent can aid the ship a great deal in obtaining a successful overhaul.

The ship's superintendent is usually on hand at the dock when the ship arrives and ties up. He checks to make certain that the required dock services are promptly furnished. He attends the arrival conference and has a list of the ship's work items. He has, as a rule, a good knowledge of the repair work that is to be accomplished on the ship during the overhaul period.

The ship's superintendent maintains liaison with all personnel involved—the ship's officers and key personnel, civilian planners assigned to the ship, shop supervisors, supervisors in charge of repair details on board ship, and other Naval Shipyard supervisory personnel. When any delays or interferences develop on a repair job, the ship's superintendent can immediately check with the responsible yard personnel and obtain detailed information or assist in overcoming any difficulties that may be present. If any information on a job is requested by ship's force, the ship's superintendent can usually furnish it. He also keeps the ship's key personnel posted on the progress of all repair jobs. The ship's superintendent is available for advice on (1) repair procedures, (2) unsatisfactory work by yard personnel, and (3) tests made by the shipyard. The primary duty of the ship's superintendent is to assist the ship in all matters regarding repair when the ship is in the yard.

Ship's Progressman

The ship's progressman is a civilian who is assigned to the production department of the Naval Shipyard and has the job of keeping a running check on the progress of all Naval Shipyard work being done on the ship. It is customary to assign one progressman to each ship but, as in case of the ship's superintendent, conditions will vary.

In addition to keeping the production department informed, the ship's progressman will keep the ship posted on the progress of each job. A good ship's progressman.

especially for a small ship, will perform most of the duties assigned to the ship's superintendent. Because of his experience and knowledge of the Naval Shipyard, the ship's progressman is in a good position to give assistance, advice, and information in regard to any repair problems that may come up.

Checking on Progress of Work

During a routine Naval Shipyard overhaul period the ship has to submit weekly shipyard progress reports in accordance with the type commander's instructions. In order to submit weekly progress reports, as well as for their own information, ship's supervisory personnel must keep an accurate check on the progress of work at all times. This should include ship's force work as well as Naval Shipyard work. One of the best methods of keeping track of the numerous repair jobs is by means of a progress chart. Any number of copies can be used as necessary. Usually one is made out for Naval Shipyard work and another one for ship's force work. The job order number and title are listed in the left-hand columns. The right-hand columns are usually marked to show the percentage of completion for each job listed. One copy of the progress chart is usually posted outside the log room where assigned ship's key personnel can keep it up to date. Many ships use the same progress chart for tender or repair ship overhaul periods.

Some Naval Shipyards hold a weekly repair progress conference. This conference is attended by ship's representatives as well as by all interested shipyard activities. Usually jobs that are encountering delays or other difficulties are discussed. When these conferences are held a good deal of knowledge can be obtained in regard to the progress of Naval Shipyard work.

The ship's superintendent and the ship's progressman are the important yard personnel who can provide information on the progress of the various shipyard jobs.

Checking on the progress of a job requires detailed information of what repair work is to be accomplished on

the job. This information can be obtained from the job orders that are issued by the planning department of the yard. The ship receives 3 or more copies of these job orders, which are usually delivered to the ship by the ship's progressman. In most cases a complete set of these job orders is kept on clipboards or file folders in the log room.

Copies of the job orders that are applicable to your division are usually kept by your division officer. It is important that you understand the details of these job orders before you start checking on the progress of the individual repair jobs.

Obtaining Additional Repair Jobs

It is sometimes necessary to prepare supplementary repair lists, which embody items arising subsequent to the submission of the original list. This additional work may be the result either of recent voyage casualties or of shipyard tests or inspections; or, in the period (approximately 3 months) between the submission of the original work lists and the ship's arrival at the shipyard, there might be some unforeseen difficulties come up which will require shipyard repairs. In cases of this kind, an added repair list is made out which is called the 1st supplement. If possible, this should be done prior to the ship's arrival at the yard.

The same procedure holds true for submitting the supplementary list as for the original, and these supplementary lists are dovetailed into the ship's priority index.

Outside of the two cases mentioned, there should ordinarily be no further need for submitting supplementary repair items. In other words, all items requiring shipyard repairs should be written up and submitted before a ship arrives in the yard and not after it has been in the yard for some period of time. A last-minute repair job is usually a headache for everyone concerned, and all possible precautions should be taken to avoid this sort of thing happening. In most cases where a last-minute job comes up, there has been an inadequate ship's mainten-

ance inspection program on board ship; or else the CSMP record-keeping has not been complete or up to date; or there has been a lack of experience or knowledge in submitting a complete list of repair items for a Naval Shipyard overhaul period. Because of the stringency of funds available for an overhaul period, many repair jobs are assigned for ship's force accomplishment. Sometimes it is found, after these repair jobs have been started, that making satisfactory repairs is beyond the capacity of ship's force, and that a request has to be made for the yard to do them. To avoid such difficulties, jobs of this type should be written up as "ship-to-shop" shipyard jobs. In this manner ship's force will disassemble, assemble, and test machinery while the yard will do the needed repair work.

Inspection Duties of Ship's Force

The inspection of work being done by a repair activity for a ship is the responsibility of both the repair activity and the ship. The repair activity should require such inspections to be made as will ensure the proper execution of the work and adherence to prescribed specifications and methods. The ship should make such inspections as may be necessary to determine if the work is satisfactory, both during its progress and when completed.

The Chief or First Class Petty Officer should schedule his work in such a manner that he is free at all times to inspect and check the progress of work being performed by naval shipyard personnel in his space or being performed on equipment for which he has the responsibility of maintenance and upkeep. A check should be made to see if any required tests are made by the shipyard before the job is considered fully completed. The Naval Shipyard job order will list any tests that have to be made by yard personnel.

If you find that unsatisfactory work has been performed by shipyard personnel, you should follow the instructions put out by your engineer officer. Talking it over in a friendly manner with the workmen will usually

solve the problem. If it doesn't, you should notify your division officer or engineer officer, who can take up problems of unsatisfactory work with the ship's superintendent. In exceptional cases the commanding officer of your ship can take necessary action in accordance with Navy Regulations.

On many ships it is customary for your division officer or engineer officer to check with you before he signs a job order as being completed. By continuous inspection of shipyard work and checking off the jobs that have been satisfactorily completed, you can furnish the required information without unnecessary delays.

Drydocking the Ship

The ship is drydocked each time it goes to the Naval Shipyard for a routine overhaul. The regular procedure is to keep the ship in drydock for as short a period as practicable. As soon as the necessary work has been completed the ship is removed from drydock.

Before the ship goes into drydock you should check the machinery history and the CSMP to make sure that you have all necessary information on the sea valves. It is a good idea to make a check-off list from the ship's blueprint of sea valves, so that no valve will be overlooked. In other words, you want to be fully prepared to inspect the fire-room sea valves because this work should be started as soon as the ship is drydocked.

A ship should enter drydock without list and without excessive trim. Trim in excess of 1 foot per 100 feet of length normally makes the docking operations hazardous. If practicable, the trim should be brought below this limit before any attempt is made to dock the ship. While the ship is in drydock, no weight, fuel oil, or water should be shifted, added, or removed, except as specifically authorized by the docking officer. Any tanks containing water or oil should be either completely full or completely empty. When permission to shift weight is given, the responsibility for keeping an accurate record of the amount and location of the change of weights rests with the ship.

Provisions should be made to ensure that the ship will lift from the block without taking a list when the drydock is flooded.

When the ship is in drydock, no fuel oil or other flammable liquid should be drained or pumped into the dock.

When ships are docked in cold or freezing weather, the valves, pipes, or similar fittings attached to the shell should be examined and any water remaining therein should be drained to prevent freezing and possible cracking of the fitting.

Whenever a ship is drydocked all sea valves must be examined and the result of the examination must be entered in the engineering log. The yoke, yoke rods, valve stem, and securing bolts, as well as the internal parts of the valve, must be examined. At least two of the bolts holding outboard valves to sea stools should be removed from each valve for inspection; the remaining bolts should be sounded with a hammer. If defects are found in a bolt, all the bolts for that valve should be removed for inspection. Where all bolts have been removed, the gasket should be replaced. All necessary repairs to place sea valves in good condition should be made while the ship is still in drydock.

While the ship is in drydock, openings in the hull caused by ship's force disassembling sea valves should be closed temporarily at the close of working hours, by replacing the valve bonnets or by blank-flanging the openings. The docking activity is responsible for openings on which they are making repairs. At the end of working hours, a report should be made to the engineer officer in regard to the status of all sea valves. The same information should be written up in the engineering log.

Before the drydock is flooded, all sea valves must be carefully inspected to ensure that they are properly secured. The result of this inspection should be reported to the engineer officer and entered in the engineering log.

While the drydock is being flooded, a continuous inspection must be made of sea valves until the ship is afloat and

all valves are under a normal operating head of water. Any unsatisfactory conditions must be reported at once to the engineer officer so that the docking officer can be notified. A report of leakage must be made in sufficient time so that the docking officer can stop flooding, if necessary, before the ship lifts from the supporting blocks.

Dock Trial

A dock trial is held whenever major repairs have been made by a Naval Shipyard to propulsion machinery. The trial is usually held as a precautionary procedure at the completion of a routine Naval Shipyard overhaul period.

At least one day prior to the dock trial all auxiliary machinery should be tested to prevent delay or interference with the testing of the main engines and associated equipment.

The ship's engineering personnel, under the direction of the engineer officer, should make such tests of boilers and machinery, with the ship properly secured to the dock, as will enable them to ascertain that this equipment is ready for operation at sea. Sufficient inspections and tests should be made to ensure that machinery and equipment have been properly repaired and are in a good operating condition. Any defect, deficiency, or maladjustment will have to be corrected either by ship's force or the shipyard. The dock trial should be repeated as often as is necessary, until conditions are satisfactory.

Post Repair Trial

The post repair trial is mandatory whenever the machinery of a vessel has undergone extensive overhaul, repair, or alteration which may affect the power or capabilities of the vessel or the machinery; it is optional whenever machinery has undergone only partial overhaul or repair. The object of this trial is to ascertain if the work has been completed and efficiently performed and if the machinery in all its parts is in all respects ready for service.

The post repair trial should be held as soon as practicable after the repair work has been completed, the preliminary dock trial made, and the persons responsible for the work satisfied that the machinery is in all respects ready for a full-power trial. The conditions of the trial will be largely determined by the character of the work that has been performed. The trial should be conducted in such manner as the commanding officer shall deem necessary. If the repairs have been slight, and the commanding officer is satisfied that they have been satisfactorily performed and can be sufficiently tested without a full-power trial, such trial may be dispensed with. A post repair trial is usually made when the ship has completed a routine overhaul period.

Unsatisfactory conditions that are beyond the capacity of ship's force should be corrected by the Naval Shipyard. Machinery should be opened up and carefully inspected if necessary, to determine the extent of any injury, defect, or maladjustment which may have appeared during the post repair trial.

Some Naval Shipyard personnel, such as technicians, inspectors, and repairmen, accompany the ship on a post repair trial. The yard personnel witness the operation of machinery that has been overhauled by the yard. If a unit of machinery is not operating properly, the yard technicians will carefully inspect it and try to determine the cause of unsatisfactory operation.

Upon the completion of the post repair trial, a report of the circumstances and results of the trial is made to CNO and to BuShips.

QUIZ

1. What kind of record should be kept of repairs which cannot be accomplished immediately, or which are beyond the capacity of the ship's force?

2. What term is used to describe an alteration which affects the military characteristics of a vessel?

3. What term would describe minor modifications which were made to a machinery unit in order to prevent recurrence of trouble?

4. What form is used to record all pending repairs?

5. What three cards make up the CSMP?

6. With what record is the CSMP kept?

7. When should the CSMP cards be made out?

8. How can you make sure that you have an up-to-date blueprint, rather than an obsolete one?

9. On board ship, how would you go about finding a particular blueprint?

10. What is the subject matter filing number for pumps?

11. What term describes a ship's availability for the accomplishment of specific items of work by a repair activity, with the ship present?

12. What type of availability is involved in the accomplishment of a specific item of work by a repair activity, with the ship not present?

13. What is the main difference between voyage repairs and restricted or technical availabilities?

14. What t<:rm describes a period of time assigned a ship for the uninterrupted accomplishment of work by the ship's force?

15. What activity performs most of a ship's routine inspections and tests?

16. What is the normal duration of a routine overhaul period alongside a repair ship or tender?

17. What is the basic difference between repair ships and tenders?

18. What is the designation of the general-duty repair ship which is designed to meet the maintenance requirements of capital ships?

19. How much notice is generally required for major work items to be performed by repair ships or tenders?

20. Who is usually present at arrival repair conferences?

21. At what point does a work request become a job order?

22. What is the best way to identify material delivered to a repair ship or tender?

23. How are repair jobs designated for such portable items as valves, gages, etc?

24. What action should be taken by a ship, if the repair ship or tender does not provide the services of a ship's superintendent?

25. What activity is responsible for the inspection of work done for a ship by a repair activity?

26. Under what circumstances are dock trials held?

CHAPTER

THE OIL KING
FUEL OIL SYSTEMS

Training an individual or a group requires that the instructor have a complete knowledge of the subject matter he wishes to pass on. It is mandatory that the Boilerman First or Chief know the systems relative to fuel oil on board the ship to which he is attached.

All of this knowledge cannot be obtained from a book or a training manual, because it is not practicable to describe and illustrate in detail the fueling systems of all types of ships. This chapter describes the fuel oil service and transfer systems for the DD445 class and DD692 class, which have the same general features as the systems on larger vessels. By careful study of these systems, you'll be better able to trace a similar system on any vessel. Then, after you have become thoroughly familiar with the system, you'll be qualified to pass the knowledge on to other men.

Service Systems

For each plant there are two fuel oil tanks designated as fuel oil service tanks, from which oil furnished to the oil burners is pumped. These two tanks are the after centerline tanks of the forward group (that is, A2F and ASF), and the forward centerline tanks of the after group (that is, ClF and C4F). The fuel oil service pumps
installed in each fireroom take suction through a common suction line.

The fuel oil service suction manifold is located on the forward bulkhead of the forward fireroom. It consists of three valves; the one on the left is a cut-out valve for the entire manifold, the center valve is for suction from tank A2F, and the right-hand one for suction from tank ASF. The suction line leads from this manifold down the port side, as illustrated in figure 14-1. The after system, illustrated in figure 14-2, is similar to the forward system in general arrangement. The service suction manifold, however, is in the engineroom in this case.

BuShips recommends the following procedure for fuel oil service tanks:

1. Take a suction from one service tank of each system at a time.

2. Maintain the fuel oil service tank levels between 95 and 50 percent, to ensure an adequate gravity head.

3. Fill the standby service tank of each system shortly after shifting suction to the adjacent service tank, for settling out water and sediment.

4. Prior to taking suction from a standby service tank take a "thief sample." If necessary,

take a low suction for a short interval and discharge to the contaminated oil tank or overboard, as operating conditions permit.

A satisfactory thief sampler can be manufactured aboard ship. One suitable for sounding-tube use is described in BuShips Plan S5501-64052-SK of 16 August 1943; illustrations of two types of samplers are given in chapter 55 of BuShips Manual.

Transfer Systems

The only pump involved in the transfer system is the fuel oil BOOSTER or TRANSFER PUMP. This pump is constructed in the same manner as the fuel oil service pump, except that it is designed to operate at a lower pressure and with a higher capacity than the service pump. The

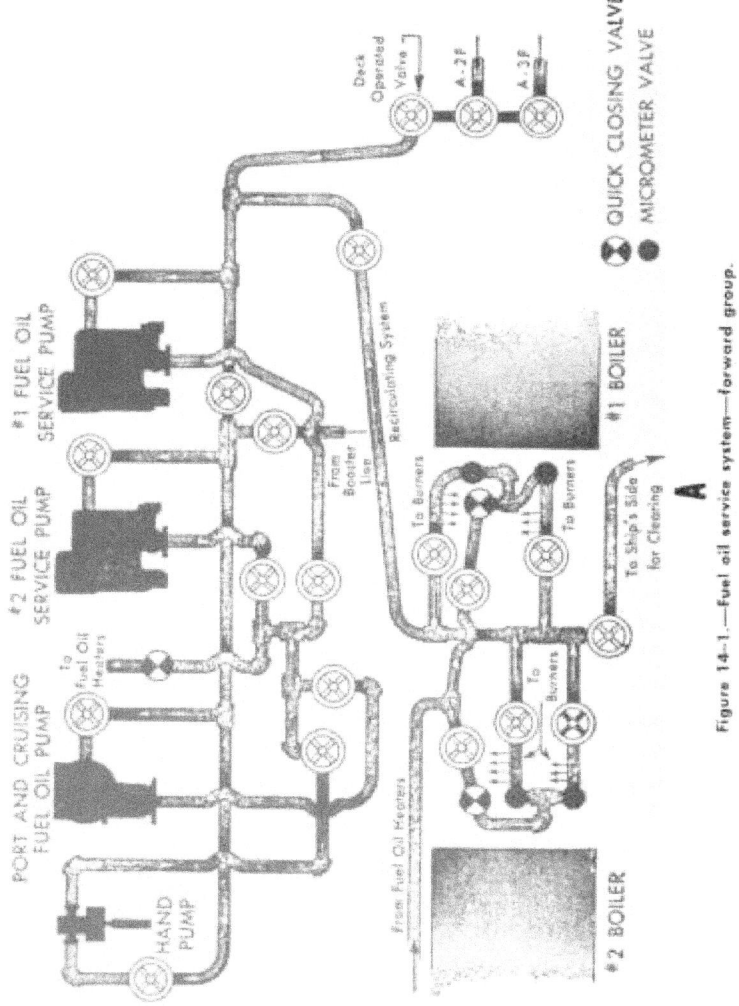

Figure 14-1.—Fuel oil service system—forward group.

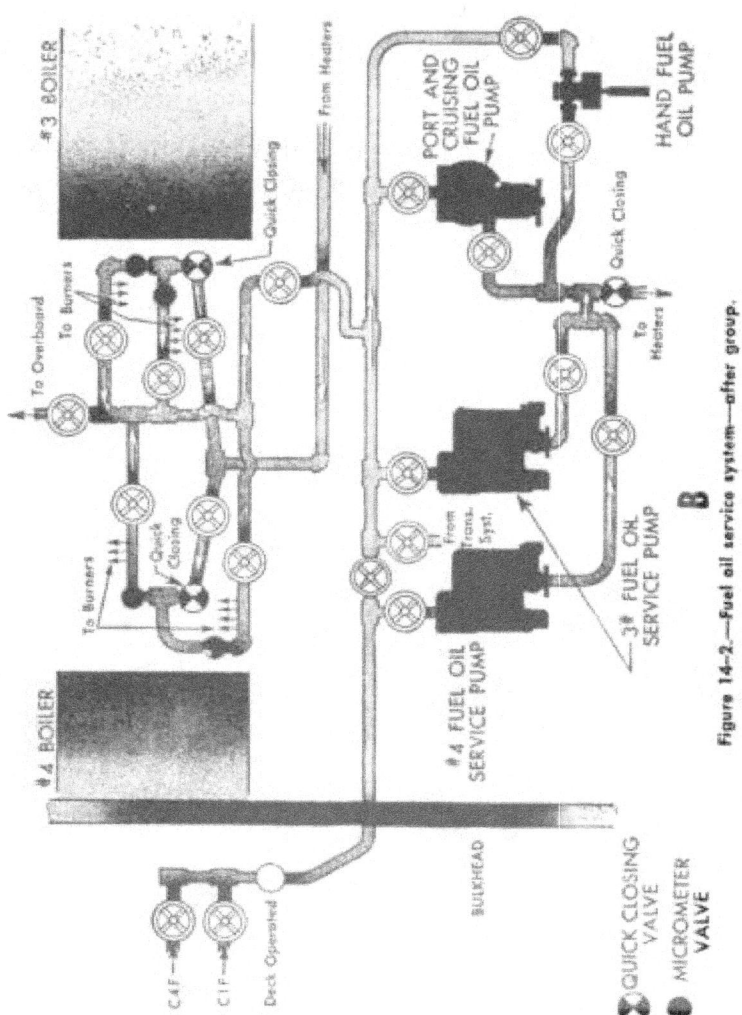

Figure 14-2.—Fuel oil service system—after group.

pump is rated to discharge 100 gallons per minute at a pressure of 100 psi. When transferring oil under normal operation, the discharge pressure will not be this high (owing to the lack of a head to pump against), and the capacity will be controlled by the speed of the pump. The transfer pump is connected to the forward fuel oil tanks, as shown in figure 14-3, by three manifolds located on the

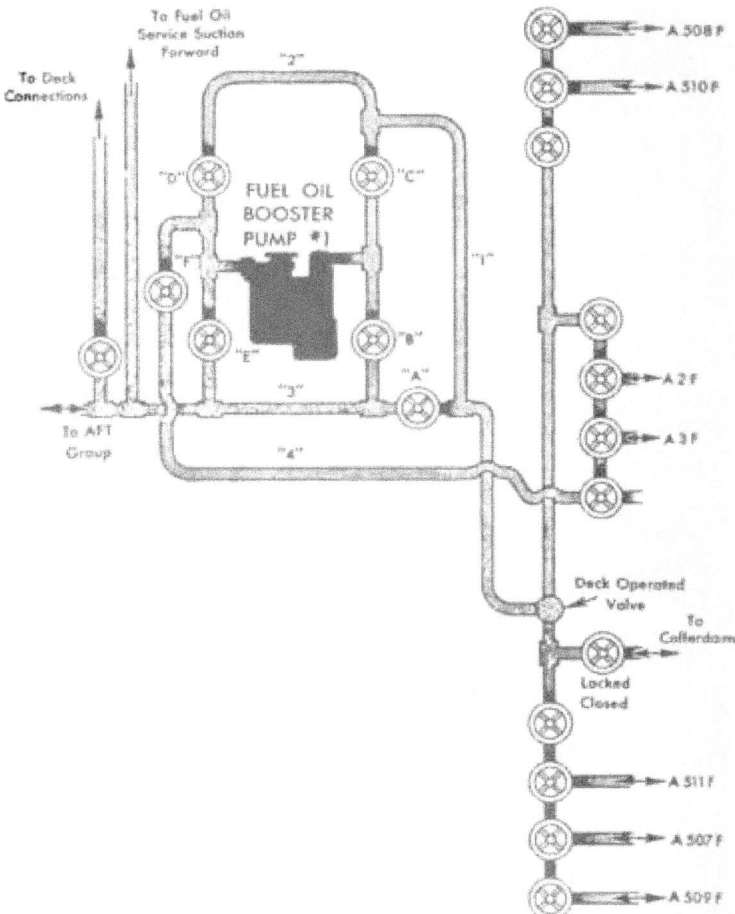

Figura 14-3.—Fu«l oil transfer tyctem, DD445 clots—forward group.

forward bulkhead of the forward fireroom. These manifolds serve as cut-outs for suction and discharge to the tanks. A portside manifold connects to the storage tanks and wing tanks. The manifolds have cut-out valves and are all joined together by a common line which leads to the No. 1 fuel oil transfer pump through a cut-out valve which may be operated either from the fireroom or from the deck. See figures 14-3 and 14-4. In addition, a single valve branches from the common line and connects to the cofferdam. This valve is locked closed.

The line leading to the fuel oil transfer pump passes through another cut-out valve and then leads on past the transfer pump to cross-connect with the after transfer system.

The fuel oil transfer system for the DD692 class differs from that of the DD445 class, as illustrated in figures 14-5 and 14-6. Note that the manifolds connecting to the forward fuel tanks are not all connected to a single line as in the DD445 class. You must bear in mind that these simple examples develop into more complicated and extensive arrangements on larger ships. A close study of the simple schematic arrangements shown here, however, will give you an idea of what to expect on any ship. It is mandatory that as oil king you know the location and purpose of every line, valve, and manifold in the entire system of your ship. You must have this knowledge before you can pass it on to other men.

It is suggested that as soon as possible after reporting aboard ship you check the fuel oil service and transfer systems, whether you expect to be assigned to the duties of oil king or not.

A proved method of training men to know and to operate a system is: (1) have them trace out the system and draw it, and (2) go over the system with them, using a schematic arrangement and showing and explaining as you go. It is not a good practice to instruct a large group of men at one time; training just a few at a time is highly

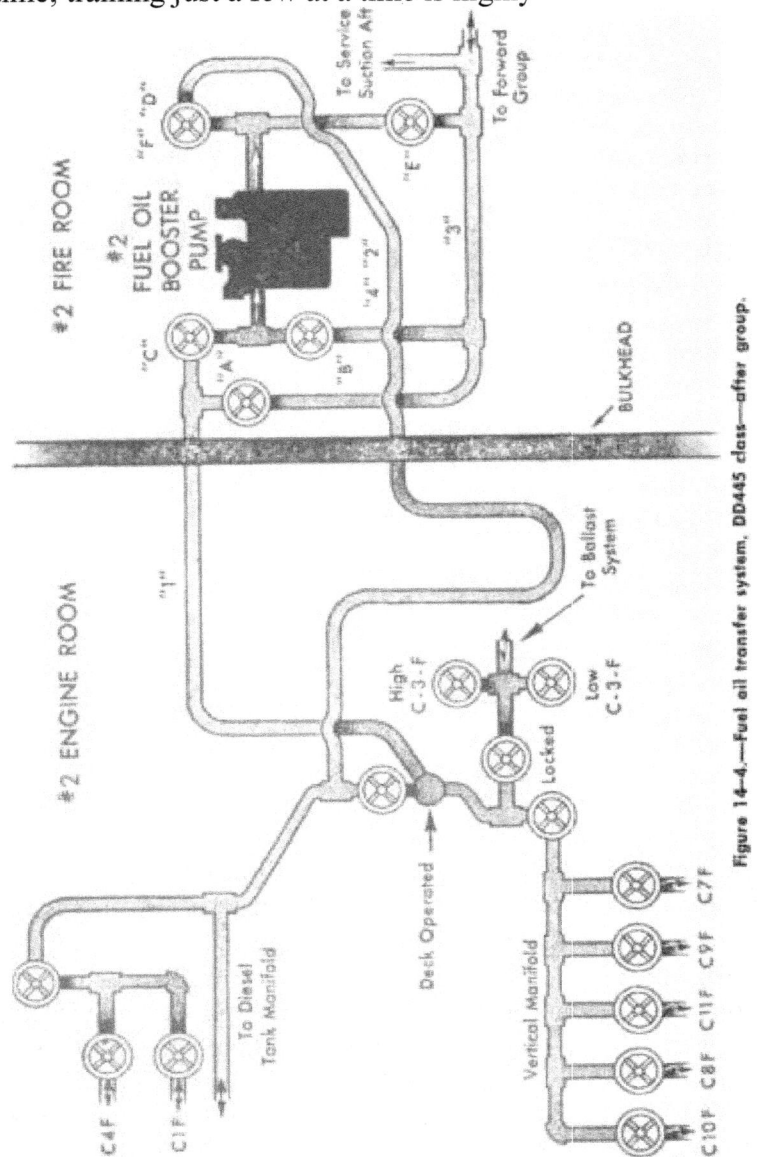

Figure 14-4.—Fuel oil transfer system, DD445 class—after group.

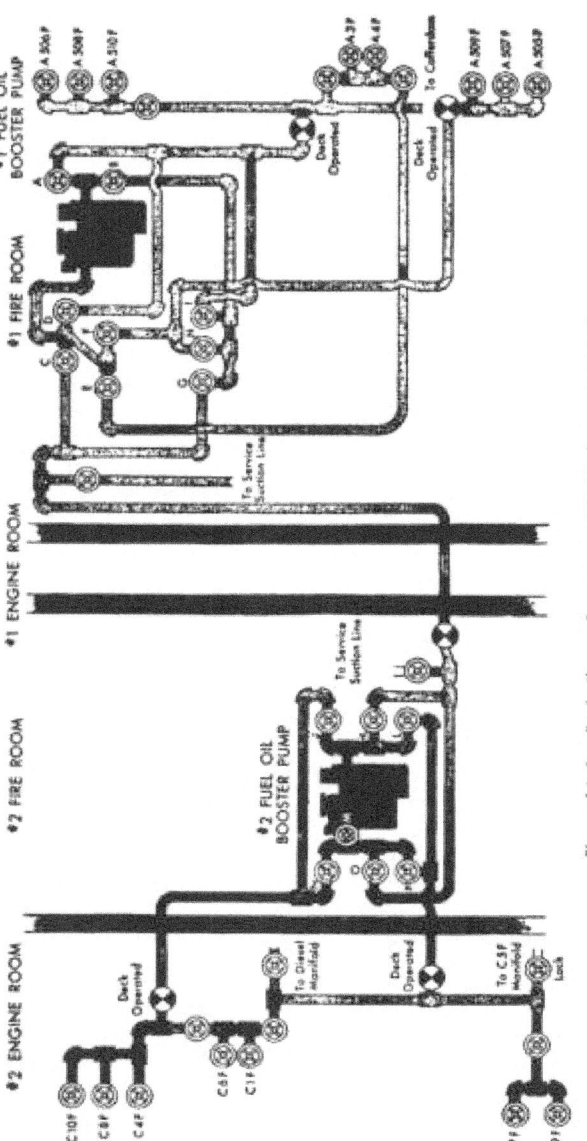

Figure 14–5.—Fuel oil transfer system, DD692 class—short hull.

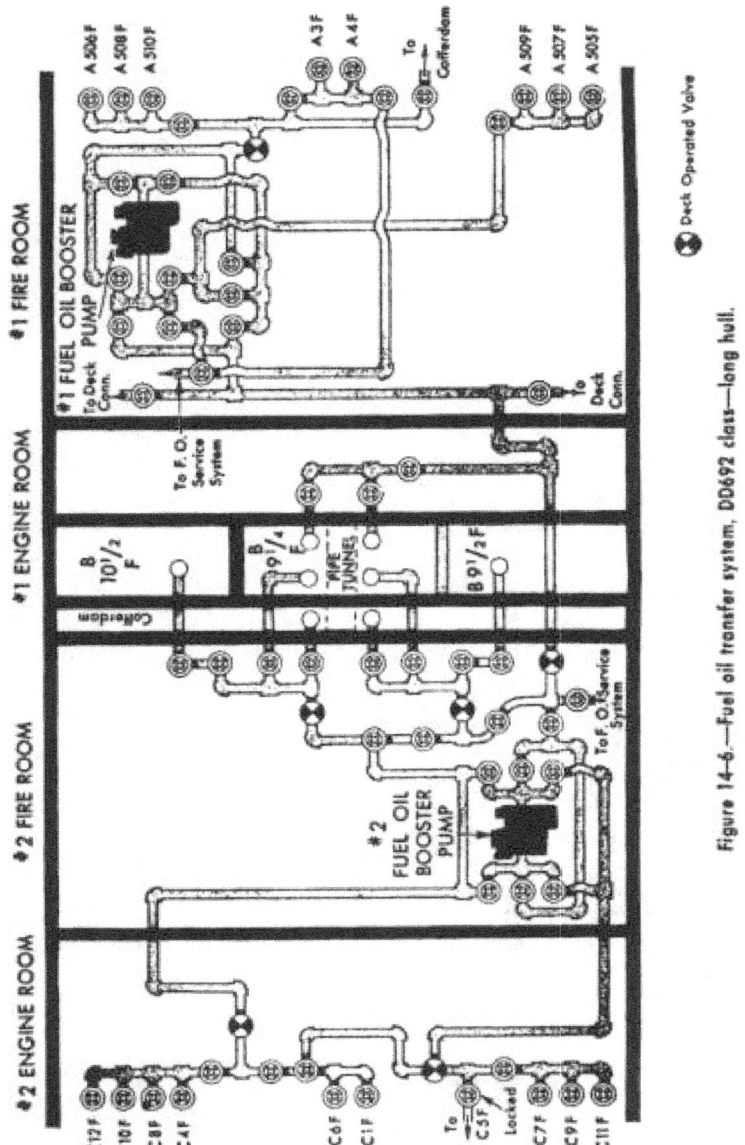

Figure 14-6.—Fuel oil transfer system, DD692 class—long hull.

recommended. When each man has a working knowledge of a system, you should bring the various training groups together and have each man explain to the others the system he has studied. Afterwards have the group criticize each man's explanation, for both the group and the individual can learn from such criticism.

FUELING SHIP

The oil and water king is required in general to receive, store, and test fuel oil, and in some cases to supply other ships with fuel from reserve storage tanks. In order to accomplish these things, the fueling gang must be well versed in the location and sequence of tanks, cross-connections, pumps, heaters, pneumercators, manifolds, and the like.

Your men must also store, test, and assure a continuous supply of boiler feed and ship's service water, to meet the daily requirements of the ship. It is your responsibility to see that they acquire the knowledge necessary to perform these duties satisfactorily.

Preparation for Receiving Fuel

Preliminary to receiving fuel there are numerous preparations for which you will be held responsible. The first is the DEBALLASTING and STRIPPING of all ballasted tanks as soon as possible after receiving word from the fuel officer that the ship will take on fuel. If the ship plans to fuel after entering port, permission must be requested, from port authorities, to deballast upon entering port. The conditions of the sea may make it impossible to de-ballast before entering port. (Some loss of stability ensues from emptying low tanks and thereby creating a high center of gravity of the ship.) It is very possible that it will be necessary to deballast into a barge after coming to anchor in port. This procedure is mandatory at most anchorages for the obvious reason that an oil slick may ruin beaches and contaminate the water. During wartime, if you plan to fuel at sea it is usually the

procedure to deballast at night (conditions of the sea permitting) so that an oil slick is not evident to enemy search planes.

The actual pumping of ballasted tanks will in most cases follow a recommended sequence table, in order to retain maximum stability and maneuverability while under way. Although such a sequence plan need not be strictly adhered to while the ship is in port, it should be followed at least in part, to retain a maximum stability in case a sudden storm or a conflagration should occur before the lost weight is restored to the ship in the form of fuel oil. After a tank has been pumped, it should be stripped by using the tank drain pump to remove all remaining oil and sludge.

Prior to fueling, the oil king should top off all the service tanks and then top off as many storage tanks as possible, in order to reduce the number of tanks to be filled while fueling.

Manning the Fueling Stations

Your first step in organizing a fuel detail will be to select men with enough experience to carry out efficientl: the individual duties required at each fueling station. If there are not suflftcient men available with experience, you will have to make up a skeleton crew, and fill in the gaps with untrained men, who will have to be trained by YOU with the help of other leading Petty Officers.

In order to effectively train a fueling detail, you must first know the fueling system of your ship and the numerous stations to be manned, in addition to your duties involving daily tests and reports as described in Boilerman 3 & 2, NavPers 10535-C.

Although fueling is classed as an *'all hands" procedure, in actual practice it is primarily an engineering operation under the direct supervision of the engineer officer, assisted by the fueling officer and the oil king and his detail. This detail will in most cases be composed of the follow-ing: (1) messenger to the fueling officer, (2) pneumer-cator men and/or tank sounders, (3) forward- and after-hose connections men, (4) forward manifold controfiman, (5) midship manifold controlman, (6) after manifold controlman, (7) forward manifold JV talker, (8) midship manifold JV talker, (9) after manifold JV talker, (10) forward- and after-fuel connections JV talker, (11) forward sounding JV talker, (12) midships sounding JV talker, (13) after sounding JV talker, (14) main control JV talker, and (15) ship-to-ship phone talkers, one stationed at the hose connection and one on the bridge. The ship-to-ship phone talkers may or may not be furnished by the Engineering Department.

The fueling stations will vary according to the type and design of the ships. For example, destroyers use sounding rods instead of pneumercators for determining the percentage of oil in the tanks. Eack tank has a sounding rod and while fuel is being received, a man should be assigned to each tank that is receiving fuel. Regardless of the system, the men should be notified of their station and/or a fueling watch list should be posted well in advance of fueling time.

Your station during fueling will normally be at the MAIN CONTROL, where you will have control of the fueling board. It will be your responsibility to see that the men of your fueling detail remain on the job and alert to their respective duties. The fueling detail is tied together with an elaborate communication system, illustrated in figure 14-7. It is imperative that you hold numerous phone talker drills to ensure close coordination of the fueling detail. These phone talker drills should be scheduled at regular intervals and often enough to ensure that every man on the detail is proficient in receiving and passing on instructions from the fueling officer or main control. You should read and study the Telephone Talkers' Manual, NavPers 14005.

You and your detail will usually man the fueling stations one-half hour prior to expected fueling time.

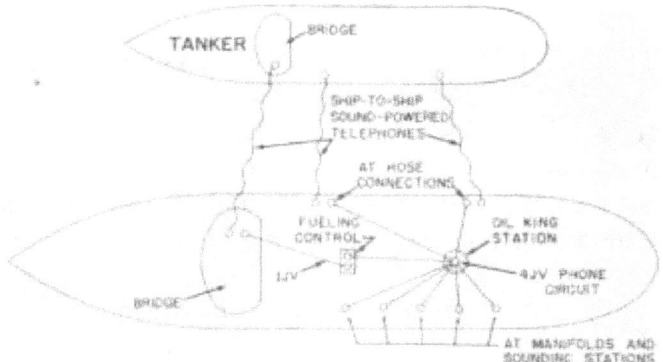

Figure 14-7.—Fueling communication tyttom.

During this period see that the phone circuits are tested, that the air hose is connected to the fueling connection, and that the thermometer and the pressure gage are screwed into the fueHng connection. After the fuel hoses have been brought aboard and the caps removed from their ends, make certain that your men clamp the quick-release coupling correctly. One type of quick-release coupling is shown in figure 14-8. When all connections are made you should immediately inform the fueling

BLANK FLANGE

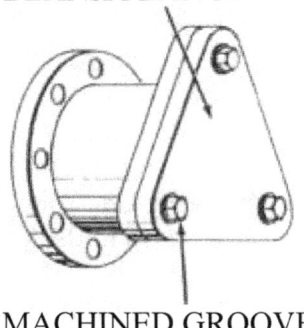

MACHINED GROOVE

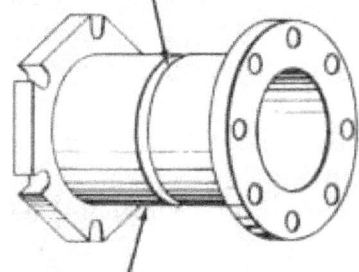

DROP BOLTS BREAKABLE SPOOL

SUPPLYING SHIP RECEIVING SHIP

Figure 14-8.—Quick-r«Uat« coupling.

officer that you are ready; he in turn will inform the bridge and request permission to notify the tanker to commence pumping.

Receiving Operation

When fuel is coming aboard, a constant check must be kept on all tanks that are receiving fuel. On large ships, in particular, the oil king must follow a systematic procedure in order to get all tanks properly filled without unnecessary loss of time. In addition, he must make sure that the vessel does not develop a list.

When there are several tanks in each overflow group, and four or more such overflow groups, one or two tanks in each group should be opened initially; when these have been filled to about 85 percent of capacity, start filling the others in the group and close down on the valves to the tanks that are almost full, topping them off slowly. The overflow tank in each group is filled last. If this procedure is followed, you will not have to request the tanker to reduce pressure until you are ready to top off the last tanks.

The percentage of capacity to which any tank has been filled is determined by sounding rods or tank-capacity indicators of the pneumercator type (fig. 14-9).

As oil king, you will be responsible for keeping the fueling officer advised as to the amount of oil received (percentage of the total to be received) and the probable time required to complete the fueling. The fueling officer, in turn, keeps the commanding officer posted on fueling progress.

When all tanks are filled, the fuel hose is emptied by either of two methods: (1) the oil in the hose is blown back to the tanker by opening the compressed air valve to the fueling connection, or (2) the tanker may take a back suction, which will also require that the air valve be opened. Which of these two methods to use is a decision that rests with the tanker or fueling station.

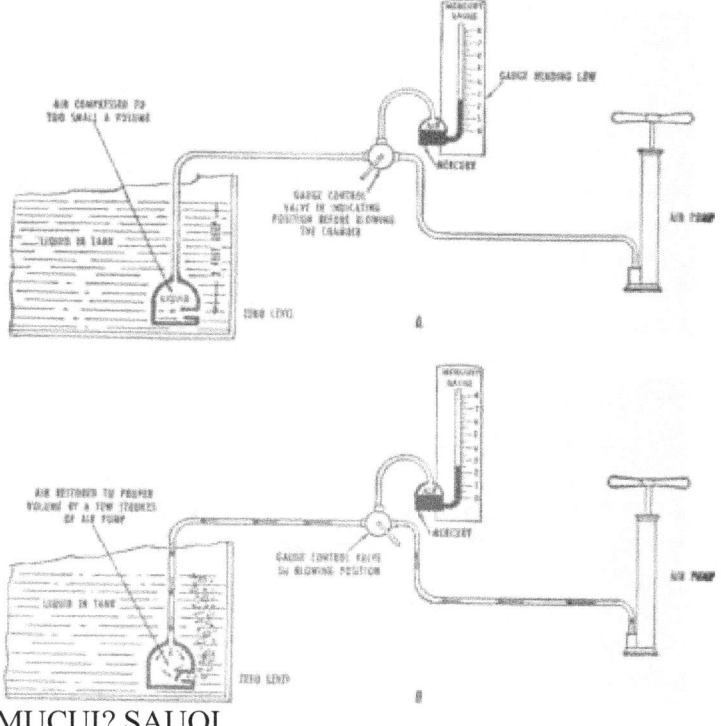

MUCUI? SAUQI

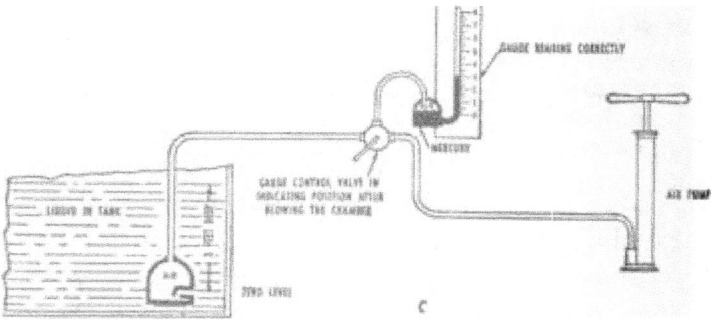

Figure 14-9.—Tank-capacity indicators.

Once the fuel line is cleared, the hose should immediately be uncoupled and returned to the tanker.

Computing Amount Received

The oil king should be familiar with the procedure for computing the amount of fuel oil received. This procedure is more than merely taking a tank sounding or reading a gage; it involves making two corrections —a temperature correction and a gravity correction.

Correction to a standard temperature is necessary because of the expansion and contraction of petroleum at different temperatures. It is impossible to accurately define the amount of oil contained in a tank, unless this property of expanding and contracting is taken into consideration. The specific gravity of the oil also varies according to its temperature. It is customary to measure volume and gravity at a standard temperature of 60° F.

Determining the specific gravity. —Specific gravity is the ratio of the weight of a given volume of a substance to the weight of an equal volume of water. This ratio varies with temperature. Petroleum oils are either tested for gravity at 60°F., or else the readings obtained are corrected to that temperature.

The American Petroleum Institute (A. P. I.) gravity, the gravity scale most widely applied in this country, is an arbitrary figure, expressed in degrees and related to specific gravity in accordance with the table in figure 14-10. This table will suffice in range for the temperatures and gravities encountered in Navy fuel oil. Here after in this discussion, the gravity of the fuel oil will be expressed as observed gravity or A. P. I. gravity.

The observed gravity of petroleum oils is determined in the same general manner as the specific gravity of storage battery solution—by floating a hydrometer in the liquid and reading on the hydrometer scale the number which coincides with the liquid level. The gravity test may be made with the sample of fuel oil at any temperature between 0° F. and 195° F. However, if it is made at a

Figure 14-10.—Tamperature reduction table (American Petroleum Institute).

temperature other than 60° F., the observed gravity must be corrected, by use of a table similar to that shown in figure 14-10, to gravity at 60° F. The temperature of the sample should not change substantially during the test. To meet this requirement the hydrometer and thermometer should have approximately the same temperature as the liquid to be tested.

The sample to be tested should be poured into the test container slowly, so as to avoid the formation of air bubbles, which contribute to inaccuracies in reading the scale. After the sample has come to rest, the hydrometer should be lowered into the liquid and then released. When the hydrometer has come to rest the reading is taken.

Using the conversion tables. —The tables shown in figures 14-10 and 14-11 are used in making the necessary conversions. As stated before, the table in figure 14-10 is used for making the gravity conversion; the table in figure 14-11 is used for determining the volume multiplier necessary for computing the amount of oil that would be in a tank, if the oil were at 60° F.

Suppose that, after a refueling operation, tank soundings indicate that 1,000,000 gallons have been received. The oil is tested to determine its temperature and observed gravity. The tests show the oil temperature to be 90° F. and the observed gravity 15 degrees. Using the table in figure 14-10, locate 90° under the column "observed temperature in °F.," and locate 15 degrees in the space allotted to "Observed Degrees A, P. I." Follow down the "15" column to a point even with 90° F. in the temperature column, and you will find 13.4 as the corresponding degrees A. P. I. at 60° F. Now use the table in figure 14-11 to find the volume multiplier. Again locate 90° F. in the observed temperature column. The column allotted to degrees A. P. I. at 60° F. is not divided into tenths, so you will locate 13 in that column. Follow down the column under 13 to a point even with 90° F. in the temperature column. At this intersecting point you will find the volume multiplier to be 0.9889.

Figur* 14-11.—Velum* muhipliar tabi* (American Petroleum Institute).

To find the correct volume taken on board, you multiply the amount indicated by the tank soundings by the appropriate multiplier from figure 14-11. In this case, you would multiply 1,000,000 gallons by 0.9889. The result is 988,900 gallons, which is the amount of oil recorded.

Emergency Procedure

Situations may arise—such as enemy action, separation of ships, and the like—which make it necessary to let go of the fuel hose immediately. To assure immediate action should this emergency operation become necessary, have

Figure 14-12.—Emergency release of fuel hose.

one man equipped with a sharp axe and one with a sledge hammer at each fueling station. First sever the securing line, and then smash the coupling at the necked portion, as illustrated in figure 14-12.

DISCHARGING FUEL

It will sometimes be necessary for your ship to fuel another ; if this situation arises, you must be ready to carry out the operation with efficiency. As in receiving fuel,

discharging fuel makes necessary certain preliminary preparations. These preliminary preparations will usually be as follows:

1. Fill and top off to 95 percent of capacity the largest tanks close to the pumps. (The fueling officer will contact the communications officer, who will inquire of the ship to be fueled what amount is needed.)

2. Heat oil, if necessary, to the temperature required to produce a viscosity of 450 Seconds Saybolt Universal.

3. Sound all tanks to be used.

4. Line up the pumping system by segregating the forward and after sections; then test and see that the pumps operate properly.

5. The first lieutenant's crew will break out sections of fuel hose for your gang to couple together, then make up the rig.

6. Place red flags over the side of the ship, at the fueling stations. (The Officer of the Deck has draft readings taken, forward and aft, both before and after fueling.)

7. Make up a fueling detail, set up the fueling board, and fill in available data on the fueling sheet for the fueling officer.

8. One-half hour before the expected approach of the vessel to be fueled, send your detail to their stations and then test the fueling circuit, as well as special circuits involved. Have the detail connect air hoses to fueling connections, screw the thermometers and pressure gages into fueling connections, warm up fuel pumps, and open valves to fuel tanks. (Be certain that the ground wire is secured to the deck.) Report to the fuel officer that the fueling detail is ready; he will inform the bridge, and request the word passed that "the smoking lamp is out."

9. When you receive word to start discharging, start the pumps and go slowly at first to build up a pressure of approximately 40 psi at fueling connections. Make sure fuel is sufficiently heated to reduce viscosity to 450 Seconds Saybolt Universal or less.

10. Maintain rated pump capacity until your ship's (the fueling ship) tanks are down to 35 percent of capacity, then shift the pump suction. Slow down the pumps when requested to do so by the receiving vessel, and stop them when ordered to do so by the receiving vessel.

11. Remove fuel oil from the hose by blowing through with air or taking a back suction. Then unfasten, unscrew, and recap the fuel hose. The first lieutenant's gang will take it from there.

12. Sound the tanks, and compute the amount of fuel discharged; report the amount to the fueling oflfi-cer. Be sure to make the proper temperature correction in computing the amount of oil.

WATER IN THE FUEL OIL

Water in the fuel oil is a common engineering casualty. When it occurs, it often results in the breakdown of the ship. This trouble is usually caused by the lack of trained personnel, or by failure of personnel to take the necessary precautions.

Oil tanks frequently develop leaks that are not immediately apparent. Hard bumps against fenders sometimes start leaks. In wartime, battle damage, "near misses," and underwater explosions may start leaks. Some of the reasons for water in the fuel oil are:

1. Failure to take samples as required by BuShips Manual;

2. Failure to test fuel on receipt from another ship;

3. Carelessly or unknowingly taking suction from tanks containing water;

4. Failure to inspect and overhaul inoperative deck vents;

5. Water from heavy seas entering through unprotected deck vents;

6. Leaky sluice valves or incorrect manipulation of valves when taking on, storing, or discharging water ballast;

7. Leaky hull seams;

8. Improper deballasting (failure to completely strip the tanks of water).

The best insurance against the presence of water in oil is to follow the procedure outlined in the next few paragraphs.

During the forenoon watch, all fuel oil tanks or bottoms which are to be used or are likely to be used for fuel oil suction during the next 24 hours should be tested for and rendered safe from water contamination. The tank stripping pump should be used to take suction on the tank and discharge to the contaminated fuel oil settling tank over a period of time sufficient to draw off about 150 to 300 gallons through the low suction. While the tank stripping pump is operating, drain off a sample from the test connection on the discharge side of the pump and examine it for water. In this test sample, water may appear as an emulsion in the oil or in readily separable form showing a sharp cleavage line, depending on the temperature of the oil, the characteristics of the oil, and the amount of mechanical agitation to which it was subjected. Either condition will lead to difficulties and accordingly requires that settling and separation of the oil must be accomplished before it is used. If any reasonable doubt exists as to the water content of the oil sample, subject the sample to a centrifuge test.

Once each week the oil carried in the storage tanks should be tested for water. From such tanks, either take samples as outlined above or use thief samplers.

Allow oil and water mixtures accumulated in the settling tank to settle as long as practicable before attempting separation of the two liquids. Use of tank heating coils will materially assist in separating the oil

from the water unier almost any condition, and where oils of the higher viscosities are involved, heating is mandatory to obtain satisfactory results. However, the temperature in any tank should never exceed 120" F. When oil tanks adjacent to magazines are heated, the commanding officer usually designates the maximum allowable temperature. In any event, care must be taken that the magazine temperature does not rise over 100" F. nor habitually exceed 90"" F.

Remove water from the low suction level in the oil settling tank by using the tank stripping pump and discharging overboard. This should not be undertaken in harbors or restricted waters near the shore, since even under the most favorable conditions traces of oil will usually be carried overboard. After the foregoing process of separation has been carried out, the oil in the settling tank is tested by centrifuge to see if any water remains. If the oil is found to be suitable for boiler use, draw it from the settling tank through the high suction and store it in a tank to be used when the ship is in port, or at any time when no danger may result from water contamination.

The presence of even a small amount of salt water in fuel oil is highly detrimental to the furnace refractories, and is believed to be one of the major causes of excessive accumulations of slag on the firesides. Give particular attention to the water ballast compensating system. Prior to filling the tanks with fuel oil, drain them completely of water, with the lowest suction installed. Follow the previously mentioned process of fuel oil inspection and separation when transferring oil to service tanks or when taking oil directly from a storage tank.

OIL POLLUTION ACT

As a Boilerman, you must be entirely familiar with the provisions of the Oil Pollution Act. A copy of this Act should be posted at every pump which can be used to pump oil overboard.

The Oil Pollution Act prohibits the discharge or escape of oil or water containing oil from any vessel into navigable coastal waters of the United States. Heavy penalties are imposed for violation of this Act, Similar laws are enforced in most countries throughout the world.

No oil or water containing oil should be discharged overboard within 100 miles of land, except in case of an emergency that imperils life or property. Only clean ballast water may be pumped overboard in port; dirty ballast water must be pumped directly ashore or pumped into barges. When pumping clean ballast overboard in port or within the 100-mile zone, it is advisable to discharge over the top rather than through a sea valve. A man should be stationed at the discharge hose to watch for any sign of oil in the discharged water.

BALLASTING

Taking on sea water to compensate for the loss of fuel oil weight is one of the many important tasks of an oil king. This process of COMPENSATION, known as ballasting, places you on the spot, inasmuch as ballasting before battle plays an important role in the toughness of your ship. It provides liquid layers at the shell to absorb fragments and minimize torpedo damage, and is also a means of retaining low weight to ensure against off-center weight and against list.

The Damage Control Book for your type of ship will outline the procedure for you to follow in ballasting tanks. Your type commander will usually issue detailed instructions for ballasting your type of ship. Figure 14-13 illustrates the recommended sequence table for the DD692 (long hull) destroyer. This sequence of liquid loading is governed by the following considerations:

1. Adequate low weight for stability;

2. Keeping off-center tanks filled;

3. Maintaining the ship in proper trim without list;

4. Early use of wing tanks dovjn to the waterline, to improve freeboard;

OOifl ClASS liON6 HUU) OiSTtOTIRS Rtcomwiemdtd ttqutmct tabU ft emptjimg fnfl oil tamki

I BALLAST C-7r AND C-iP

Noic—Whm il ■« known iKal llx DicmI oil or Dwl will haw to be burned, mix ihew oib with al Vat an tqful qtoality of furl oil Enirtnw caution mmt br nercited lo prevent contaimnation o(Dietd Of Diol iwikl with Mack oil when IcaM-icrring Ibex oiU. Do not kalUtt C-50IF or C-S02F

Figwr* I4~13.—S«qw«nc« table for •mptying fw*l oil tanks.

5. Emptying first those tanks which are not to be ballasted, in order to improve freeboard and reserve buoyancy;

6. Ability to maintain "split plant";

7. Ability to refuel rapidly;

8. Maintaining a minimum of free surface in the ship's tanks at all times.

Some of these requirements may conflict; however, their relative importance depends on the study of each individual ship.

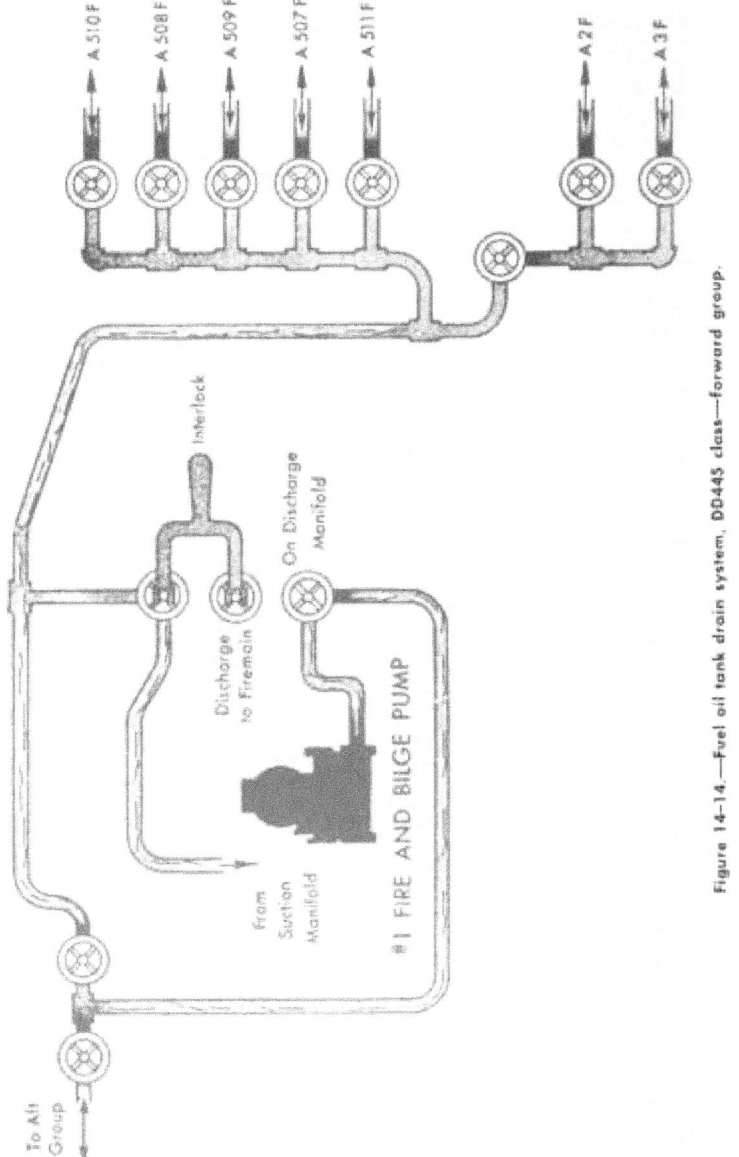

Figure 14-14.—Fuel oil tank drain system, DD445 class—forward group.

Ballast Systems

In order to accomplish this ballasting, it is necessary to provide a system for filling empty fuel oil tanks with SEA WATER and draining them of OIL. This is done by means of separate manifolds which connect to the fuel oil tanks and to two of the fire and bilge pumps. For example, in the DD445 class the manifolds to the forward tanks connect to No. 1 fire and bilge pump, and the manifolds to the after tanks connect to No. 4 fire and bilge pump. The ballast system forward has two manifolds located on the forward bulkhead of No. 1 fireroom. The 5-valve manifold leads to all tanks forward except the service tanks, and has no cut-out valve installed. The manifold for the service tanks is separate, with a valve for each service tank and also a cut-out valve. Both manifolds lead into a common line which acts as a discharge as well as a suction line. This line leads to the No. 1 fire and bilge pump suction manifold through a valve and also branches off, as indicated in figures 14-14 and 14-15, through a cut-out valve, to the cross-connection valve and fire and bilge pump discharge manifold. From the cross-connection valve, the pipe line leads through the plant to the after ballast system. The fire and bilge pump can take suction from the sea, and discharge through the discharge valve and the cut-out valve into the manifold line, to fill any tank.

The fuel oil tank drain and ballast system for the DD692 class is a revision of that installed on the DD445's. The revised system permits the pumping of the forward group of tanks with Nos. 1 and 2 fire and bilge pumps, and the pumping of the after group of tanks with Nos. 3 and 4 fire and bilge pumps. A cross-connection is provided between the forward and after systems.

Drainage eductors are connected to the ballast system to provide an alternate emergency dewatering facility in the event that one or more of the fire and bilge pumps is inoperative or in use for another purpose.

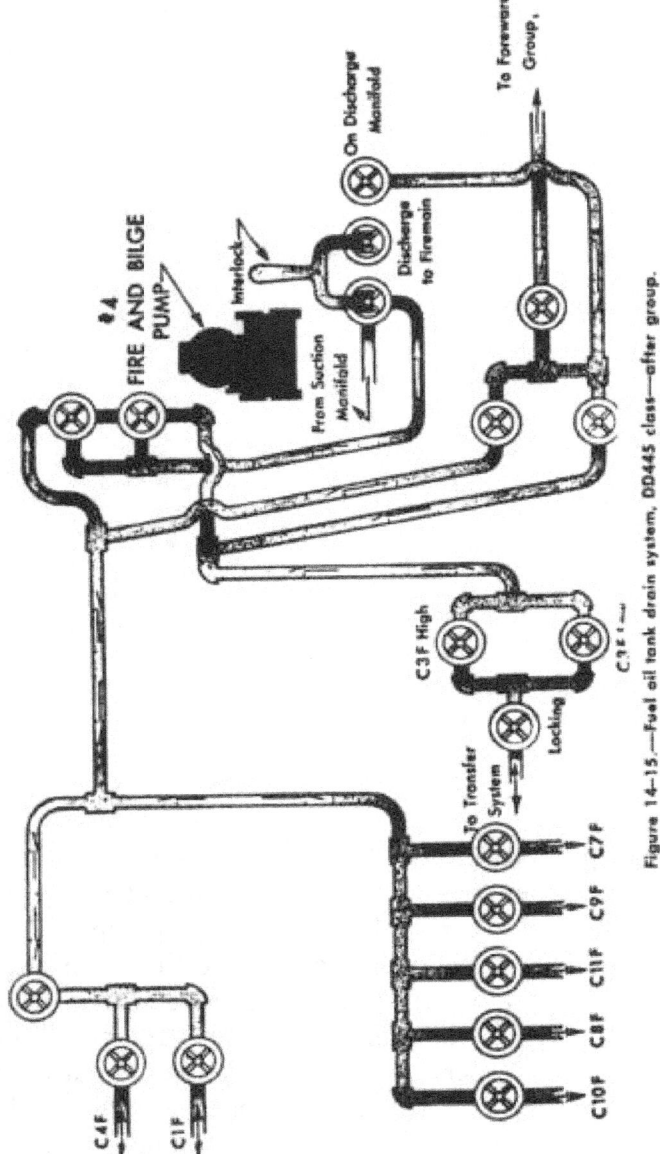

Figure 14-15.—Fuel oil tank drain system, DD445 class—after group.

A swing check valve is installed in the piping connecting the eductor to the main drain system, to prevent contaminated water or fuel oil from entering this system.

The interlocks previously provided between the fuel oil tank drain suction valve and the firemain discharge valve have been omitted in this revised arrangement.

The diagrammatic arrangements of this system in the DD692 class (short and long hull) are shown in figures 14-16 and 14-17.

Stability

In the preceding paragraphs the terms "buoyancy," "stability," "center of gravity," and "off-center weight" have probably raised the question in your mind: "What have they to do with me?" As oil king these terms have a lot to do with you, and you with them. The success of your ship in battle or in a rough sea may well depend on your help in the practical application of the principles of stability.

The oil king is usually required to submit a daily report to Damage Control Central containing information on the amount of fuel and water in each tank so that the damage control assistant can make his daily stability calculations. Damage Control Central will also keep an account of ballasted tanks, empty tanks, and voids.

Before we discuss the reason for the importance of this information, let's define the terms used in the theory of stability.

Buoyancy. —When an object is immersed in water, it displaces a volume of water whose weight is equal to that of the object. The displaced fluid exerts a pressure on the under and side surfaces of the immersed object, tending to force the object out of the water. This upward force is equal to the weight of water which the object displaces, and is called the force of buoyancy, or just "buoyancy." The buoyancy of a ship is, therefore, an upward force equal to the weight of water displaced by the ship.

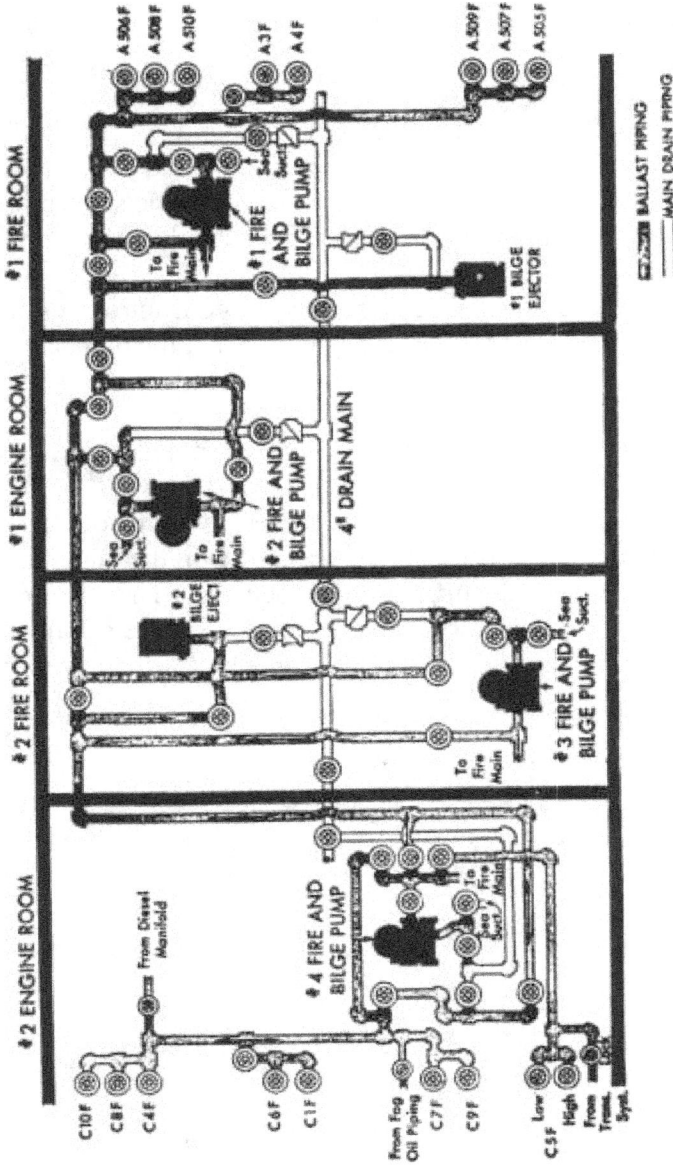

Figure 14-16.—Fuel oil tank ballast system, DD692 class—short hull.

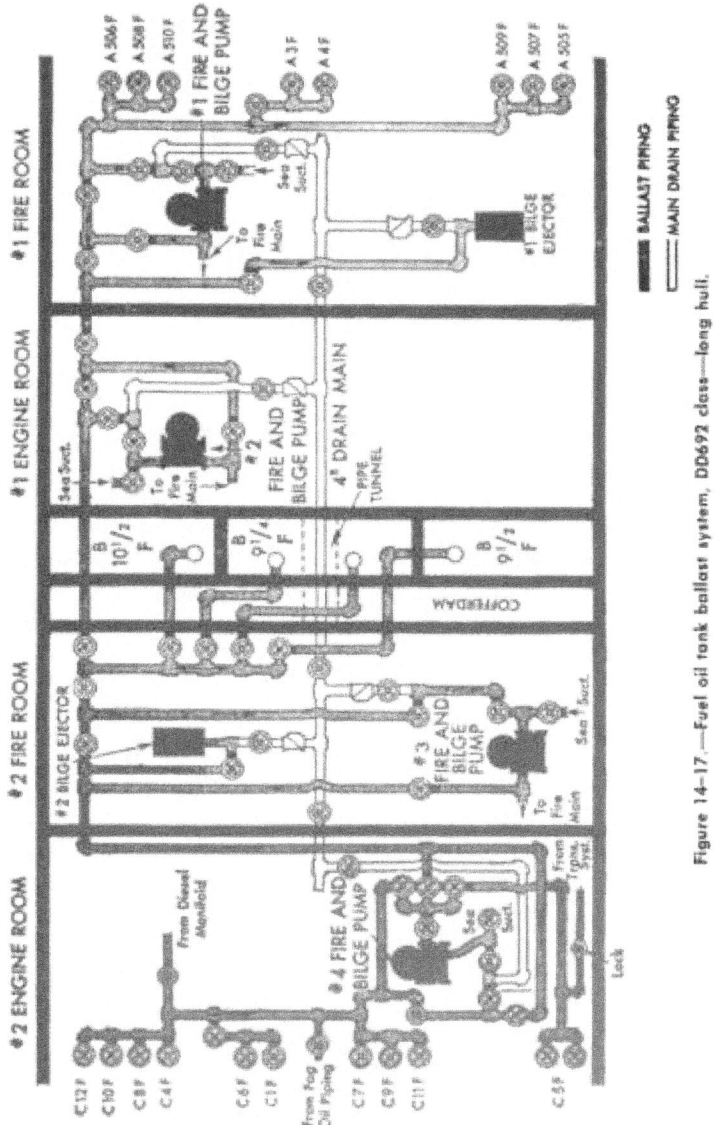

Figure 14-17. —Fuel oil tank ballast system, DD692 class—long hull.

Displacement. —A ship will float if the force of buoyancy is equal to her weight. Thus, the weight of a ship may be expressed as its "displacement," meaning the weight of the volume of water displaced by the hull. Displacement is expressed in long tons (2,240 pounds).

Stability. —When a ship is tilted by some disturbing influence, she tends either to return to her upright position, or to overturn. This tendency to rotate one way or the other is referred to as the stability of the ship.

Center of buoyancy. —The upward force of buoyancy acts in a vertical line through the geometric center of the volume displaced. This center of the ship's underwater body is called the "center of buoyancy" and is designated by the letter B, as shown in figure 14-18.

Center of gravity. —In a ship the center of gravity is the point at which the weight of the ship's structure and contained load may be regarded as acting vertically downward. The center of gravity is designated by the letter G, as shown in figure 14-18.

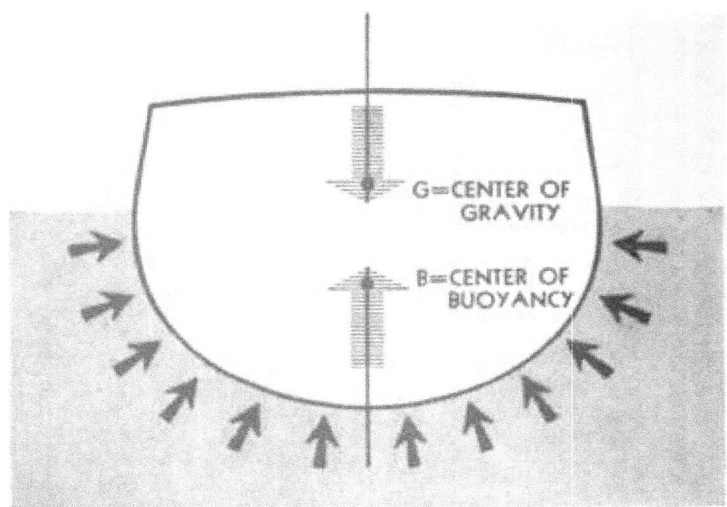

Figwra 14-1S.—Forcts of buoyancy and gravity.

Added weight. —This is the weight added to the ship's load either by necessity (such as a topside load of cargo) or by casualty (such as sea water flooding a portion of the ship).

Freeboard. —This is the perpendicular distance from the main deck to the waterline. (See fig. 14-19.)

List. —This term is used to describe inclination caused by a rotation of the ship about her longitudinal axis.

Trim. —This is an inclination of the ship around her transverse axis.

tUOtMKf AND TRANSVERSE STAMUTy

Figur* 14-19.—Diagram showing relationships of rosorvo buoyancy, froobeard, draft, and depth of hull.

We have discussed keeping the ship on an even keel by compensating for the loss of weight resulting from the daily expenditure of fuel oil and water. However, when a ship sustains damage in battle, by collision, or by running aground, a certain portion of the ship may take on sea water; the weight of this water will cause an undue list or trim, no matter how well the ship was ballasted before this damage. Such added off-center weight makes necessary further compensation to return the ship to a stable condition. This compensation is known as counter-flooding and is carried out under the supervision of the engineer officer with the help of damage control parties and the oil king.

The term counterflooding refers to the practice of deliberately taking sea water into the ship on the side opposite the damage, to correct list caused by off-center weight. In some cases counterflooding is also used for the correction of trim, either simultaneously with, or independent of, the correction of list.

Ships fitted with torpedo-protection systems are susceptible to off-center flooding in wing voids. Consequently, they are fitted for counterflooding the opposite wing voids. Remote-control sea valves, or in some cases damage control pumping systems, are used. In order to flood such tanks, the air escapes must be opened before sea water is turn(3d in. After the voids are filled, the sea valves are closed to prevent progressive flooding in the event of further damage.

In some cases, counterflooding of the wing voids of a torpedo-protection system is carried out on the diagonal principle, correcting for both list and trim at the same time. In this procedure, if underwater damage on the starboard bow had caused the ship to list to starboard and trim by the head, counterflooding would be undertaken aft and to port.

Some large ships have adopted the waist principle of counterflooding; this counterflooding is done in two steps: (1) for most of the list (the major menace), and (2) for trim and the remainder of the list. This requires establishment, prior to action, of a counterflooding doctrine, to meet an anticipated list of 7 or 8 degrees. When the engineer officer issues the order "Counterflood port" (or starboard, as the case may be), the appropriate assigned repair party promptly opens counterflood valves to a predesignated group of voids in the waist of the ship. This corrects the list. As more exact information on the location of the hit is received and analyzed, secondary orders are issued to counterflood additional wing voids near the undamaged end of the ship, to compensate for

trim and the remainder of the list. The net result is compensation for both list and trim, with minimum confusion in communications (such as might be created by lengthy orders as to numbers of voids to be flooded) and rapid initiation of list removal.

It is very possible, in case of serious damage, that the predetermined plan will not be sufficient to bring the ship back to an even keel. In this case the oil king must work with Damage Control Central in shifting oil or water from one tank to another. The damage control assistant can by means of stability calculations advise which tanks to use in order to obtain the best results.

There are certain precautions necessary in correcting list and trim in order to alleviate the possibility of bodily sinkage. For example, if a ship has a list due to negative GM (GM is a measure of stability), and if an attempt is made to remove list by correction for off-center weight instead of restoring positive GM (by dewatering or lowering liquid or other weights), off-center weight will be created where none existed. In fact, the ship will lurch to an even greater angle of list on the other side and may even capsize. With this in mind, it is evident that the oil king cannot flood tanks or transfer fluids on his own initiative. He must work with Damage Control Central.

If the list is due to a combination of negative GM and off-center weight (or if such a condition is suspected), it is most important that measures for improvement of GM be carried out vigorously until positive GM is assured. Thereafter the correction of only the identified amount of off-center weight is undertaken.

If measures must be taken to correct list and trim and to relieve stress in the hull girders, confine the measures to those that will not reduce an already critical stability. Likewise, if extreme trim and severe structural damage indicate that the ship is in danger of plunging or breaking in two, measures to improve off-center weight (transfer of liquids or counterflooding) must be confined

to those which will not aggravate trim or hull stress. For example, in counterflooding forward to correct trim and relieve some of the stress on the hull girder, you may, by losing freeboard, cause the ship to sink bodily. On the other hand, if you remove weight (by dewatering tanks, compartments, etc), you may aggravate the stress on the hull girder. A decision, then, to flood or to de-water or to shift liquids must take into account all possibilities and should attempt to strike a medium between the two main dangers—namely, breaking up and bodily sinkage.

The duties of the oil king are indeed most important, so you must learn all you can in preparation for the time when your knowledge and experience may mean the difference between success or failure of an operation. This knowledge can only be attained through diligent study and actual experience. You will find a discussion of stability in Damage Controlman 1 and C, NavPers 10572-A. For additional information, see chapter 88 (section I) of BuShips Manual.

QUIZ

1. To ensure an adequate gravity head, what is the minimum level that should be maintained in a fuel oil service tank?

2. Why is a sequence table followed in pumping ballasted tanks?

3. Who is responsible for obtaining ship's draft, fore and aft, both before and after fueling?

4. How long before fueling time will the fueling stations normally be manned?

5. What operations are performed during this preliminary period?

6. When tanks receiving fuel have been filled to 85 percent of capacity, what must the oil king do?

7. Should tank capacity be exceeded, what will happen to the excess oil?

8. What are the two corrections involved in computing the amount of oil received?

9. What is the standard temperature at which specific gravity is measured?

10. How is the specific gravity of a petroleum oil usually determined?

11. What temperature control must be exercised during a test for specific gravity of oil?

12. If you are discharging fuel to another ship, the oil in your tanks should be brought up to (at least) what temperature?

13. When you are discharging fuel to another ship, about what pressure should you maintain at the fueling connections?

14. How often should the oil in the storage tanks be tested for water?

15. Where will you find an outline of the procedure to be followed in ballasting?

16. What is meant by the "displacement" of a ship?

17. What is the "center of buoyancy" of a ship?

18. What is freeboard?

19. What is the difference between list and trim?

20. What is meant by **counterflooding?

21. What is the waist principle of counterflooding?

CHAPTER

AUTOMATIC COMBUSTION CONTROL

From previous study and experience you are familiar with the construction and operating principles of the automatic boiler feed water level regulators. The automatic combustion control equipment is a further application of automatic control. There is scarcely any limit to the amount of automatic equipment which could be utilized for boiler control; and such automatic equipment is, in fact, extensively used for boilers in civilian power plants. It must be borne in mind, however, that the more automatic a boiler or fireroom becomes, the more complicated the machinery becomes, and that the training of operators and technicians must be correspondingly greater. On naval ships it is necessary to decide just how much automatic control is necessary for good operation. Naval ships present many problems in regard to design and practical operation of automatic equipment. The equipment must be reliable, easy to operate, and foolproof. It must serve a useful purpose sufficient to offset the disadvantages of added weight and extrs maintenance problems.

Automatic combustion control has been installed on many of the Navy's auxiliary ships, and on some combat vessels. In the future it may be installed on all ships. As a Boilerman, therefore, you will need to understand the principles of automatic combustion control. The general

discussion of the Bailey system presented in this chapter is not intended to qualify you either as an operator or as a maintenance man on automatic combustion control equipment. But you will learn from this discussion something about the general features of the system as well as the fundamentals of its operation. The primary purpose of this chapter is to give you sufficient information to enable you to study, without undue difficulty, the manufacturer's instruction books for automatic combustion control systems; and to find in these books such detailed information and technical data as you may need.

The function of an automatic combustion control system (or equipment) is to maintain the fuel oil and the air input into the boiler furnace in accordance with the demand for steam, or the load on the boiler, and to proportion the amount of air to the amount of fuel oil in such a way as to provide maximum combustion efficiency.

There are three basic types of automatic combustion control systems: electrical, hydraulic, and pneumatic. For naval use, the pneumatic system has certain distinct advantages over the other two types; and for that reason 90 percent of the control systems used on board ship are of the pneumatic type. The air operated control system has the following advantages:

1. There is no danger from fire.
2. The system is easily adjusted.
3. The system is easily installed.
4. The equipment is less complicated.
5. The equipment requires less weight.
6. Compressed air is easy to obtain.
7. Less maintenance is required.
8. The system is not easily affected by moisture. Pneumatic automatic combustion control systems now

installed on naval ships of one class or another are manufactured by Bailey, Hagan, and Mason-Neilan. At the present time, most of the equipment in naval use is made by Bailey; and for this reason most of the discussion in this chapter will be on the Bailey control system. Other

pneumatic control systems are more or less similar in nature. (Major differences of the other systems will be mentioned.) Knowing the construction and operating principles of one system should enable you to understand something of the other types of control systems as well.

THE BASIC COMBUSTION CONTROL SYSTEMS

The basic operation of the automatic combustion control system is dependent upon the measurement of steam pressure in the boiler steam drum or steam header, as indicated in figure 15-1. In accordance with this steam pressure, the automatic system will regulate the amount of air and fuel supplied to the furnace. The four methods of accomplishing this are:

1. Series control in which the steam pressure adjusts THE FUEL RATE. The measurement of the fuel rate establishes a metered air flow.

2. Series control in which the steam pressure adjusts THE COMBUSTION AIR FLOW. The measurement of the air flow establishes a metered fuel flow.

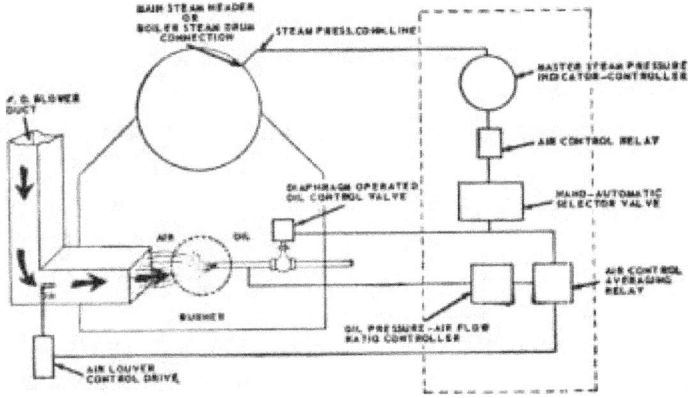

COWTttOL rAMEL

Figure 15-1.—Principal parts of the Bailey combustion control tyttem.

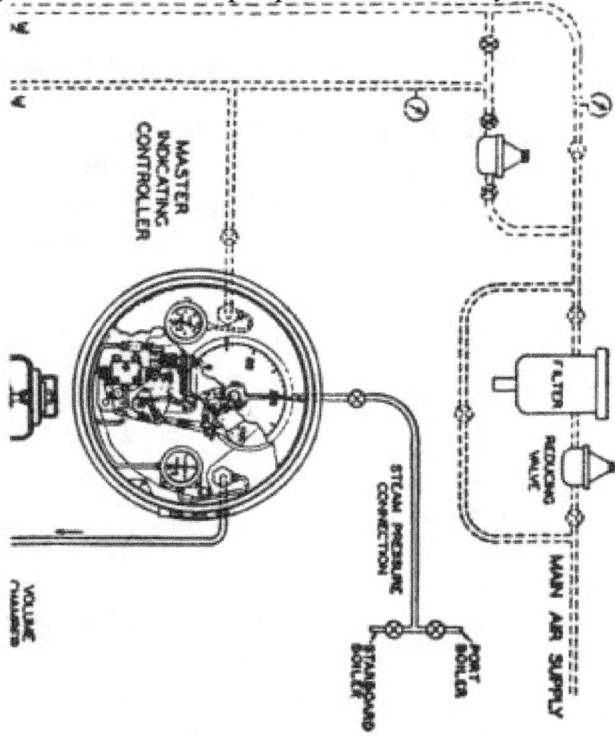

3. Parallel-series control in which both the fuel rate and the combustion air flow are adjusted simultaneously by the steam pressure, and the fuel flow is measured and readjusted by a ratio controller on the basis of the measured combustion air flow

4. Parallel-series control in which both the fuel rate and the combustion air flow are adjusted simultaneously by the steam pressure, and the combustion-air flow is readjusted by a ratio controller on the basis of the measured fuel
RATE.

The first two methods are know as series control because the elements of combustion enter the furnace in series. When both the elements of combustion are increased or decreased by the master controller for a change in rate and the ratio controller simultaneously repositions the power drive which realigns either the fuel flow or air flow (as in the latter two methods), the control is described as parallel-series. The Bailey system of automatic combustion control, described in this chapter, employs the fourth of the four methods listed in the preceding paragraph.

It should be noted that the Bailey system can also be used for the measurement and control of steam temperatures, but this aspect of automatic combustion control will not be discussed in this chapter.

It is obvious that a pneumatic control system will depend upon compressed air for the operation of the system and its component parts. Compressed air is used as a source of power to operate the fuel oil control valve and the forced-draft blower air supply control mechanism. Compressed air is used also, in a somewhat different way, as the controlling or balancing force of the system's control instruments. Figure 15-2 shows the compressed air supply to the different units or instruments of the automatic combustion control system.

In order to give you a basic understanding of an automatic combustion control system, the function or purpose of each component part of the system will first be discussed briefly in general terms. More detailed information on each unit will be given later in the chapter, after you have some idea as to what each part or unit of the system does.

The first unit or component part of the control system is the master steam pressure indicator-controller. It is connected, by means of high-pressure tubing, to the main header or steam drum, as shown in figure 15-1. The master steam pressure indicator and controller (master controller) has two functions: first, to register the steam pressure in the boiler in a manner similar to that of a main steam pressure gage; and second, to operate an air pilot valve which sets up an air loading pressure which controls the operation of the air control relay.

The air control relay (standatrol relay) controls or regulates the amount of air (under pressure) that is allowed to flow from the compressed air supply line to the hand-automatic selector valve. The function of the STANDATROL RELAY is to instantly reproduce the incoming loading air pressure changes in the outgoing loading (or control) air pressure; and, by means of a slow regenerative action, to amplify the change in outgoing loading air pressure until the incoming loading air pressure is returned to a predetermined standard value.

The primary function of the hand-automatic selector valve (selector valve), as its name indicates, is to select either the hand or the automatic method of control of the system.

The loading air pressure from the air control relay is transmitted to the selector valve, from which it is relayed to the oil control valve and to the air control averaging relay.

The fuel oil control valve (figs. 15-1 and 15-2) is operated by means of a diaphragm mechanism. This dia-
phragm arrangement operates in a manner similar to that of a steam reducing valve, except for the fact that the loading air pressure is variable. In accordance with the loading air pressure on top of the diaphragm, the valve will control the amount of fuel oil fired in the boiler furnace. The rate of oil flow (as determined by the diaphragm-operated oil control valve)

controls, in turn, the oil pressure in the line between the fuel line and the ratio controller; and this oil pressure determines the loading air pressure in the ratio controller. Another type of fuel oil valve, the piston-operated control valve, is used in many new installations.

The purpose of the oil pressure-air flow ratio controller (ratio controller), as its name indicates, is to control the amount of air flow (from the forced-draft blowers) to the boiler furnace in proper ratio to the amount of oil that is being burned. As you may remember from previous discussion, the first step is to obtain the proper amount of oil flow; the second step is to match the oil flow with the right amount of air, in order to obtain effficient combustion. In order to accomplish this control or regulation, the ratio controller sends a loading air pressure to the air control averaging relay.

The air control averaging relay (averaging relay) receives loading air pressures from two sources—from the air control relay (relayed by the selector valve) and from the ratio controller. The pressures from these two sources of controlling air (loading air pressures) are balanced or averaged out in the averaging relay. The averaging relay then controls the operation of the damper control drive mechanism.

The damper control drive (also called receiving regulator or fan control drive) operates the louvers in the forced-draft blower supply duct to control the amount of air being supplied to the boiler furnace, in order to maintain proper combustion.

The automatic combustion control equipment is designed so that the boiler can be operated manually, in case

of derangement or faulty operation of the automatic equipment. By shifting over to manual control the boiler could be operated practically as though no automatic equipment had been installed.

The units described above constitute the basic parts of the automatic combustion control system. Any amount of additional equipment such as gages, thermometers, recording devices, and alarm circuits, can, of course, be added to the automatic combustion control system. This type of equipment will vary somewhat, depending upon the installation, and will not be taken up in this training course.

COMPONENT PARTS OF THE BAILEY SYSTEM

The component parts or units and the diagrammatic arrangement of the Bailey automatic combustion control system are shown in figures 15-1 and 15-2. The major component parts of the system, as shown in figure 15-2, are as follows:

1. Master indicator-controller
2. Standatrol Relay (air control relay)
3. Selector valve
4. Oil control valve
5. Ratio controller
6. Averaging relay
7. Damper control drive

Before we can describe the operation of the system it would be well to discuss in detail the operation of each of the above component parts and associated equipment.

AIR SYSTEMS

In a pneumatic automatic combustion control system (such as the Bailey), the air systems are extremely important. Since all the component parts of the combustion control system operate by means of air, under different pressures and conditions, the air system must be thoroughly understood before it is possible to understand the operation of the system as a whole.

Compressed Air Systems

The MAIN AIR SUPPLY is usually obtained from the ship's service low-pressure air compressor and system, at approximately 100 psi. This pressure is reduced to the required operating pressure by means of a reducing valve. Usually a bypass line and valve is installed around the reducing valve and air filter, as shown in figure 15-2. If required, a separate air compressor may be installed for the automatic combustion control system.

The 40-POUND AIR SYSTEM is used to operate the forced-draft blower damper or damper control drive unit. This 40-pound air is obtained by means of a reducing valve from the main air supply line. This compressed air is used only as a source of power to operate the louver control drive unit; it has no other use in the automatic combustion control system.

The 28-POUND AIR SYSTEM is the fundamental compressed air supply to the different units of the automatic combustion control system. These units develop the air pressures required for the actuation of the system's control devices.

The 28-POUND AIR SYSTEM is used to furnish the operating power or medium to the different major units of the system, as shown in figure 15-2. This compressed air is kept at a constant value of 28 psi. The source of this air is a separate reducing valve which drops the air pressure from 40 psi to 28 psi.

Loading Air Pressure

The term loading air pressure refers to any variable pneumatic pressure developed directly or indirectly from the 28-pound air supply, between each piece of control equipment and the last intervening unit before a control valve or drive. The air pressure between the last intervening unit and the control valve or drive is called control air pressure. Variable air pressures are used to actuate the control units of an automatic combustion control system. Many of the units have a supply of 28

pounds of compressed air which is used by the individual units to produce or develop air pressures to control other units in the system.

The major units of the automatic combustion control system are interconnected by air pressure only, since no other control—electrical, hydraulic, or mechanical—exists between these units.

The pneumatic system uses air-actuated control devices which are positioned from air pressures set up by the different controlling units of the system. The controlling units measure the variable factors such as steam pressure, combustion air flow, and oil flow. The results of these measurements determine the loading air pressures which in turn are used to maintain the automatic control of the pneumatic system.

In figure 15-2 you can see the solid lines connecting the various units of the automatic combustion control system. These lines (tubing) carry air pressures between the respective units of the system. As you can see, there are no other means of control between the units.

THE MASTER INDICATING CONTROLLER

In taking up the master indicating controller we will first describe the construction of the unit and its different major parts. Then, with an understanding of the construction of the various parts, it will be easier to understand the operating principles of the master controller. Some of the major parts are linked together; from the operational viewpoint, therefore, we must consider several parts as a complete operational unit instead of a various individual parts with their own functions.

Construction of the Master Controller

The master steam pressure indicating controller, as shown in figure 15-3, is a rather complicated instrument, when considered as one unit. The approximate over-all size of the instrument is 14 inches in diameter and 4

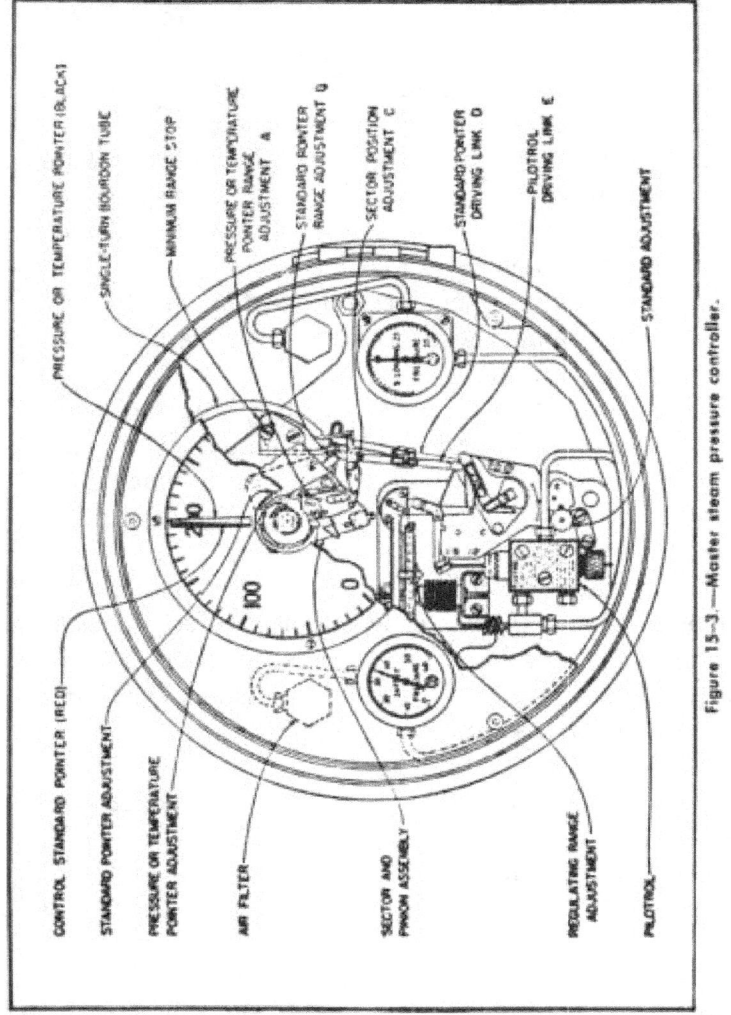

Figure 15-3.—Master steam pressure controller.

Inches in depth. The master controller consists essentially 01 5 different parts, as follows:

1. The air supply pressure gage,
2. The loading air pressure gage,
3. The pressure pointer (black),
4. The control standard pointer (red),
5. The PILOTROL, a small air throttle or pilot valve. Each of these assemblies will be described.

Air supply pressure gage —The air supply pressure gage, shown on the left side of figure 15-3, is an ordinary small air pressure gage, with a scale of 0—60 psi. It shows the pressure on the compressed air supply line from the ship's low-pressure air compressor or a special air compressor installed for the automatic combustion control system. The gage pointer indicates 35 psi compressed air supply. This should now read 28 psi, because all the later automatic combustion control systems use 28 instead of 35 psi compressed air supply. The top connection leads to an air filter. The compressed air supply line is con nected to the air filter on the back side of the controller. The indicated connection on the left side of the air gage (fig. 15-3) leads to the PILOTROL. This line is actually the compressed air supply line because it connects into a fitting, similar to a tee fitting, in back of the gage.

Loading air pressure gage. —The loading air pressure gage, shown on the right in figure 15-3, is a small air gage, like the supply pressure gage, but with a range of 0—35 psi. The indicated reading of the gage is 15 psi, which is the normal or static condit'on of the system when no changes are occurring. The gage shows the loading air pressure from the PILOTROL of the master controller to the STANDATROL RELAY. The normal range of this loading air pressure is from 5 to 25 psi. The bottom connection of the gage is for the line that comes from the PILOTROL. The right hand connection leads to the air filter and then to the STANDATROL RELAY. Here is, also, the same type of tee connection, which allows the gage to be connected to the line.

Steam pressure gage or mechanism (black pointer) . —The black (steam pressure) pointer and its control mechanism is, for all practical purposes, a Bourdon-type steam pressure gage which indicates the steam pressure in the boiler steam drum or header. (If necessary, review your previous study of the construction and operating principles of the Bourdon-type pressure gage.) A steam pressure connection is made from the boiler to the back side of this unit of the controller. The range of the scale will be according to the steam pressure used where the equipment is installed.

The assembly (gage) consists of a sensitive single turn Bourdon tube, which has the movable end connected through a pinion and sector assembly to the indicating (black) pointer in the conventional manner. The added device is a mechanical driving link which connects the black pointer to the air pilot valve of the pilotrol assembly. In other words, any force that moves the pointei (of the gage) will also move the air pilot valve.

The control standard pointer (red). —The control standard (red) pointer and its mechanism is attached to the black pointer assembly as indicated in figure 15-3. The red pointer remains in a stationary position, except that a change in adjustment may be made when the equipment is installed or when repairs are made. The red pointer setting is called the standard. It is the desired— or standard—steam pressure required for the satisfactory operation of the boiler. There is a connecting mechanical driving link between the red pointer mechanism and the pilotrol assembly. When the red pointer has been set for the desired steam pressure, the driving link will control the operation of the pilotrol unit.

The reading on the scale, as shown in figure 15-3, is 200 psi steam pressure. This scale and reading will, of course, depend upon the steam pressure of the particular installation. In other installations, for example, it might be 400 or 600 psi, depending upon the steam pressure used.

The pilotrol. —The pilotrol unit, which is a part of
PRINCIPAL PARTS
1- BELLOWS TRAVEL LIMIT SCREW
2- REGULATION RANGE ADJUSTABLE PIVOT
3- RESTORING BELLOWS
4- ADJUSTABLE MOUNTING BRACKET
5- PILOT VALVE STEM
6- ADJUSTABLE CAP AND LOCK NUT
7- PILOT VALVE BODY
8- LOADING AIR PRESSURE LINE
9-DRIVING LINK LENGTH ADJUSTMENT
10- REGULATING RANGE SECTOR ("D" NARROW RANGE "C" WIDE RANGE)
11- STANDARD SECTOR

12- LOCK SCREW
13- 50?^ POSITION OF STANDARD ADJUSTMENT
14- SUPPLY AIR PRESSURE LINE
15- STANDARD ADJUSTMENT (USE LOWER EXTENSION WHEN PROVIDED)

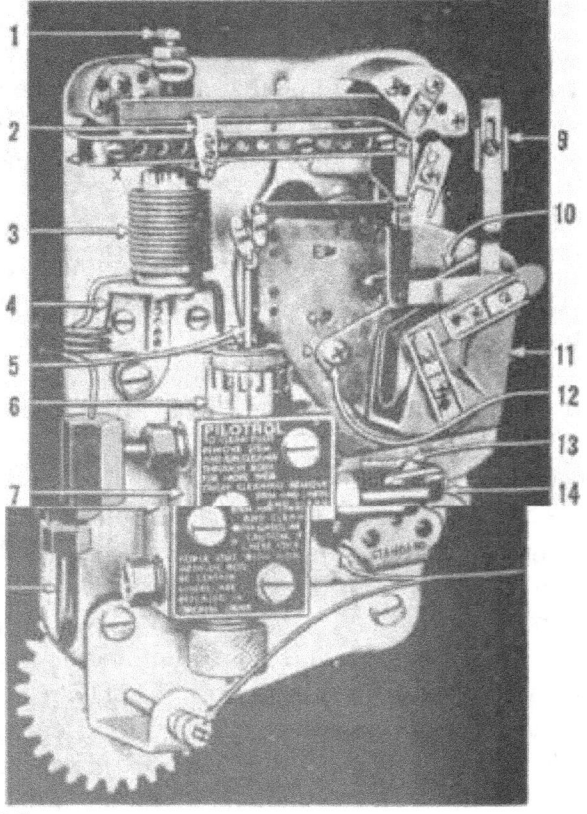

15

Figure 15-4.—PilotreL

the master controller, is shown in figures 15-3 and 15-4. The PILOTROL is primarily an air control pilot valve. The PILOTROL is a compact assembly consisting of a pilot valve, a restoring bellows and interrelated linkages which are mounted on a base plate. The movement of the steam pressure gage mechanism is transformed, by means of the linkage arrangement and pilot valve, into loading air pressures. The PILOTROL linkage is arranged for con-

RCCULATINC RANGE ADJUSTMENT PIVOT
miJ ADJUSTNENT^^^^
RESTORING • ELLOVS
ADJUSTABLE MOUNTING —i 9RACKET
LOADING AIR PRESSURE
jpL DRIVING LINK IJP^ LINGTH ADJUSTMiNT
.DRIVING LMK
REGULATING RANGE SECTOR

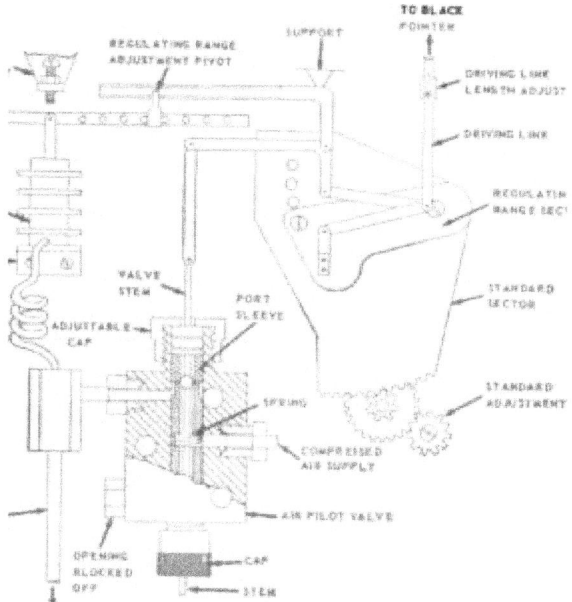

Figure 15—5.—Diagrammatic arrangement of Pilotrol.

venient adjustments. The restoring bellows is used to combine greater speed in loading air pressure changes with stability in the control system operation.

Figure 15-5 is a diagrammatic arrangement of a PILOTROL control system. The linkage for the red (standard) pointer (not shown) connects to the back side of the standard sector plate; it will move in accordance with the

standard sector when its position is changed by means of the standard adjustment gearing.

The PILOTROL assembly allows numerous adjustments. These adjustments are more or less permanent in nature. They are made at the factory, when the equipment is installed, or when repairs are being made. Should it become necessary to change these adjustments, carefully follow the instructions outlined in the manufacturer's instruction book.

From figures 15-4 and 15-5, the following adjustments can be seen:

1. The bellows travel limit adjustment. The screw and locknut are set so that the bellows travel is limited or stopped when the loading air pressure is 28 pounds. (No. 1, as shown in figure 15-4.)

2. The regulating range pivot adjustment (No. 2). The regulating range of the PILOTROL is adjusted to produce the desired change in loading air pressure for any given change in steam pressure.

3. The restoring bellows is installed by means of an adjustable mounting bracket (No. 4). The bellows unit is set at the midpoint of its designed limits of travel.

4. The adjustable cap and locknut (No. 6). When other adjustments have been made, the cap of the air pilot valve is turned for alignment of the port sleeve with the valve stem land (disk) until 15 pounds of loading air pressure is obtained.

5. The driving link length adjustment (No. 9). The black pointer link adjustment is made to align the movements of the steam gage with the linkage system of the PILOTROL. The red pointer link adjustment (shown only in figure 15-3) is made to align the position of the red pointer with the pilotrol mechanism.

6. The regulating range sector (No. 10). The left end of the range sector plate can be set in any one of 4 positions. These are marked D, C, E, and F in figure 15-4. The positions D and C make the instrument

direct-acting in regard to loading air pressures; the positions E and F make the instrument reverse-acting. The position D (or F) gives a narrow range of performance of the pilotrol. A narrow range produces a wide change in the loading air pressure for a small change in the steam pressure. The position C (or E) gives a wide range. A wide range gives a small change in loading air pressure for the same (as above) change in steam pressure. You should note that the regulating range sector is set in position D (figure 15-4) with a lock-screw (No. 12).

7. The standard adjustment (No. 15 in figure 15-4). By observing figure 15-5, you will see that the standard adjustment will rotate the standard sector plate around a pivot point. This adjustment will set the standard or the desired steam pressure which is to be maintained by the boiler. The red pointer (not shown) is connected to the standard sector plate by a linkage system. The red pointer will indicate the steam pressure for any setting of the standard adjustment.

No changes are made in any of the above adjustments, during normal operation of the automatic combustion control system.

The interior construction of the air control pilot valve is indicated in figure 15-5. The compressed air (28 pounds) supply line is attached to the right side of the valve. There are two air line connections on the left side of the valve. The top connection is used for loading air pressure from the air pilot valve. The bottom connection is blanked off or capped. The valve is of double construction with two ends (top and bottom) which are the same. When installed in the master controller, the upper half, or end, is used. The valve stem runs all the way through the valve body. It has two ball-shaped valve disks (only the top one is shown) which are permanently attached to the valve stem. The spring, in the center of the valve, separates two port sleeves. The outside ends of these sleeves

have air ports, which can only be closed by the valve stem and its valve disks. Besides the two inner (port) sleeves, there are two outer sleeves. These plain outer sleeves are secured in place by means of the lower cap and the upper adjustable cap with its associated lock nut. An annular air space is provided around the valve stem.

When the valve stem moves upward the air valve is opened in proportion to the movement of the stem. Air from the incoming compressed air line passes through the valve to the outgoing loading air pressure line. The air in the incoming supply is at a constant pressure of 28 pounds. The air in the outgoing loading air line is at a variable pressure, the pressure depending upon the opening and closing action of the pilot valve. This variable air pressure is maintained at 15 pounds when the system is in a normal position and no changes are taking place. When the system starts to function or move, as in taking care of a drop in steam pressure, the pilot valve may produce a loading air pressure varying between 5 and 25 pounds, approximately, depending upon the valve stem position.

The upper end of the valve stem is connected through a system of mechanical linkages to the black pointer mechanism of the steam gage.

The loading air pressure line runs from the air pilot valve to the standatrol relay, which is the next unit in the system. The restoring bellows is connected to the loading air pressure line.

The bellows unit operates in a conventional manner. The action of the restoring bellows is reflected back to the valve stem through a system of linkages. It tends to restore the valve stem back to a closed position before it is closed by the movement of the steam gage mechanism. This action tends to stabilize the system and minimize hunting.

The air pilot valve is fastened to the supporting plate of the PILOTROL assembly by means of three screws, as shown in figure 15-4. Also, all the other parts of the PILOTROL assembly are mounted on the same base plate.

Operation of the Master Controller

As previously stated, the master steam pressure indicating controller is primarily composed of an air pilot valve operated by a Bourdon tube, which is responsive to changes in steam pressure. The loading air pressure set up by the master controller is sent to the standatrol RELAY for further action and control. When the value of the steam pressure is the same as the operating standard (with the black and red pointers together), the loading air pressure is 15 pounds. Variations in the steam pressure from the operating standard will cause the master loading air pressure to vary. A decrease in steam pressure will increase the loading air pressure above 15 pounds. An increase in steam pressure will decrease the loading air pressure below 15 pounds. The boiler steam pressure is transformed by the master controller into corresponding loading air pressures which can operate the pneumatic devices of the automatic combustion control system.

The master controller has four visual means of showing the operator its condition of operation. The first two—the red pointer and the supply air pressure gage—show constant readings which do not change. The black pointer shows the steam pressure in the boiler. The loading air pressure gage gives a reading which should correspond with the change in steam pressure. At the moment things are normal (or no change is taking place), the black pointer should be in line with the red pointer and the loading air pressure gage should show a reading of 15 pounds. It must be remembered that the 15 pound reading of the loading air pressure is not a constant reading. The reading will vary from 5 to 25 pounds, depending upon corrective action taking place in the automatic control system. If the system is raising or lowering the steam pressure of the boiler, this action will be indicated by the value of the reading (below or above 15 pounds) on the loading air pressure gage.

PRINCIPAL PARTS
.SPRING COVER .SPRING GUIDE .SPACERS
.ADJUSTMENT NUT .THROTTLING VALVE .CLAMP BOLT .INLET VALVE
I-ADJUSTING SCREWS 9-SPRING
10- SEALING BELLOWS
11- OPERATING BELLOWS
12- OPERATING ARM
13- EXHAUST VALVE

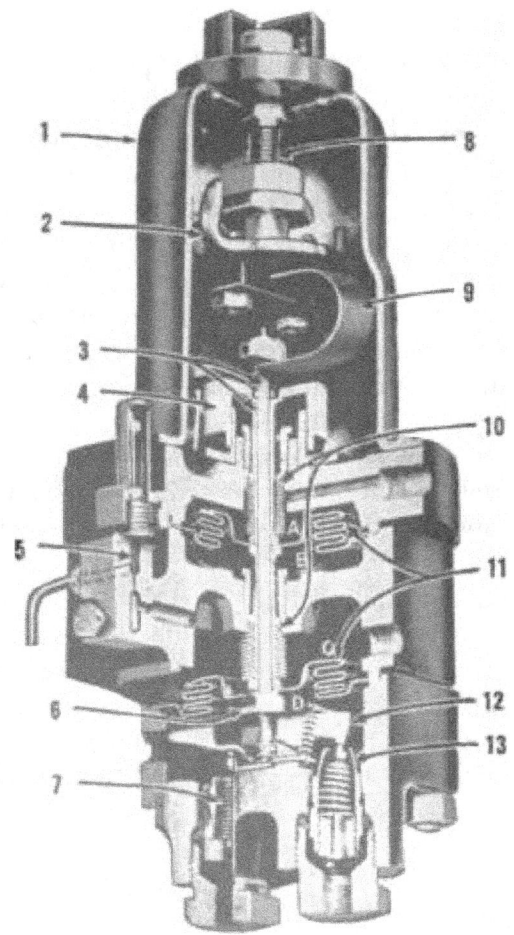

Figur* 15-6.—Sectional view of the Standatrol and avoraging relay.

THE STANDATROL RELAY

After the master controller, the STANDATROL relay is the next unit in the automatic combustion control system. The STANDATROL RELAY instantaneously changes the incoming loading air pressures (from the master controller) into stronger and amplified outgoing control or loading air pressures. In performing this function, the STANDATROL acts essentially as a mechanical relay— an apparatus which, when actuated by a relatively feeble force, exerts a greater force which is used to control a powerful appliance or instrument. Thus the STANDATROL, actuated by the loading air pressure received from the main controller, furnishes the necessary control and loading air pressures to the next units in the system.

Construction of the Standatrol Relay

The construction features of the standatrol relay are shown in figures 15-6 and 15-7. By looking over these figures, you will observe that the relay has four pressure chambers, indicated as A, B, C, and D in both figures 15-6 and 15-7. Operating bellows separate chamber A from chamber B, and Chamber C from Chamber D. A metal sealing bellows is used between chambers B and C. A similar sealing bellows is installed as a packing gland around the clamp bolt (the long main center bolt) assembly where the bolt extends upwards through the A chamber body.

The operating bellows and the sealing bellows are secured (in an airtight manner) to the clamp bolt assembly. The clamp bolt assembly is free to move up or down 1^2 iiich in each direction. The travel is limited by the adjustment nut.

A loop spring is attached to the upper end of the clamp bolt assembly. The upper part of the spring is attached to an adjusting screw in such a manner that the spring pressure can be applied in either direction.

The clamp bolt extends down into chamber D where the lower end positions an operating arm which operates the inlet and exhaust valves.

(l^fASTCN iMt SPIMiC COVER TO CHAMBER A* BOOT w»TM rOVR BOtTS TuRN TmE ADJUSTMENT SCREW COUNTCd CtOC"wiSE INTO NUT AS COVCM IS ASSEMBLED

(m) measure OiMENS«(1 V*

ANO Turn the nut ano uDCit NuT UNTIL tmc Dimension'x iS^'kESS Than OmtNSiON'y

(5 :^ER ANOifPRfR

OPER^TlNC BEllo#»s

i2) CHiivBERSV* C BOOT COM^LCrC mhTh SEAt.NC SElUMS CLAMR BOLT AND LO^ER OPERATING BELLOWS.

(7) ADJUST ThC positions Of TmC »^ET ANO t*HAoST vAlwC'. unTc TMtT BA«tuT CONTACT ThC NiflS ON THf O^CRATiNG ASM, MHEN ThC .IRERATINC ARM iS PARALLEL TO ThC /0» SuRfACC Of ChamBER'O boot

/X.ET VALVE

(f) CONNtCT 35"AIR FHtSboHt TOTH£ INLET vA.vE

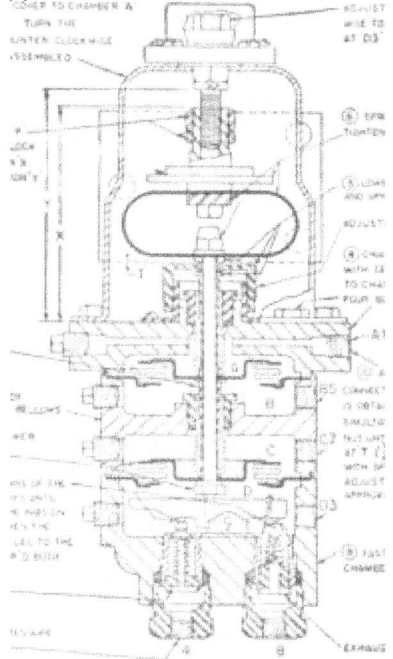

ADjoSTmcnT SCREm- Turn CLOCK* M SE TO INCREASE AIR PRESSURE AT oy

© SPRING. SMIMC CLAMR AND CuCC

Tighten nut on clampbolt assV

(?)LO«tR SPACER. AOJUSruiCNUt ANO UPPER SPACER ADJUSTMENT NUT

0CHAMBER a'bOOT complete WITH SEALING BELLOWS. ASSEMBLE TO CHAMBERS B AG' MOT W.rl BOLTS.

ADMIT AIR PRESSURE TO 65 COW^CCTlON Al'uNTlL l4*PRESSURE
IS OBTAINED IN Chamber *0* SiMojANEOuSur Turn the adjustment
C7 NUT until ^ CLEARANCE IS OBTAINED k AT T (J TURN FROM POWTOf
CONTACT WfTH SPACER) i«EEP EXHAUST vALvE ADJUSTED fOR ;>L»GHTAIR
DISCHARGE. i^03 '^^^^^^^^'^^^ j CU.FTPERMiNuTt
^® FASTEN CHAMBER O BOOT TO THt ^ CHAMBERS 8 AC'ftOOT
EXHAUST VAiVC

Figure 15-7.—Assembly and adjustment of Standatrol and averaging relay.

Operation of the Standatrol Relay

In studying the operating principles of the STANDATROL RELAY, refer to figure 15-8.

Chamber B of the relay is not used for the standatrol RELAY and is left open to atmospheric pressure. The volume chamber is also shown in figure 15-2. A bleedei valve, shown as item number 5 in figure 15-6, is installed in the line between chambers D and C. Two valves are installed at the bottom of the relay, an inlet valve and an exhaust valve. The spring is adjusted to 15 pounds pressure in the upward direction.

The basic principle of operation of the relay is that the admission or discharge of air from chamber D balancp-s

the variations in loading air pressures which may be applied to chambers A and C. A compressed air supply, normally 28 pounds pressure, is connected to the inlet valve. The outlet valve exhausts to atmosphere. The loading air pressures, from 5 to 25 pounds, are used in chambers A and C.

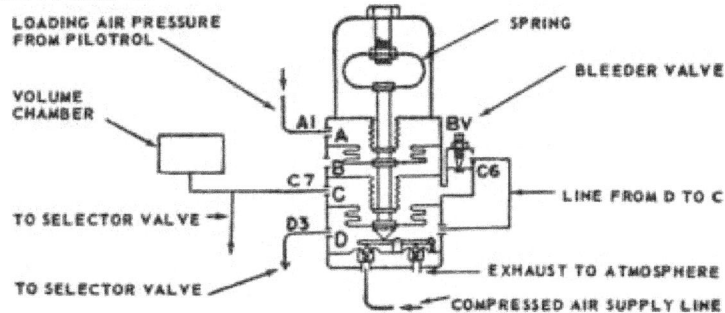

Figur* 15-8.—Diagrammatic arrongsmant of the Standotroi Relay.

If the loading air pressures are increased to chambers A and C, the air pressures acting on the operating bellows will force the clamp bolt assembly down and open the inlet valve in chamber D. Air pressure from the compressed air line will be admitted to chamber D through the inlet valve until the force upward on the lower operating bellows (with the assistance of the spring tension) is sufficient to balance the force on both operating bellows from the air pressures in chambers A and C, thereby causing the clamp bolt assembly to move upward until the inlet valve is closed.

If the loading air pressures are decreased to chambers A and C, the force of the chamber D air pressure on the lower operating bellows will move the clamp bolt assembly upward and cause the exhaust valve to be opened, thereby decreasing the air pressure in chamber D and returning the exhaust valve to a closed position.

The forces within the relay are thus kept in balance through the operation of the inlet and exhaust valves.

which admit or discharge air pressure as required. The air pressure in chamber D may be made to assume any value by varying the incoming loading air pressures to chambers A and C.

The spring loading is set at 15 pounds to balance the incoming loading air pressure. The balance establishes a normal control point of 15 pounds air pressure for the incoming loading air pressure. Any variations in loading air pressure from this control point (15 psi) will produce variations in chamber D air pressure.

Another function of the standatrol relay is to amplify the change in outgoing loading air pressure, until the incoming loading air pressure is returned to 15 pound standard value. When the loading air pressure from the master controller varies above or below the standard value of 15 pounds, the air pressure in chamber D may be made to assume any desired value between zero and the maximum pressure of the compressed air supply line. Since a slight change in the incoming loading air pressure can be transformed into a large variation in the outgoing control or loading air pressure, the standatrol relay is used in the automatic control system to maintain the steam pressure within close limits. Chamber C of the relay performs this function of amplification of loading air pressures.

The loading air pressure from the PILOTROL in the master controller is brought into chamber A. When the black and red pointers of the controller are matched, no changes are taking place in the system, so the incoming loading air pressure to the relay will be 15 pounds. With this 15 pounds of air pressure in chamber A, the incoming air loading pressure tends to push the operating bellows down. This force is opposed and held neutral by the 15 pounds force of the spring in the upward direction. With the two forces balanced there is no movement of the clamp bolt assembly of the relay. (We can speak of pressures as total forces because the relay operating bellows have equal areas.)

It will be noted that in this discussion the pressures in chambers C and D were assumed to be equal. However, these pressures are actually equal only when the system is in a neutral condition.

Suppose, for example, the steam pressure in the boiler drops, thus causing the master controller to bring the loading air pressure up to 17 pounds. This increase of 2 pounds of air pressure on the operating bellows in chamber A will move the clamp bolt assembly down and open the compressed air supply valve. The immediate action would be to increase, by 2 pounds, the air pressure in chamber D. If this air pressure in chamber D was 20 pounds (it could have been anywhere from approximately 5 to 25 pounds), it now becomes 22 pounds. The immediate increase of 2 pounds of air pressure in chamber D causes two simultaneous actions.

First, the outgoing control air pressure to the fuel oil valve diaphragm will be immediately increased by 2 pounds. This action will open the fuel oil valve wider to get more oil fired into the furnace so that the steam pressure will be increased. When the steam pressure has been returned to normal, the loading air pressure from the master controller will also return to a normal pressure (15 psi).

Secondly, the other action takes place simultaneously in the STANDATROL RELAY, due to the function of chamber C. Although chamber D air pressure immediately increases to 22 pounds, the air pressure in chamber C will increase at a very much slower rate which is determined by the setting of the bleeder valve. There is a volume chamber attached to chamber C, which also will require more air pressure. Suppose, for the purpose of explanation, that the air pressure in chamber C and the volume chamber reaches 22 pounds and is equal to that of chamber D. The forces of the air pressure in chamber C and D will be in balance.

But let's take a look at what is happening between the pressure of the spring and the air pressure in chamber A.

These pressures are not balanced because the spring pressure is 15 pounds and chamber A air pressure is 17 pounds. This unbalance will push the clamp bolt assembly down and open the compressed air inlet valve. More air will enter chamber D and build up more air pressure. In the above assumed case it would be 2 more pounds. This air pressure is also transmitted to the fuel oil valve diaphragm and other units of the system. It will be seen that as long as air pressure leaks from chamber D to C the compressed air inlet valve will be opened.

This regenerative action can keep going indefinitely. The air pressure in chamber D keeps gradually building up at a rate allowed by the setting of the bleeder valve. It is stopped when the boiler steam pressure returns to normal, which causes the loading air pressure from the controller to drop from 17 pounds back to the normal pressure of 15 pounds. The forces of the spring and chamber A will be equal. The compressed air inlet valve will be closed and the forces in chambers D and C will be equal. The system is in balance or equilibrium again. The outgoing loading air pressure of the STANDATROL relay is now at the new value which is needed to hold the load oh the boiler.

If the steam pressure rises above the control standard, the PILOTROL decreases its loading air pressure. This action causes the standatrol relay to reverse its operation, which has just been described. Let's assume that the loading air pressure in chamber A of the standatrol drops down to 13 pounds. This will cause the clamp bolt assembly to move upward due to the action of the spring. This action of the clamp bolt assembly will cause the exhaust valve to open in chamber D. This will release the air pressure in chamber D to the atmosphere, until the clamp bolt assembly lowers and closes the exhaust valve. Here again we have the regenerative function of chamber C. The air pressure in chamber C, and the volume chamber, will be higher than chamber D, which causes air to leak from chamber C to D. This upsets the temporary

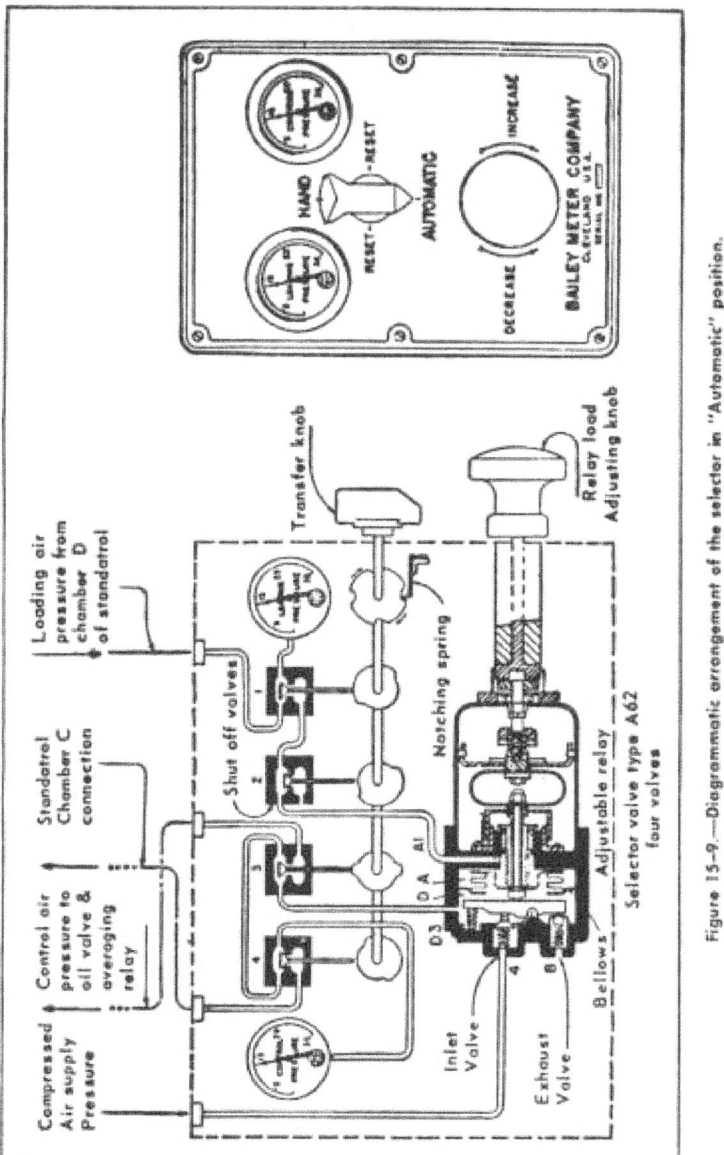

Figure 15-9.—Diagrammatic arrangement of the selector in "Automatic" position.

balance of the relay and the exhaust valve opens up. This progressive procedure keeps on going until the loading air pressure to chamber A has returned to its normal pressure of 15 pounds. As you will remember, the boiler steam pressure controls the value of the loading air pressure from the master controller.

Study of the action of the standatrol shows that the setting of the bleeder valve determines the speed of reactions of the control to the changes in the loading air pressure produced by the pilotrol. If the bleeder valve is opened too much, the loading air pressure reproduced by the STANDATROL will change rapidly and tend to cause hunting. If the bleeder valve is not open enough, the reaction will be too slow and the steam pressure will vary too far from the control standard. The result will be unsatisfactory control.

SELECTOR VALVE

The selector valve, mounted on a control panel, provides a means of operating the combustion control system either on "Automatic" or on "Hand" control.

The selector shown in figure 15-9 consists of an adjustable relay, four cam-operated valves, and pressure gages to indicate loading and control pressures. The connections to the other control units are shown in the control system piping diagram in figure 15-2.

The assembly of the selector is shown in figure 15-10.

By observing figure 15-9 you will see that the selector valve is made up of four different parts or assemblies, as follows:

1. The loading pressure gage. This gage measures the value of the incoming loading air pressure from chamber D of the standatrol relay.

2. The control pressure gage. This gage records the value of the outgoing control air pressure from chamber D of the adjustable relay of the selector.

3. The transfer valve assembly. This assembly consists of the four separate air valves, which are operated by the cam arrangement.

in the vm-i is

on, of es-lay to ive ng by or

^he

:he 1. ow 2n, ;he Ive on of be :h-as let ;x-lir re

be he To. ill ch

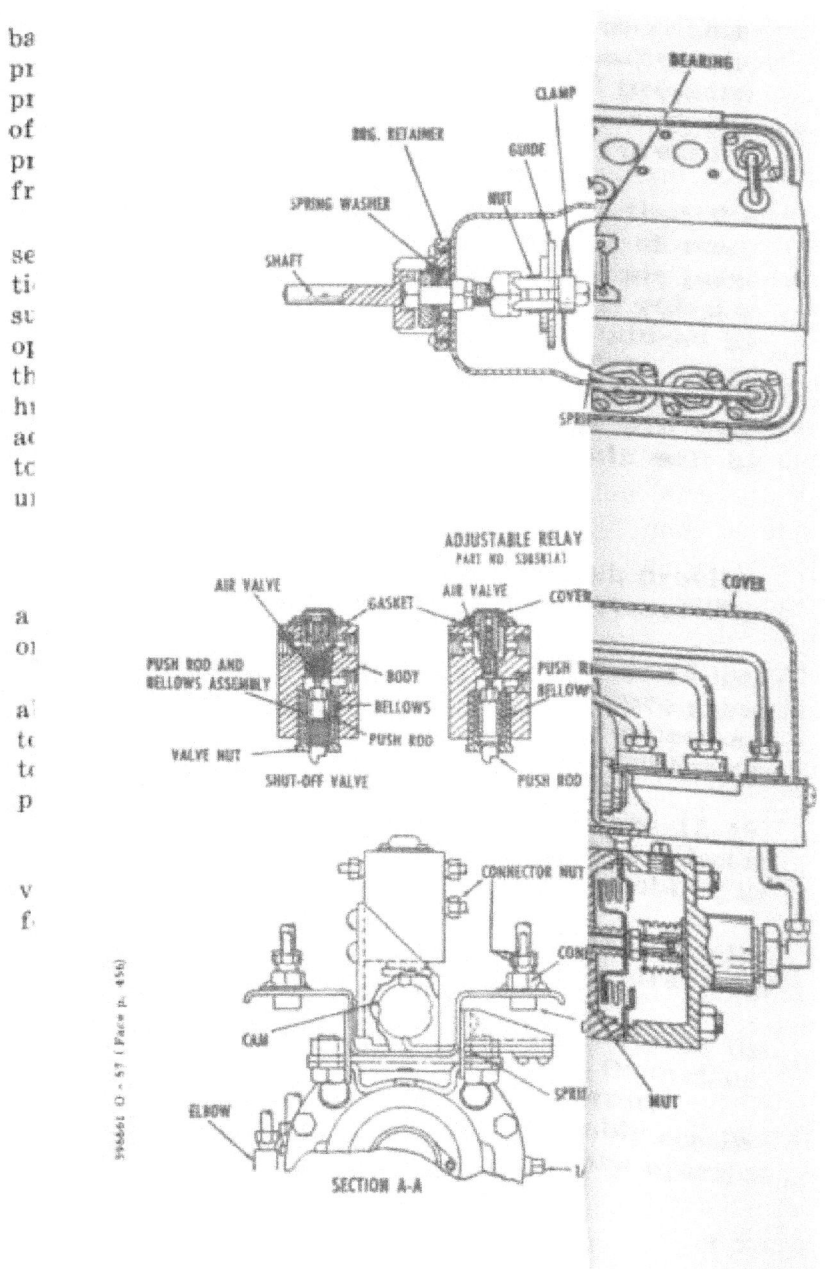

ba
pı
pı
of
pı
fr

se
ti
su
oₚ
th
hı
aₒ
to
uı

a
oₙ

aₜ
to
to
p

v
fₒ

ADJUSTABLE RELAY
PART NO S385N3A1

AIR VALVE   AIR VALVE   COVER   COVER

GASKET

PUSH ROD AND   PUSH R
BELLOWS ASSEMBLY   BODY   BELLOWS

BELLOWS

VALVE NUT   PUSH ROD

SHUT-OFF VALVE   PUSH ROD

CONNECTOR NUT

CON

CAM

ELBOW   SPRI   NUT

SECTION A-A

4. The adjustable relay. This air relay is similar in construction and basic operating principles to the STANDATROL RELAY. This air relay has two chambers, A and D. In this case the spring pressure is controlled by the relay load adjusting knob.

Automatic Operation of Selector

When the transfer knob is in the automatic position, the loading air pressure is received from chamber D of the STANDATROL RELAY. This incoming loading air pressure is reproduced or transformed in the adjustable relay so that the outgoing control air pressure is always equal to the incoming loading air pressure. However, the relative value of the outgoing control air pressure to the incoming loading air pressure may be increased or decreased by turning the load adjusting knob, which will apply more or less spring loading pressure.

Now, let us trace the air through the selector valve. The incoming loading air pressure from chamber D of the STANDATROL RELAY goes to the Upper part of valve No. 1. (See fig. 15-9.) The loading air pressure gage will show the pressure of this incoming air. Valve No. 1 is open, the selector knob being in the automatic position, and the loading air pressure will pass on to valve No. 2. Valve No. 2 being closed, the loading air pressure will pass on into chamber A of the adjusting relay. The pressure of the air on the operating bellows of chamber A will be transmitted by the clamp bolt assembly to the valve mechanism in chamber D. It will be noted that this is as far as the incoming loading air goes. The operation of the inlet valve, from the compressed air supply line, and the exhaust valve, to atmosphere, will maintain a control air pressure in chamber D similar to the loading air pressure in chamber A.

Any difference between these two air pressures will be due to the setting of the relay load adjusting knob. The control air pressure passes out of chamber D to valve No. 3. Valve No. 3 being open, the control air pressure will go in two directions. First, it goes to valve No. 4, which is closed, and on to the control pressure gage. (Note that the two gages in figure 15-9 read the same.) Second, it goes to the outgoing control pressure line of the selector valve. From this point the air goes to the air diaphragm of the fuel oil valve and to the averaging relay.

Hand Operation of Selector

By observing figure 15-9 let us now trace the air through the selector valve with the transfer knob on the hand position. Then we can check the difference in operation of the selector valve for hand and automatic operation.

The incoming loading air pressure to the upper part of valve No. 1 will only go to the loading pressure gage because the valve will be closed, for hand operation. That is the end of the incoming loading air pressure. Valve No. 2 will be in an open position, which will allow any air pressure in chamber A to flow out to atmosphere. Valve No. 3 will be the same, in the open position.

Since there is no loading air pressure in chamber A of the adjustable relay for the hand position, the relay load adjusting knob must be used to adjust the outgoing control air pressure. The load adjusting knob is turned clockwise to increase the flow of fuel oil and air to the boiler, regardless of whether the control air pressure increases or decreases.

The control air pressure will flow out at chamber D of the adjustable relay, through valve No. 3, just as it does when the selector valve is set on automatic, as before. However, valve No. 4 is now open. This allows control air pressure from the lower part of valve No. 4 to go back to chamber C of the standatrol relay. The air pressure in chamber C of the standatrol relay is maintained at the same value as the control air pressure during hand operation so that transfer to automatic controls can be made without delay when desired.

It should be noted that the standatrol relay has no control of the automatic combustion control system when the selector valve is in the HAND position.

With the selector valve in the hand position, the operator has now taken over control of the automatic combustion control system. He observes the black pointer of the master controller, and raises or lowers the fuel oil and air input to the boiler by turning the relay load adjusting knob clockwise (raise) or counterclockwise (lower).

FUEL OIL CONTROL VALVE

The construction of the fuel oil control valve is shown in figure 15-11. Since you are already familiar with various types of valves, the details of this fuel oil valve will not be discussed.

The diaphragm control unit is of the conventional type. The diaphragm moves down in accordance with the

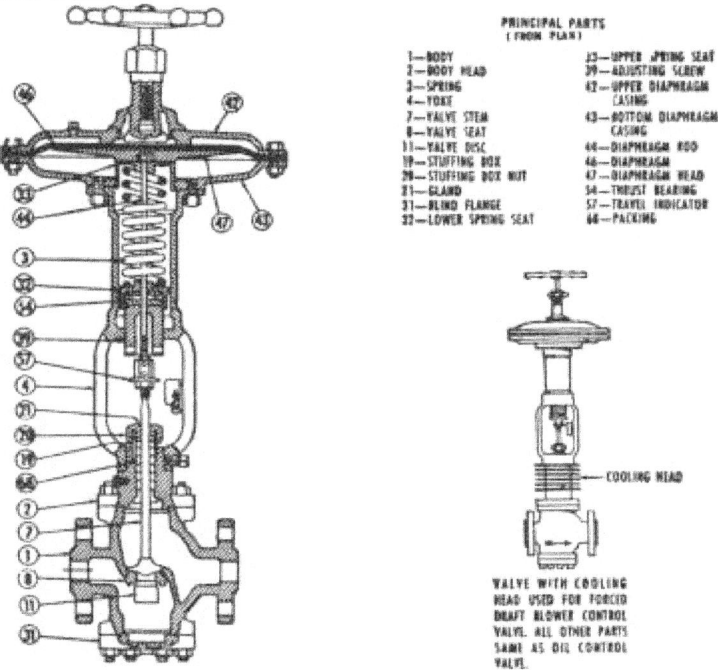

PRINCIPAL PARTS
(FROM PLAN)

1—BODY
2—BODY HEAD
3—SPRING
4—YOKE
7—VALVE STEM
8—VALVE SEAT
11—VALVE DISC
19—STUFFING BOX
29—STUFFING BOX NUT
31—GLAND
31—BLIND FLANGE
32—LOWER SPRING SEAT

33—UPPER SPRING SEAT
39—ADJUSTING SCREW
42—UPPER DIAPHRAGM CASING
43—BOTTOM DIAPHRAGM CASING
44—DIAPHRAGM ROD
46—DIAPHRAGM
47—DIAPHRAGM HEAD
54—THRUST BEARING
57—TRAVEL INDICATOR
66—PACKING

COOLING HEAD

VALVE WITH COOLING HEAD USED FOR FORCED DRAFT BLOWER CONTROL VALVE. ALL OTHER PARTS SAME AS OIL CONTROL VALVE.

Figure 15—11.—Assembly of oil control valves.

amount of control air pressure on top of it, and this action regulates the amount of fuel oil that flows through the valve.

In some installations the valves are actuated by a piston-operated control drive instead of the diaphragm-operated drive. The Bailey piston-type control drive (not shown) is basically a double-acting air cylinder and piston assembly. The operation of the air piston assembly is somewhat similar to that of the damper control, described later in this chapter.

The fuel oil control valve is provided with a manual control wheel or hand jacks, in case the automatic combustion control system is not used or fails.

RATIO CONTROLLER

The purposes of the ratio controller are (1) to measure the fuel oil and the forced-draft air supplied to the burners, (2) to readjust the air supply as necessary to maintain the correct proportion of air to fuel oil at all firing rates, and (3) to provide a quick means of readjusting the control for each of the various sizes of sprayer plates to be used in the burners.

The location of the ratio controller in the control system can be seen in figure 15-2. A diagram of the ratio controller is shown in figure 15-12. The assembly and details are shown in figure 15-13. By observing figure 15-12 we can see that the ratio controller can be roughly divided into 7 parts or assemblies as follows:

1. The diaphragm unit;
2. The bellows unit;
3. The two beam assemblies;
4. The fulcrum assembly including the chain drive, pointer, scale, and adjustment screw "A";
5. The loading air pressure gage;

6. The compressed air supply gage;

7. The PiLOTROL.

NEUTRAL POSITION INDICATOR,

BELLOWS PRESSURE

BEAMS

DIAPHRAGM UNIT

ROLLER ASSEMBLY

DirrERENTlAL (HIGH Smml PRESSURE 1 LOW

BELLOWS UNIT

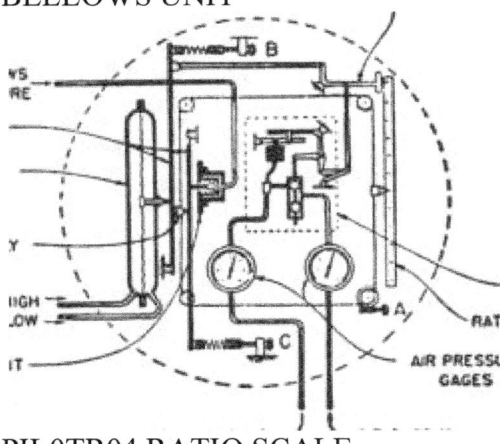

PIL0TR04 RATIO SCALE

AIR PRESSURE CAGES

AIR LOAOING^-^ PRESSURE

^ AIR SUPPLY PRESSURE

Figure 15-12.—Diagram of the ratio controller.

The DIAPHRAGM UNIT has two sides. The "high" side is connected to the boiler air casing by a special pipelike fitting. The "low" side is connected to the furnace (passing through the brickwork) with a similar fitting. The diaphragm unit thus measures the drop in the forced-draft air pressure as it passes from the windbox into the furnace. This draft loss across the burners is a direct measure of air flow.

If the forced-draft air pressure is too high, the outer diaphragm chamber air pressure will push the diaphragm control needle against the beam. This action (described later) will cause the air pressure to the furnace to be decreased. If the forced-draft air pressure is too low, the air pressure in the inner diaphragm chamber will back the diaphragm control needle away from the beam. This action will finally cause the air pressure to the furnace to be increased.

The BELLOWS UNIT is connected to the manifold of the burner oil supply as shown in figure 15-2. Since the capacity of the various sizes of sprayer plates in relation

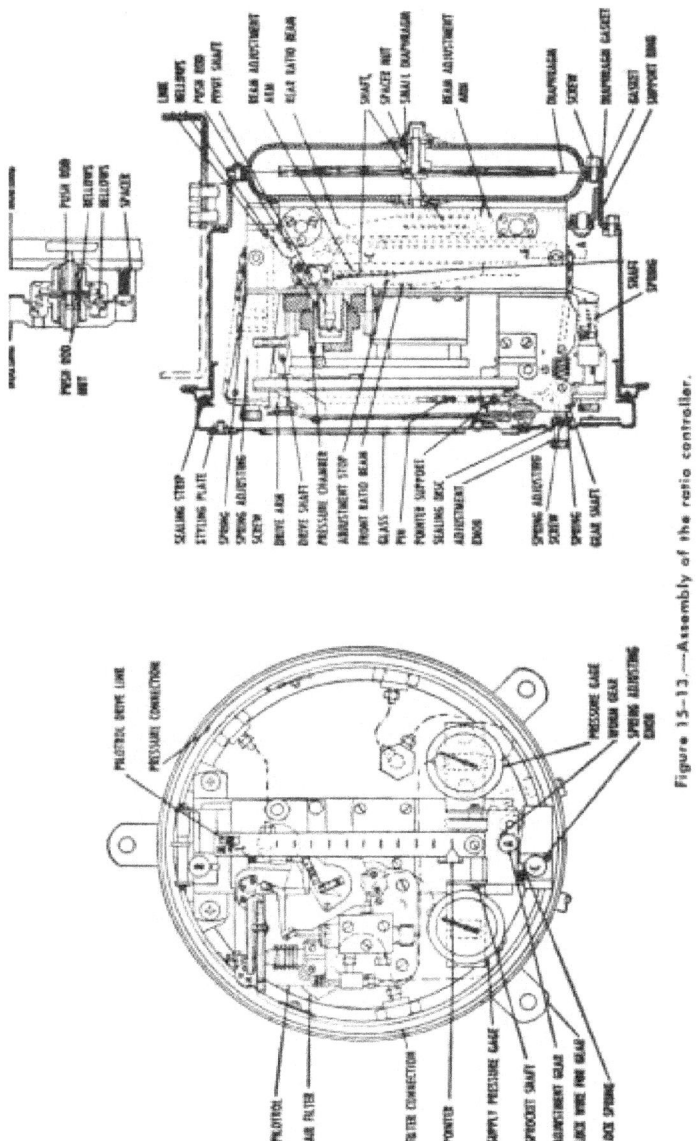

Figure 15-13.—Assembly of the ratio controller.

to oil pressure is known, the oil manifold pressure is a direct measure of the oil supply to the burners. A pressure is maintained on the (inner) beam by the bellows unit in accordance to the oil pressure.

The two BEAMS are similar in that one end is attached to a pivot and the other end to a spring and adjusting screw (B and C). It should be noted that the control needles of the diaphragm and bellows units oppose each other. When the equipment is installed, the adjusting screws, B and C, are set to give proper performance of the ratio controller.

The FULCRUM ASSEMBLY is located between the two beams. The position of the fulcrum assembly can be readily changed by means of the chain drive and the adjustment screw "A." Also, the position of this fulcrum assembly is shown by pointer on the ratio scale. This is necessary because different sizes (capacities) of sprayer plates are used for different boiler loads. A pointer position on the ratio scale is determined, at the time of installation, for each size of sprayer plate to be used in the operation of the boiler. The fulcrum assembly is set for the size of sprayer plate being used. When the size of the sprayer plates is changed, the position of the fulcrum assembly must also be changed. The relation between the fuel oil pressure and the forced-draft air pressures required to hold the ratio controller in a neutral position depends upon the position of the fulcrum assembly.

The PILOTROL unit is connected by a mechanical linkage system to the rear (outer) beam. Any movement of this beam will be transmitted to the pilotrol.

The operating principles of this pilotrol are the same as for the one in the master controller, which is shown in figures 15-4 and 15-5. It would be a good idea to look over these figures and to review, if necessary, the previous discussion of the pilotrol.

When the fuel-air ratio is correct, the normal loading air pressure set up by the oil pressure-air flow ratio con-

troller is 15 pounds. If the proportion of fuel and air varies from the predetermined adjustment, the loading air pressure will vary from the 15 pounds by an amount sufficient to modify the control pressure from the averaging relay and reposition the forced-draft damper control drives to re-establish the correct proportion of air to oil.

When a burner is taken out of service and the register closed, the relationship between the fuel oil flow and boiler air flow, as measured by the ratio controller, is not changed.

THE AVERAGING RELAY

The location of the averaging relay in the control system and the associated piping is shown in figure 15-2. The construction features of the averaging relay are shown in figures 15-6 and 15-7. As you will notice, the

LOADING Alii PRESSURE , FROM SELECTOR VALVE I
CONTROL AIR PRESSURE TO DAMPER DRIVE

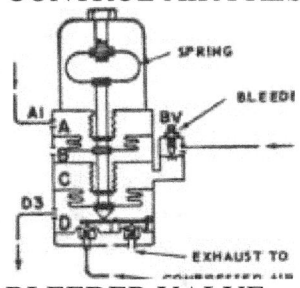

BLEEDER VALVE
LOADING AIR PRESSURE PROM RATIO CONTROLLER
EXHAUST TO ATMOSPHERE COMPRESSED AIR SUPPLY LINE

Figure 15-]4.—Diagrammatic arrangement of the averaging relay.

construction and operating principles of the STANDATROL and averaging relays are identical.

By observing figure 15-14 we can study the operation of the averaging relay in the automatic control system. The loading air pressure from the adjustable relay of the selector valve enters chamber A. The loading air pressure from the pilotrol of the ratio controller enters chamber C through the bleeder valve. The function and operation of the air valves and chamber D is the same as

the STANDATROL RELAY. Chamber B is left open to atmospheric pressure. It should be noted that the air pressures in chambers A and C both act downward.

The loading air pressures from the two separate controllers are combined in the averaging relay to produce a control air pressure which operates the air pilot valve; this, in turn, by means of the 40-pound air pressure, positions the louver or damper in the forced-draft blower duct to the boiler.

The bleeder valve is used to introduce a delaying action between the ratio controller loading air pressure change and the ultimate effect of this change on the air pressure in chamber C. This delaying action can be increased or reduced as desired by adjusting the bleeder valve.

The spring pressure is normally set to balance 15 pounds air pressure in chamber C, so that the control air pressure in chamber D will follow the loading air pressure changes in chamber A. However, should the relation between the fuel oil flow and the forced-draft air flow vary, so that the ratio controller loading air pressure is less or greater than 15 pounds, the air pressure in chamber C will change to produce a like change in the air pressure in chamber D. Therefore, the control air pressure in chamber D will vary from the loading air pressure admitted to chamber A in the amount and direction that the chamber C loading air pressure varies from 15 pounds.

The loading air pressure changes which come into chamber A of the averaging relay from the adjustable relay in the selector valve (originally from the master controller) are immediately effective. The changes in the loading air pressure from the ratio controller to chamber C are kept from being immediately effective by action of the bleeder valve.

The forced-draft blower damper is positioned by the loading air pressures from one controller (the adjustable relay of the selector valve) and is readjusted in position by means of the loading air pressures from a second controller (the ratio controller). The two controllers or relays are adjusted to operate in proper relationship with each other.

THE DAMPER CONTROL DRIVE

The air-operated forced-draft blower air damper control drive is shown in figure 15-15. It consists of a double-acting air cylinder, and the assembly is mounted on a pivoted base. The control drive is provided with a bellows-actuated double-acting air pilot valve and restoring spring, which causes the drive to assume a definite position for every incoming control air pressure.

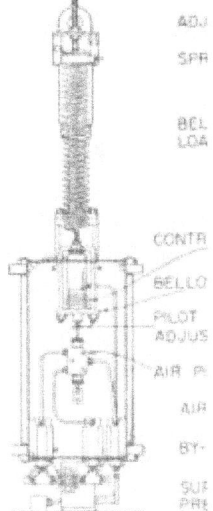

ADJUSTING SCREW SPRING HOLDER
BELLOWS LOADING SPRING
CONTROL BELLOWS
BELLOWS PLATE
PILOT STEM ADJUSTING SCREW
AIR PILOT VALVE

AIR FILTER
BY-PASS VALVE
SUPPLY
PRESSURE CONNECTION

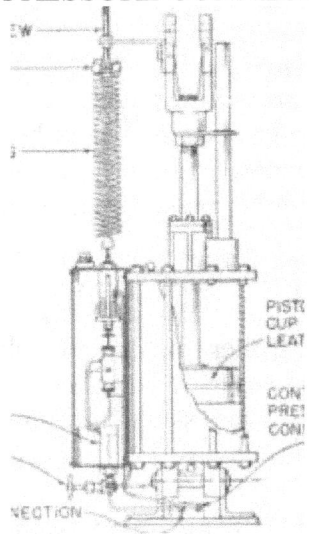

PISTON CUP
LEATHERS
CONTROL PRESSURE CONNECTION
Figure 15-15.—The damper control drive.

The control bellows is operated by the control air pressure from the averaging relay. The air pilot valve is the same (only larger in size) as the PILOTROL in the master controller, which is shown in figures 15-4 and 15-5. In this case both ends of the air pilot valve are used. The line from one end of the air valve goes to one side of the

air cylinder piston, and the line from the other end goes to the other side of the piston.

Air from the 40-pound compressed air line is used to operate the piston of the control drive.

The bellows unit is attached to the air pilot valve stem, and is used to operate the air valve. The air pilot valve operates in the same manner as the pilotrol. The position of the piston of the control drive is controlled by the air pilot valve, which in turn is controlled by the bellows unit.

The piston rod is connected to a louver or damper mechanism, which is not shown.

The bypass valve is provided for hand operation of the piston. The hand lever for manual operation is not shown. When the bypass valve is open it allows air to flow from one side of the piston to the other.

In some installations a steam control valve is used instead of the forced-draft damper and control drive. The damper mechanism is usually used with electrically driven forced-draft blowers. When steam turbine-driven forced-draft blowers are used, a steam throttle valve is used. This throttle valve is similar in operation to the fuel oil control valve. The steam valve will have cooling fins attached to it. The speed of the forced-draft blower, hence the supply of air to the furnace, is controlled by means of this steam throttle valve.

OPERATION AND CONTROL OF THE SYSTEM
Starting a Boiler
When a boiler is being started, the control valves must be handled manually by means of

the hand jacks. The other boiler, if in service and carrying a load, can be operated on HAND or automatic control, as desired.

When the boiler being started is at normal operating pressure and has been cut in on the load, it should be synchronized with the boiler in service. (By "synchronizing" is meant the equalization of firing rates of the two

boilers so that the load will be equally divided between them.) The procedure for synchronizing is as follows:

1. Slowly adjust the fuel oil pressure, using the hand jack on the oil control valve until the fuel oil pressure is the same on both boilers. Do not forget that if you are synchronizing boilers when one is operating on AUTOMATIC control, a change in the firing rate of the boiler being controlled manually will change the firing rate on the other boiler, due to the distribution of the load.

2. Close the three-way cock in the control air pressure line to the oil control valve and then turn the hand jack back to allow the control air to operate the valve.

3. Adjust the ratio controller fuel-air ratio knob to suit the size of the sprayer plate in service.

4. Adjust the forced-draft pressure with the blower rheostat (electric blower installation) to agree with the blower pressure of the other boiler.

5. Cut in the air supply to the damper drive, unlock the lever, and close the bypass valve.

Both boilers are now on either hand or automatic control, depending upon the selector valve setting.

Changing from "Hand" to "Automatic" Control

Before changing from hand to automatic control with either one or two boilers in service on HAND control, be sure the steam pressure connection valves and the air supply valves to the master controller and the air supply valve to the standatrol relay are open. Then transfer from HAND to AUTOMATIC control by the steps given below, using the selector unit on the control panel (fig. 15-16).

1. Observe the value of the control pressure gage.

2. Turn the transfer knob to the reset position. This will cause a drop in the control pressure.

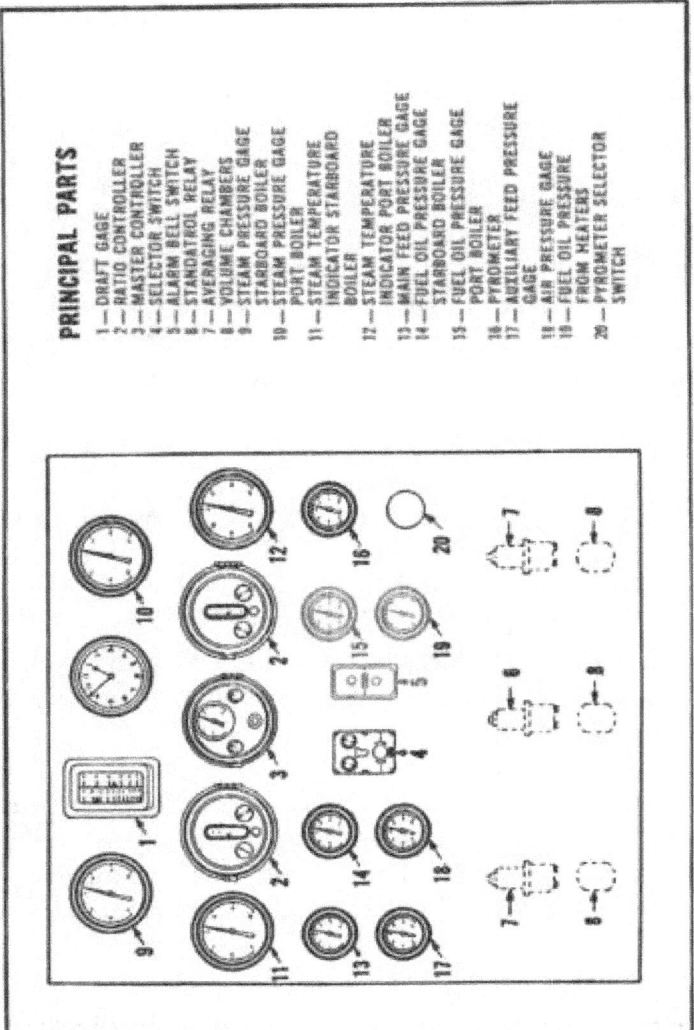

PRINCIPAL PARTS

1—DRAFT GAGE
2—RATIO CONTROLLER
3—MASTER CONTROLLER
4—SELECTOR SWITCH
5—ALARM BELL SWITCH
6—STANDATROL RELAY
7—AVERAGING RELAY
8—VOLUME CHAMBERS
9—STEAM PRESSURE GAGE
STARBOARD BOILER
10—STEAM PRESSURE GAGE
PORT BOILER
11—STEAM TEMPERATURE
INDICATOR STARBOARD
BOILER
12—STEAM TEMPERATURE
INDICATOR PORT BOILER
13—MAIN FEED PRESSURE GAGE
14—FUEL OIL PRESSURE GAGE
STARBOARD BOILER
15—FUEL OIL PRESSURE GAGE
PORT BOILER
16—PYROMETER
17—AUXILIARY FEED PRESSURE
GAGE
18—AIR PRESSURE GAGE
19—FUEL OIL PRESSURE
FROM HEATERS
20—PYROMETER SELECTOR
SWITCH

Figure 15–16.—Arrangement of control panel.

3. Rotate the relay load adjusting knob until the control pressure is again the same value as observed in operation 1. Do not leave the transfer knob in the reset position longer than is necessary to adjust the control pressure.

4. Turn the transfer knob to the AUTOMATIC position.

5. If any difference exists between the loading pressure and the control pressure, the pressures should be equalized by slowly turning the relay load adjusting knob.

When the transfer knob is in the AUTOMATIC position, the loading pressure is received from the standatrol, and is reproduced in the adjustable relay so that the control pressure is always equal to the loading pressure. However, the relative value of the control pressure to the loading pressure may be increased or decreased by turning the load adjusting knob.

Changing from "Automatic'^ to "Hand'^ Control

To change from AUTOMATIC to HAND control, proceed as follows;

1. Observe the value of the control pressure, then:

2. Turn the transfer knob to the reset position.

3. Rotate the relay load adjusting knob until the control pressure is the same as observed in the first operation. Do not leave the transfer knob in the reset position longer than is necessary to adjust the control pressure.

4. Turn the transfer knob to the hand position.

GENERAL CARE AND MAINTENANCE

The automatic combustion control system will require very little maintenance or readjustment if a few simple precautionary measures are followed, most important of which are maintaining a supply of clean, dry compressed air, and prohibiting inexperienced personnel from tampering with the control devices.

Calibration

Once the control system has been calibrated the original adjustments ordinarily should not be changed. If changes are necessary they should be made only after a careful study of the system and the manufacturer's calibration instructions.

When it is necessary to disassemble any part of the control equipment for cleaning or repair, it should be returned to the original adjustment shown on the calibration instructions.

Compressed Air Supply

The most common source of trouble in a combustion control system is the fouling of pilotrol ports with dirt, oil, or water originating in the supply of compressed air.

The compressor lubrication instructions should always be followed carefully. Avoid excessive lubrication. It does the compressor no good and is certain to result in trouble with the automatic combustion control.

The receiver tank and the air filters should be blown out every 4 hours to remove any accumulation of water or oil.

The main air line filters should be inspected once every 6 months, and the filter medium should be renewed when necessary.

General Operation of the Control System

With the adjustment properly made and coordinated, the control system should maintain the steam pressure, stabilize the firing rate to correspond to the steaming load, and readjust the forced-draft damper to maintain the correct fuel-air ratio.

Bear in mind that the control can function properly only when the firing rate is within the range of the sprayer plates and the number of burners in service. It is the duty of the operator to select the size of the sprayer plate or vary the number of burners in service to allow

the control to increase or decrease the firing rate. The fuel oil pressure at the burners must not be allowed to fall to less than 120 psi or rise to more than 300 psi.

The control system, if adjusted to react too quickly to changes in steam pressure, will tend to hunt rather than to stabilize at the correct point. If adjusted to react too slowly, steam pressure will vary beyond acceptable limits.

These adjustments are made in the "range adjustment" of the master controller pilotrol, and in the bleeder valve of the standatrol. Similar adjustments to obtain proper action of the forced-draft dampers are made in the ratio controller pilotrol and in the bleeder valve of the averaging relay.

For each installation the original settings of the PILOTROL slides in both the master controller and ratio controller and the original settings of the standatrol and averaging relay bleeder valves have been recorded on a calibration card. (It is a good idea to make a copy of this card and to file it in a safe place as a precautionary measure should the original be misplaced.)

If the entire control system hunts, the trouble will lie in either the master controller PILOTROL or the standatrol. Check the position of the bleeder valve in the STANDATROL to be sure it is in the position shown on the calibration card (usually about four to five turns open). If hunting continues with this setting of the STANDATROL bleeder valve, move the

PILOTROL range adjustment slide to the right in small steps until the control system stabilizes.

If hunting occurs only in the forced-draft damper control drive, check the bleeder valve in the averaging relay to see that it is set as shown on the calibration card (usually four to five turns open). If hunting continues with the correct bleeder valve setting, move the range-adjustment pivot in the ratio controller pilotrol to the right far enough to eliminate the hunting.

The forced-draft damper drive will require 15 seconds to change from FULL open to closed position. This lag

must be allowed for when making the range adjustments in the master controller and ratio controllers, and also when setting the standatrol and averaging relay bleeder valves.

Connecting Piping

If repairs or replacements are made to the air supply lines, blow them out thoroughly before making the final connection.

All air piping should be installed in a neat, workmanlike manner and supported as necessary to hold it in line. Piping should be pitched towards the settling chambers or filters. Blow out the settling chambers or filters each watch (every 4 hours). In addition, it is necessary to observe the precautions listed below.

1. At regular intervals inspect all JOINTS in the air lines for leaks. Brushing the air-line joints with a soapsuds solution will readily show up any leaks.

2. Once every month disconnect the steam pressure CONNECTION at the master controller and give the STEAM LINE a free blow of 5 to 10 seconds to clear out any sludge.

3. At each boiler overhaul, disconnect the draft CONNECTIONS at the ratio controller and blow out the DRAFT LINES towards the boiler with air at a pressure of 50 pounds or more.

4. Check the separating chambers in the fuel oil pressure lines to the ratio controllers at least every 6 months and renew the prestone seal.

More detailed instructions for maintaining the various units making up the control systems are given in the following pages.

Control Standard Adjustment of Master Controller

The standard pressure can be changed, if desired, by merely turning the standard adjustment until the standard pointer indicates on the pressure scale the pressure to be maintained.

If the standard is changed while the controller is in service, move the standard adjustment slowly and in small steps, allowing the control to stabilize each time before taking another step.

Calibration of Master Controller

It is not necessary to alter the adjustments of the master controller except on replacement of the sector and pinion assembly or the Bourdon tube. The proper sequence in which such adjustments should be made is indicated by the letters A to E in figure 15-3.

The pressure pointer is adjusted at the hub and by the connecting linkage so that it accurately indicates steam pressure through the scale range.

The standard pointer (red) is adjusted at the hub and by the connecting linkage so that when the red and black pointers are together, the loading pressure developed by the PILOTROL is 15 psi. This adjustment should also be such that when the control standard pointer is at the middle of the pressure scale, the standard adjustment scale pointer on the pilotrol (see item 13, fig. 15-4) is at the midpoint of the scale.

Regulating Range Adjustment of "Pilotrol"

Move the regulating range adjustable pivot toward the stop screw Y (fig. 15-4) to increase the regulating range. The slide on which the pivot is carried should be loose enough to permit easy adjustment, yet firm enough to hold the adjustment after it has been completed. The adjustment is made by inserting the sharp end of a pencil into the hole beside the slide to pry it along the beam.

Maintenance of the "Pilotrol"

All pivots used in the assembly of the PILOTROL are made of stainless steel and require no lubrication.

The air connections to the pilot valve and filters must be kept air-tight. Check for leakage with soapsuds.

Files, reamers, or abrasives must never be used on the valve stem lands, sleeves of the pilot valve, or the PILOTROL pivots and bearings.

If the pilot valve is cleaned, or if repairs are made to the PILOTROL linkage, a recalibration of the pilotrol linkage should be made. For the procedure, see manufacturer's instruction pamphlet.

Maintenance of the "Standatrol" and Averaging Relays

Figure 15-7 shows the method of assembling and adjusting the relay with each step indicated in numerical order. To disassemble the relay, follow this procedure in reverse order from step 12 to step 1.

To adjust the bleeder valve, remove the cover and turn the stem clockwise to close or counterclockwise to open.

Caution: Never use a wrench on this stem. When replacing the cover be certain that gasket is in place.

It is essential that the assembly have no air leaks. To check the operating and sealing bellows, apply an air pressure to any one chamber, open the adjacent chamber connections and note if any air is being discharged. If leakage is detected, disassemble the relay to inspect and replace the faulty bellows. In reassembling, the joints between the bellows, the clamp bolt, and the body should be made up with a non-hardening compound such as FOSTORIA TITE SEAL. (Caution: Do not use Copalite or Glyptol.)

To clean the inlet and exhaust valves, remove the chamber D section. Remove, clean, and replace one valve at a time to simplify replacing the valve so that the operating arm is level and the nibs are just touching each valve. Clean the valves with carbon tetrachloride.

The spring adjustment of the relay is checked after reassembly in accordance with the following procedures.

Spring Adjustment of the "Standatrol"

To calibrate the standatrol in place, the board must be taken out of service, since the tie-back from the selector to chamber C must be disconnected and the chamber

C plugged. The selector should be placed on HAND, the bleeder valve stem removed, and the cap cover and gasket replaced. Adjust the standard pointer of the master controller to send a pressure of 15 pounds to chamber A. Adjust the spring to obtain a pressure of 15 pounds in chamber D as indicated on the loading pressure gage of the selector. Insert the bleeder valve and screw it closed, and open four or five turns.

To check adjustment, drop the master loading pressure to 14 pounds. The selector loading pressure should drop to zero. Raise the master loading pressure to 16 pounds, and the selector loading pressure should rise to 25 pounds. If this action is obtained, reconnect the tie-back line from the selector to chamber C and the standatrol is ready for service.

Spring Adjustment of the Averaging Relay

Adjust the spring loading by turning the adjusting screw to give an air pressure of 15 psi at D3, with the bleeder valve in its full open position and with an air loading pressure of 15 psi applied at connections A1 and C7. Then adjust the throttling valve to give the desired rate of change in the control pressure at D3 when the loading pressure to C7 changes. Be sure the air supply is at 28 psi, for this adjustment. (See fig. 15-8.)

Air pressure at A7 is indicated by the control pressure gage on the selector, and the pressure at C7 by the loading pressure gage on the ratio controller. Connect the spare gage at D3,

Selector Maintenance

The air pressure gages should give accurate readings or should be replaced. If air leakage occurs through shut-off valves, replace the air valve assembly. The shut-off valves may be checked for air leakage by connecting an air supply pressure of 28 psi to the lower connection and testing with soapsuds.

To clean the inlet and exhaust valves, remove the chamber D section and clean the inlet and exhaust valves by washing with a common solvent such as carbon tetrachloride. After the valves have been cleaned, it is essential that you return the inlet and exhaust positions to the same relation with the operating arm, in order that the travel adjustment will be correct. (See adjustment No. 5, in fig. 15-17.) Adjust the exhaust valve for a very slight air leakage and check the operation of the relay by applying air loading pressures to chamber A. Any adjustments which may be required should be made as described in figure 15-17. The leakage at the exhaust port should be about cubic foot per minute.

The cams must not contact the push rods when the shut-off valves are in the closed position. In open position, the push rods should be lifted Vi^ inch by the cams. The shut-off valve bodies may be adjusted up or down to

U CUMP MLT AND lOXOWS

t CMAMtCI A MOY COMPini WITH ICAUMC laiOWS

1 lOWCM SPACER A0JIST1M6 NIT AND IfPCI SPACER

4 SMtNtt. SPRIM CLAMP AMO MIOL T16NTCN MVT ON CUMP lOlT ASSCMRT. AOilSTMOIT tCRCV

I. AMNT All fUSMM TO COMICCTm

M iNTR H maos ftosm it oi-mm AT comifCTVM w. pmm.-

TAMCOIUT TVRfl TNC AOJISTMCMT N«T 0ITII \/n IHCN ClCARAMCt IS ORTAINCO AT T(1/4 TIR« FROM POINT Of CONTACT WITN SPACER). KECP ONAfST

mn AOjfSTco roR uiont air ois-

CNAROC AffROIIMATUT 1/1 CO. FT/

QNAtST VAin

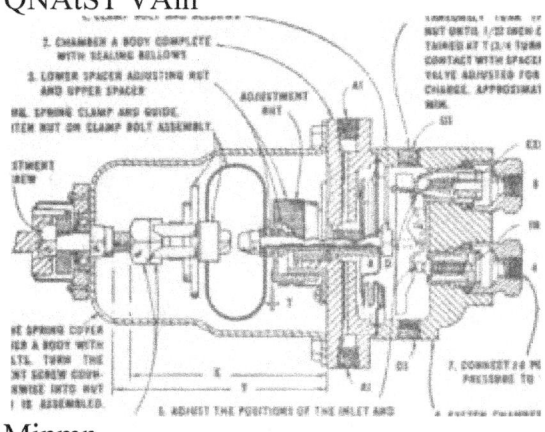

Minmn

Nl FASTCN TNC SfRINO COVER TO CNAMIER A ROOT WITH root ROiTL TfRN THE AAlf HMENT SCREW COVN-TU aOCKWIiC INTO NUT AS COVU IS AUEMIIEO.

I. MCASORC DIMENSION Y AND TVRN THE NIT AMO LOCK NUT UNTIL THE DIMCIIIIOII I IS •/» INCH lUS THAN 0IMEK90II T.

UNA0ST VUVES UNTIL THIT lAREIT CONTACT TNI NIRS ON TNI OPfRATMO ANM, WHEN TNI OPCUTINO ARM IS fAAAUXL TO TNC TOP SORTACC Of CNARWCi 0 ROOT.

7. CONNCCTril

rRESSf RC TO TNC INIH VAlVt

i FASTEN CHAMRER 0 ROOT TO THE CHAMRCR A ROOT.

Figure 15-17.—AcJjutfment of tolector.

Transfer knob position

Hand

Reset •

Automatic

Reset »

No. 1

Closed. Open.-Open. -Closed.

Shut-off valve position

No. 2

Open

Closed

Closed „-Open

No. 3 I

Open -. Closed-Open_. Closed-

• No. 3 shut-off valve operates before other valves, > Do not leave the transfer knob in reset position.

No. 4

Open. Closed. Closed. Open.

Figure 15-18.—Positions of shut-off valves.

obtain the necessary clearance and lift after loosening the mounting screws. The open and closed positions of the shut-off valves may be obtained from figure 15-18.

Maintenance of the Ratio Controller

Refer to instructions for the maintenance of the PILOTROL unit.

The operating mechanism of the controller requires no lubrication or care other than preventing excessive accumulation of dust or dirt within the case.

Break the unions at the diaphragm and blow compressed air through diaphragm connecting lines at each boiler overhaul. Do not apply compressed air pressure to the diaphragm because the fabric is easily ruptured.

Keep the bellows, the connecting line, and the separating chamber filled with sealing fluid (Prestone or Ethylene Glycol). Vent the air from the bellows connecting line after adding the sealing fluid, and at regular intervals thereafter.

Adjustment of Control Valves

Both oil control valves and blower turbine control valves are adjusted to be closed when the loading pressure is 5 psi and FULL OPEN at 25 psi. The loading pressure at which the diaphragm starts to move can be varied by turning the adjusting screw beneath the spring. Turning the screw in a right-hand direction increases the spring tension to increase the loading pressure required to start moving the diaphragm.

It is necessary that the diaphragm head be absolutely tight to prevent air leakage at this point. This can be tested by applying a soapsuds solution to the joint while the unit is under pressure.

In packing the stuffing box, the use of cotton wick packing is not recommended. Use either a formed ring packing or a good grade of twisted graphited asbestos packing. When properly packed, the stuffing box should hold the pressure by being tightened with only the thumb and forefinger. Never use a wrench or pliers.

If it is necessary to replace a diaphragm, remove the air lines and upper diaphragm casing; then the dia-

Maintenance of Control Valves

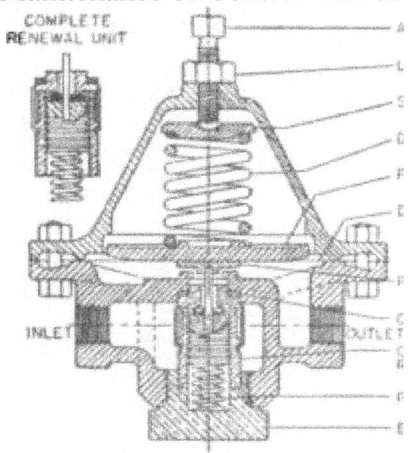

ADJUSTING SCREW
CYLINDER CASKET
LOCK NUT
COMPLETE R6NEWAL UNIT
DIAPHRAGM SPRING
PISTON SPRING
PRESSURE PLATE
PUSHER PLATE
BODY CAP
DIAPHRAGM
SPRING BUTTON

Figure 15-19.—Assembly of an oir reducing valve.

phragm may be removed. Before removing tne diaphragm it is best to relieve the tension on the spring by backing off the spring adjusting screw. This can easily be readjusted after the diaphragm has been changed in the manner described under Adjustment of Control Valves.

Combustion Control Piping, Reducing Valves, and Filters

Reducing valves, similar to those shown in figure 15-19, are furnished with the control system. They should be installed in a horizontal line with the adjustment screw on top. The valves furnished are suitable for air at an initial pressure of 80 to 100 psi and an outlet pressure of 20 to 60 psi. The size of the valve in this case is % inch.

The valves should be adjusted to carry the supply pressure indicated, 28 psi. Turning the adjusting screw clockwise increases the outlet pressure that the valve will maintain.

The main air line filters are shown in figure 15-20. You should open drain cocks once each watch to remove any water collected. Replace the charge of "Nuchar" every 6 months, or more often if fouling of the air pilot valves is noted.

Disconnect the filter from the air line, remove the cover, filter felt, and bronze wool, and dump the old charge. The air supply lines to the pilot valves of tht controllers and drive unit have small, felt, cartridge-typr filters that can be removed and cleaned.

Separating Chambers

In the oil pressure lines to the ratio controllers a separating chamber (see fig. 15-21) is used to prevent fouling the line with heavy fuel oil, since this would cause sluggish operation.

Check the level of the sealing liquid (Prestone or Ethylene Glycol) by means of the try cocks provided. Keep the level of the sealing liquid above the low try cock. Check at least once every 3 months and refill as necessary.

OUTLET-(H INCH PIPE TAP)
FILTER FaT (BETWEEN WIRE SCREEN)
IRONZE WOOL
6RANUUR NUCHAR (4-10 MESH
APPROXIMATaV ^ POUND)
FILTER FELT
MECHANICAL SEPARATOR
INLET-(li INCH PIPQ-

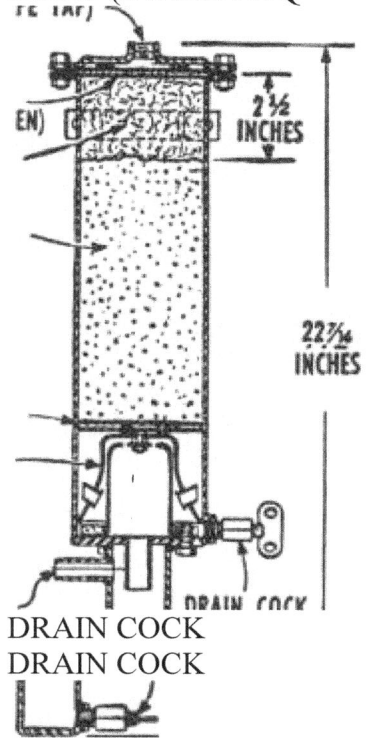

DRAIN COCK
DRAIN COCK

Figurs 15-20.—Attambly of BaiUy filter.

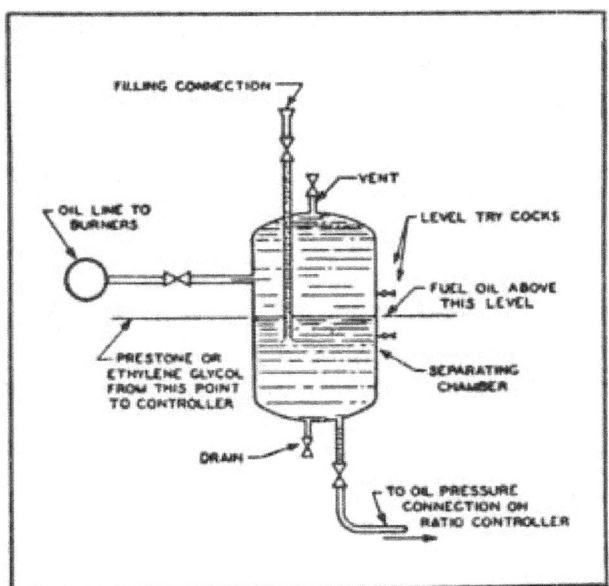

Figure 15—21.—Diagram of separating chamber.

HAGAN AUTOMATIC COMBUSTION CONTROL EQUIPMENT

It is quite possible that in the future you may encounter Hagan automatic combustion control equipment on board combat vessels. Because of its similarity to the system previously discussed, only a brief description of it is presented here.

Compressed air at a pressure of 60 pounds is used to operate the Hagan automatic combustion control system.

An air-operated power unit is provided to operate the forced-draft damper to each boiler and to control the air flow. The firing rate is controlled by an air-operated valve in the return oil line from each boiler.

Air-Flow Control

The power units controlling the forced-draft dampers, called the receiving regulators, are positioned according to a loading air pressure received from a master sender.

The master sender has a pressure connection to the main steam drum and regulates the loading air pressure according to the steam pressure in the steam drum. If the steam pressure falls, the master sender changes the loading air pressure; this causes the receiving regulators to open the forced-draft dampers. If the steam pressure rises, the master sender changes the loading air pressure, causing the receiving regulators to close the dampers.

Oil-Flow Control

A variable ratio regulator on each boiler positions the fuel oil regulating valve by controlling an air loading pressure to the fuel oil regulating valves. The variable ratio regulator has draft connections to the air casing and the furnace to measure the draft loss across the burners, which is a measure of the air flow through the boiler. The variable ratio regulator sends out a loading air pressure to position the fuel oil control valve in relation to the combustion air flow through the boiler.

The operation of the principal parts of the Hagan combustion control system is illustrated in figure 15-22.

If the steam pressure in the main steam line falls, the master sender changes the loading air pressure on the receiving regulator; this causes the damper to open and increases the air flow to the furnace.

The increase in air flow increases the draft loss across the burners, causing the variable ratio regulator to change the loading pressure on the oil control valve to raise the oil flow to correspond to the new draft loss.

If the steam pressure in the main steam line rises, the combustion control operates to close the forced-draft damper and decrease the firing rate.

A single master sender operates the receiving regulators on both boilers, the loading air pressure line being divided. In the individual circuit to each receiving regulator is a transfer valve and a compensating relay.

The transfer valves are used to switch either or both boilers to MANUAL or automatic control as desired.

Under normal steaming conditions both boilers will be operated on automatic control.

A boiler should be operated on manual control while it is being brought up to operating pressure, while the tubes are being blown, or during any emergency operation.

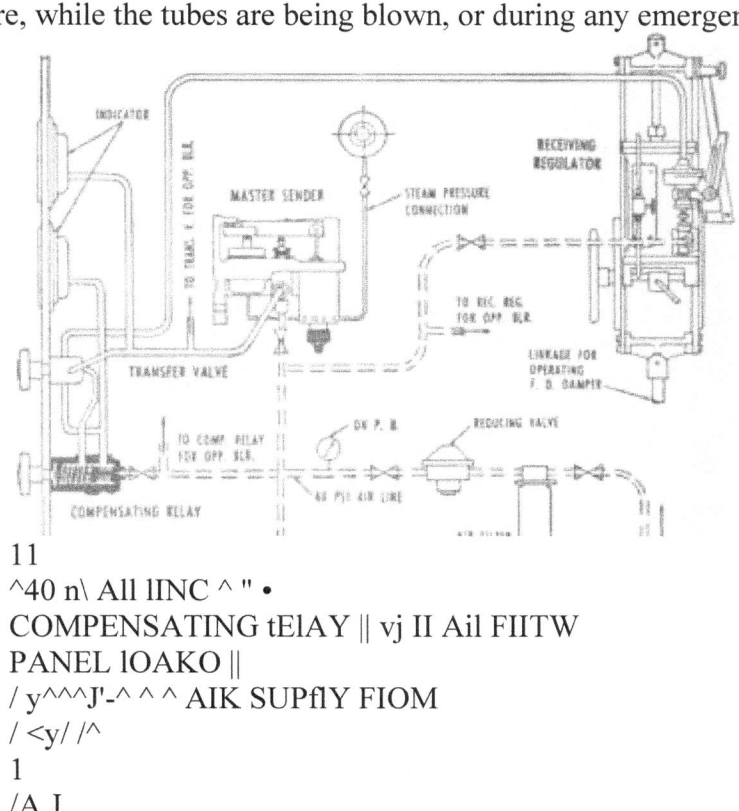

11
^40 n\ All lINC ^ " •
COMPENSATING tElAY || vj II Ail FIITW
PANEL lOAKO ||
/ y^^^J'-^ ^ ^ AIK SUPflY FIOM
/ <y/ /^
1
/A J
TO VAK. lATIO IE6. POI OPP. III.
VAIIABlE RATIO REGULATOR
r
TIP ENDED PnCOCKS - FOR TESTING
FUEL on CONTROL yAlYE
TO lUKNEI All CASING C0NNECT10II
TO fUINACE PIESSUIE CONNEaiOM

Figure 15-22.—Diogram of Hagan control piping (ono boilor).

The compensating relay transmits the master loading pressure to the receiving regulator. Its design makes it possible to vary the transmitted loading pressure to any desired proportion of the master loading pressure while on AUTOMATIC control. On manual control the loading pressure from the master sender is cut off at the transfer valve. The loading pressure to the receiving regulator depends entirely on manual adjustment of the compensating relay. Under the manual condition, the forced-draft damper position is controlled from the boiler control panel.

With both boilers on automatic control, the compensating relays make it possible to divide the load between the two boilers in any desired proportion.

Ordinarily the compensating relays should be adjusted so that each boiler will be operated at the same rating, each carrying 50 percent of the total load; that is, both forced-draft indicators reading alike.

QUIZ

1. What is the primary function of an automatic combustion control system?

2. What are the three basic types of automatic combustion control systems?

3. Which type of automatic combustion control system is used for about 90 percent of all shipboard installations?

4. What is the basic measurement which determines the operation of the automatic combustion control system?

5. What measurement determines the amount of fuel oil which will be fired in the boiler furnace, when the automatic combustion control system is used?

6. What factor determines the amount of air which is provided fc< combustion?

7. In what two different ways is compressed air used in an automatic combustion control system?

8. What are the two functions of the master steam pressure indicator-controller?

9. What unit in the automatic combustion control system has these two basic functions: (1) to instantly reproduce the incoming loading air pressure changes in the outgoing loading (or control) air pressure, and (2) to amplify the change in the outgoing loading air pressure?

10. What determines the loading air pressure in the ratio controller?

11. By what means does the ratio controller regulate the amount of combustion air?

12. From what two sources does the averaging relay receive loading air pressures?

13. What is the only way in which 40-pound air is used in the automatic combustion control system?

14. At what pressure is the basic compressed air supply maintained, for the Bailey automatic combustion control system?

15. What two terms are used to describe the variable pneumatic pressure which is developed directly or indirectly from the 28-pound air supply?

16. What type of interconnections exists between the major units of the automatic combustion control system?

17. Under normal or static conditions, when no changes are occurring in the system, what loading air pressure does the PILOTROL send to the standatrol relay?

18. What is indicated by the setting of the red pointer of the master controller?

19. What is the function of the restoring bellows in the pilotrol?

20. If the steam pressure decreases (so that the black pointer is below the red pointer), what will happen to the loading air pressure?

21. What is the normal range of loading air pressures shown on the loading air pressure gage of the master controller?

22. In the standatrol relay, what is used to balance the variations in loading air pressures which are applied to chambers A and C?

23. What is the spring-loading pressure in the standatrol relay?

24. In what part of the standatrol relay are the loading air pressures amplified?

25. What determines the standatrol's speed of reaction to changes in the loading air pressure produced by the PILOTROL?

26. At what setting of the selector valve does the standatrol relay have no control over the automatic combustion control system?

27. What unit in the automatic combustion control system allows for quick readjustment of the controls, according to the size of sprayer plates being used?

28. What is the normal loading air pressure set up by the ratio controller, when the fuel-

air ratio is correct?

29. In the averaging relay, what device is used to delay the effect of the loading air pressure changes received from the ratio controller?

30. In which part of the averaging relay are the incoming loading air pressure changes immediately effective?

31. In order to ensure proper operation of the pneumatic automatic combustion control system, what precautions must be observed concerning the supply of compressed air?

32. What is the most common cause of trouble in the automatic combustion control system?

33. What are the allowable limits of fuel oil pressure at the burners?

34. What difficulty will be encountered if the automatic combustion control system is adjusted to react too quickly to steam pressure changes?

35. What difficulty will be encountered if the automatic combustion control system is adjusted to react too slowly to steam pressure changes?

36. What are the two most likely locations of trouble, if the entire automatic combustion control system hunts?

37. What method should be used to check air-line joints for leaks?

38. What is the proper amount of air leakage at the exhaust port of the selector valve?

39. How frequently should the main air-line drain cocks be opened?

40. How frequently must you check the level of the sealing liquid used in the ratio controller?

41. In the Hagan automatic combustion control system, what force is used to position the power units which control the forced-draft dampers?

42. How should the Hagan automatic combustion control system be set while a boiler is being brought up to operating pressure?

CHAPTER
16
PROPULSION TURBINES

The purpose of this chapter is to familiarize you to some extent with the propulsion turbines used aboard naval vessels. As a Boilerman, you are not normally-responsible for the operation or upkeep of this machinery. However, a general understanding of the ship's propulsion plant as a whole will enable you to work more efficiently with other personnel of the Engineering Department; and it may be of assistance to you if, in an emergency, you are called upon to operate engineroom equipment.

Although several different types of drives are used on naval vessels, we will here be concerned only with installations using geared-turbine drive. As a Boilerman, you are most likely to be assigned to vessels having this type of propulsion machinery. It is important to remember, however, that other types of drives are used. You should make a point of studying the arrangement of the propulsion plant on your own ship.

GEARED-TURBINE INSTALLATIONS

A typical geared-turbine installation such as the one shown in figure 16-1 consists of a cruising turbine, a high-pressure turbine, a low-pressure turbine, and an astern turbine, together with a cruising turbine reduction gear and the main reduction gears. Some vessels do not have a cruising turbine; in this case, the installation con-

sists of a high-pressure turbine, a low-pressure turbine, an astern turbine, and the main reduction gears.

In an installation such as the one shown in figure 16-1, the cruising turbine is connected through a rigid coupling to the cruising gear pinion; the cruising gear low-speed

ASTERN ELEMENT

7f

LOW PRESSURE TURBINE

CRUISING TURBINE REDUCTION GEAR

CRUISING TURBINE

HIGH PRESSURE TURBINE

MAIN

REDUCTION GEAR

MAIN SHAFT

Figure 16—1.—6*n«ral arrangament of a g«ar«d-turbin« installation.

gear is connected to the high-pressure turbine through a flexible coupling. The high-pressure turbine and the low-pressure turbine are connected through flexible couplings to high-speed pinions of the main reduction gear. The astern turbine usually consists of two elements, one mounted on each end of the low-pressure turbine shaft, in the low-pressure turbine casing.

Where a cruising turbine is installed, two distinct modes of operation are possible. For most economical operation, all steam is first admitted to the cruising turbine. It then goes, by way of a crossover pipe, to the high-pressure turbine. After passing through all stages of the high-pressure turbine, the steam exhausts through another crossover pipe to the low-pressure turbine. After the steam has been utilized in the low-pressure turbine, it is finally exhausted to the main condenser. The arrangement of valves whereby the cruising turbine is used is sometimes referred to as the CRUISING turbine

COMBINATION, or the CRUISING COMBINATION.

When higher speed or greater maneuverability is required, the MANEUVERING COMBINATION is used. In this arrangement, all steam is admitted to the high-pressure turbine. The cruising turbine control valves are shut, the cruising turbine exhaust is connected to the condenser, and cooling steam taken from the first stage of the high-pressure turbine is admitted to the cruising turbine. Since the cruising turbine exhaust is connected to the condenser, the cruising turbine is maintained under vacuum. This is of utmost importance when the cruising turbine is not being used, since the cruising turbine rotors idle at a speed which is generally in excess of 10,000 rpm when the vessel is operating at full power,

TYPES OF PROPULSION TURBINES

As we saw in chapter 3, the basic distinction to be made between turbines has to do with the manner in which the steam causes the rotor to move. Impulse turbines are moved by a direct push or "impulse" from the steam impinging upon the rotor blades. Reaction turbines are moved by the reactive force produced on the moving blades when the steam expands through the blades and thereby increases in velocity, together with the reactive force produced on the moving blades when the steam changes direction.

The staging and compounding of impulse turbines was described in chapter 3. As you will remember, a simple impulse stage (Rateau stage) consists of one set of nozzles and one row of moving blades. When a number of Rateau stages are arranged in sequence, we have a Pressure-compounded or Rateau turbine. An impulse stage which consists of one set of nozzles and two or more rows of moving blades is known as a Velocity-compounded or Curtis stage.

In dealing with propulsion turbines, we must now consider another impulse turbine arrangement. An impulse turbine which consists of one velocity-compounded (Curtis) stage plus a series of pressure- compounded (Rateau) stages is generally referred to as a Pressure-velocity-co m pounded turbi ne.

In reaction turbines, one row of fixed blades and its succeeding row of moving blades are taken as constituting one stage. Since the fixed blades in a reaction turbine are roughly comparable to the nozzles in an impulse turbine, this definition of a reaction stage might sound pretty much the same as the definition of an impulse stage. However, there is this important difference: a reaction stage includes two pressure drops, whereas an impulse stage includes only one.

Each reaction turbine consists of several stages arranged in sequence, so that the pressure drop between inlet and exhaust is divided into a number of steps.

Many reaction turbines and some impulse turbines are built as DOUBLE-FLOW TURBINES. A double-flow turbine consists essentially of two single-flow units installed on one shaft, in the same casing. The steam enters at the center, between the two units, and flows from the center toward each end of the shaft. The main advantages of the double-flow arrangement are: (1) the blades can be shorter than they would have to be in a single-flow turbine of equal capacity, and (2) in reaction turbines, axial thrust is avoided by having the steam flow in opposite directions. (Impulse turbines develop relatively little axial thrust, in any case.)

Both impulse turbines and reaction turbines are used in naval propulsion plants. On many vessels, the cruising turbine, the high-pressure turbine, and the astern turbine are of the impulse type. The low-pressure turbine may be of the impulse type or of the reaction type.

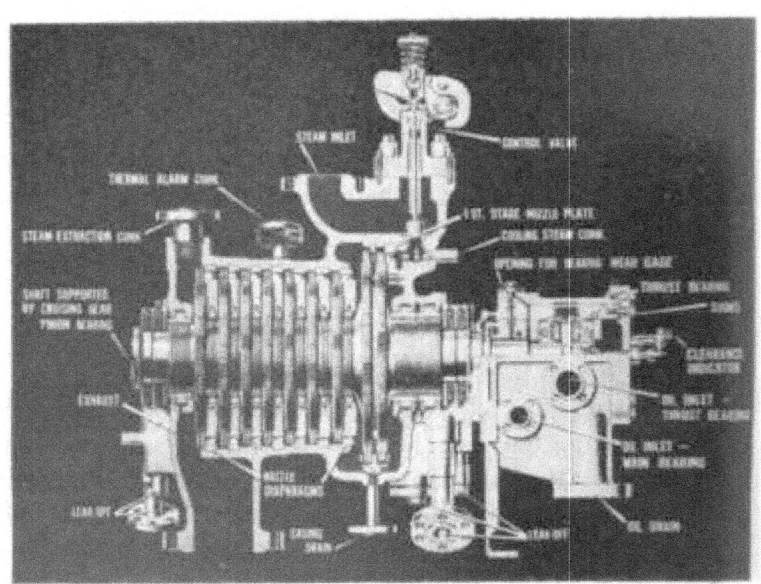

Figur* 16-2.—Cruising turbin*.

A typical cruising turbine is shown in figure 16-2. This is an eight-stage, impulse turbine. The first stage is a velocity-compounded (Curtis stage); the remaining seven stages are pressure-compounded (Rateau) stages This turbine is, therefore properly referred to as a

PRESSURE-VELOCITY-COMPOUNDED IMPULSE TURBINE. The

pressure stages are separated by nozzle diaphragms. Labyrinth packing is used between the diaphragms and the turbine shaft to minimize steam leakage between the stages. The cruising turbine is throttled by means of cam-operated nozzle control valves, which open successively as the throttle handwheel is turned. All the nozzle control valves admit steam to the first stage of the turbine.

Figure 16-3 shows a typical high-pressure turbine. This turbine has one velocity-compounded impulse stage followed by eleven pressure-compounded impulse stages; it is thus a pressure-velocity- compounded turbine. When

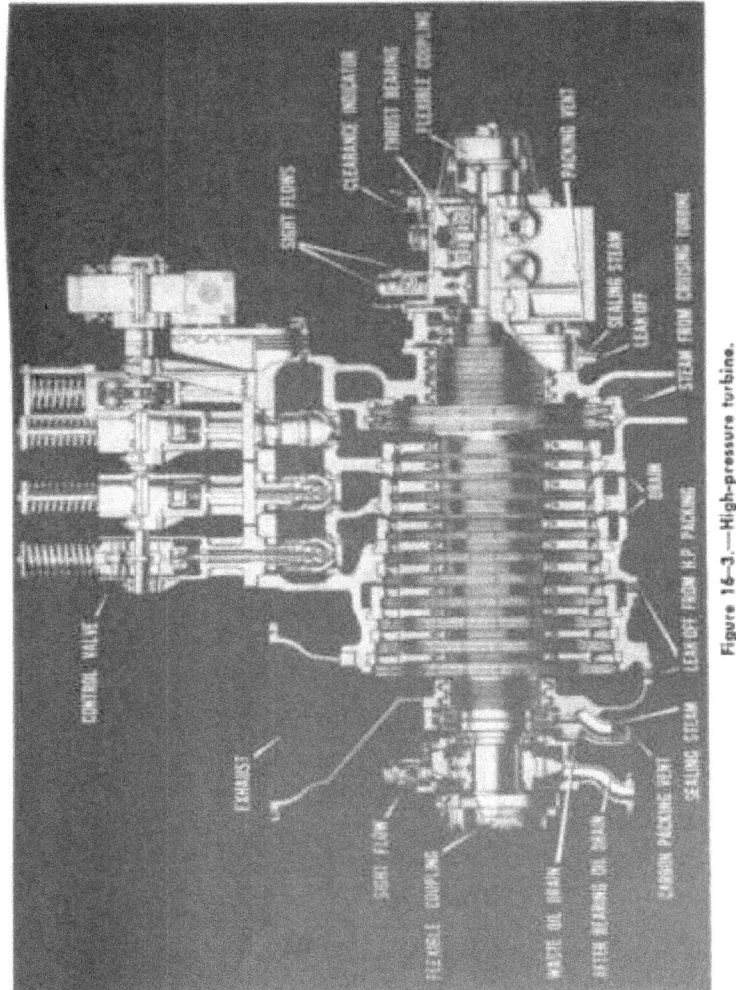

Figure 16-3.—High-pressure turbine.

the cruising combination is being used, steam from the cruising turbine is admitted to the first stage of the high-pressure turbine. The steam flows through all stages of the high-pressure turbine before being exhausted to the low-pressure turbine. When the cruising turbine is not being used, the high-pressure turbine is throttled by means of several (often five or six) nozzle control valves. As a rule, the first two nozzle control valves admit steam to the first stage, and the later nozzle control valves admit steam to later stages—the second, fourth, and sixth, for example. The nozzle control valves are cam-operated, and open in sequence when the hand-wheel is turned.

Figures 16-4 and 16-5 show the arrangement of a low-pressure double-flow reaction turbine and the astern ele-

um STUM an MUinunnii unu stum mh

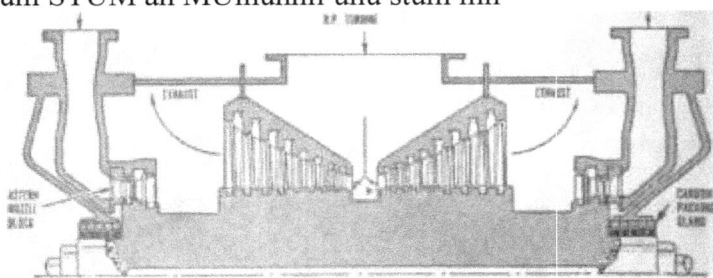

Figure 16-4.—Arrgngement of low-pretsure turbine and astern elements.

ments. The low-pressure turbine is a straight reaction turbine. In some vessels of recent design, double-flow impulse turbines of the Rateau type are used as low-pressure turbines.

The astern elements are usually velocity-compounded impulse stages (Curtis stages). On a few vessels, however, you will find pressure-velocity-compounded astern elements; that is, each element will consist of one Curtis stage followed by a Rateau stage.

Each astern element has its own steam inlet but the admission of steam to both elements is controlled by the

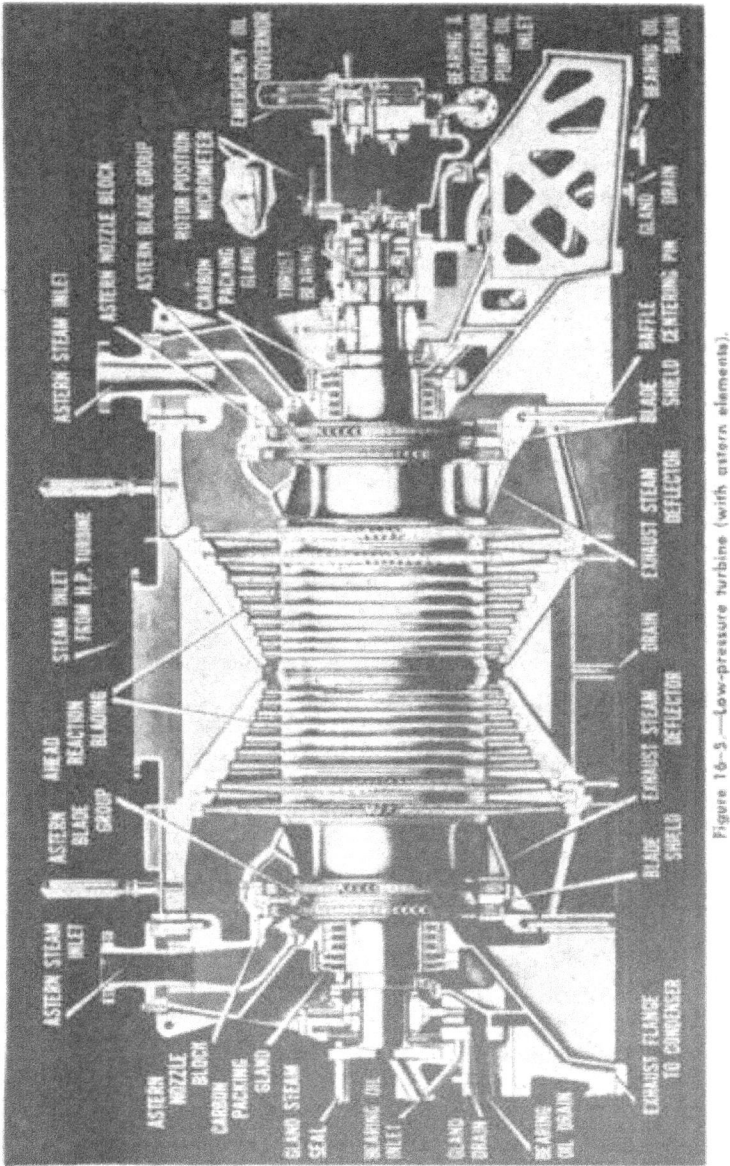

Figure 16-9.—Low-pressure turbine (with astern elements).

one astern throttle. The astern elements exhaust through the low-pressure turbine exhaust chamber to the condenser.

TURBINE OPERATION

Engineroom preparations for getting under way consist of starting the necessary auxiliaries, putting the forced lubrication system into operation, jacking over the turbines, cutting in the gland sealing steam, bringing up the vacuum on the condenser, and warming up the main steam line. All preparations for getting under way must be made in proper sequence, as prescribed by the check-off list on y4?ur own ship.

Warming Up the Main Steam Line

Before propulsion turbines are warmed up and made ready for use, the main steam line must be warmed up slowly and carefully. The steam line is usually warmed by opening bypass valves around the steam stop valves and allowing steam pressure to build up in the line. Drain lines must be opened to allow removal of condensate from the main steam line. Throttle drains and turbine casing drains must also be opened.

Warming Up the Main Turbines

During the warming-up period, turbine rotors must not be permitted to remain at rest while steam (including gland sealing steam) is being admitted to the turbine. Most ships are provided with jacking or turning gear to turn the turbine rotors; where this gear is provided, the turbine must be turned before the steam is admitted to the glands. After jacking the rotors for a sufficient period of time, disengage the jacking gear and spin the rotors ahead and astern by alternately opening the ahead and astern throttles. When spinning a turbine rotor by steam, you must be sure to admit enough steam to start the turbine rolling immediately, so as to prevent uneven heating and consequent distortion.

As a rule, the astern turbines are spun before the ahead turbines. If any condensate has been left in the piping, it will be discharged more readily through the relatively large nozzles and the relatively few rows of blading than it would be through the ahead turbines.

Standing By For Getting Under Way

If standing by for getting under way with a half-hour's notice, engineroom personnel must perform the following duties:

1. As necessary, open the recirculating valve to allow recirculation from the deaerating feed tank to the main condenser, to prevent overheating of the de-aerating feed tank and to ensure the removal of oxygen from the system. If there is insufficient condensate to maintain proper temperatures in the air ejector condenser by means of the automatic thermostatic recirculating valve, the manual recirculating valve in the line between the vent condenser and the main condenser should be opened.

2. Use the second-stage air ejector to maintain vacuum on the main condenser.

3. Use one lubricating oil pump for each engine to maintain the required oil pressure at the bearings.

4. Maintain the lubricating oil at the required temperature.

5. Keep the gland steam at the proper pressure.

6. Crack the turbine and throttle drains.

7. Operate the main circulating pumps slowly, so that the supply of water is just sufficient to maintain the required vacuum. If necessary, throttle the overboard discharge valves.

8. Keep the turbine rotors turning continuously.

Preparing to Get Under Way

When word is received to prepare to get under way, engineroom personnel must perform the following duties:

!• Cut in the first-stage air ejector.

2. Close the recirculating line from the deaerating feed tank to the main condenser.

3. Start up the auxiliaries that have been secured or placed in a standby condition.

4. If the overboard discharge valves have been throttled down, open them wide.

5. Disengage the turning gear, and spin the turbines with steam, taking care not to put way on the ship.

6. Report readiness for getting under way.

After Getting Under Way

When the ship is under way, and orders relative to standard speed have been received, engineroom personnel will:

1. Shift to the proper turbine combination for the indicated standard speed.

2. On ships with scoop injection, stop the main circulating pump when the ship reaches a speed of 5 knots or more.

3. Adjust the gland steam to the required pressure.

4. Close the turbine and throttle drains,

5. Regulate the speed of auxiliaries, as required.

6. Check all bearings; be sure that they are properly supplied with lubricating oil.

7. See that the astern valves and their bypasses are tightly closed.

8. If the vacuum is low, locate and correct the cause.

9. Check the shaft revolutions.

10. Take the rotor position reading, and log it as hot clearance.

Coming to Anchor and Securing

The preparations required for coming to anchor differ from one vessel to another. As a general rule, however, it is important to remember that the maneuvering combination MUST be used when coming to anchor. The

maneuvering combination provides sufficient power to allow immediate response to signals from the bridge.

When the ship has completed anchoring or mooring, the main turbines are secured. The throttle valves and main steam stops are closed, and the turbine and throttle drains are opened. Immediately after the throttle valves are closed, the jacking gear should be used to turn the turbine rotors continuously for a cooling-ofF period until the exhaust temperature reaches approximately 140° F. During this period, the lubricating oil pressure must be maintained on all bearings. The main steam line must be thoroughly drained before the drains are closed. The turbine drains are closed after the turbine has completely cooled, within 24 hours of the time of securing.

Maintaining a High Vacuum

A low exhaust pressure (high vacuum) permits a greater expansion of the steam, and thus makes available more energy per pound of steam. In order to obtain the greatest operating economy, the vacuum for which the turbine was designed must be maintained. The following measures will help to maintain an adequate vacuum:

1. Keep the turbine gland packing in good condition.

2. Maintain the gland sealing steam at the proper pressure at all times.

3. Be sure that there are no air leaks in the condenser, exhaust trunks, throttles, gage lines, lines to air ejectors, makeup feed lines, and idle condensate pump packing and valves.

4. Be sure that there is an adequate supply of water in the feed tank from which makeup feed water is being taken by vacuum drag.

5. Be sure that there is an adequate supply of cooling water flowing through the condenser and the air ejector condenser.

CASUALTIES WITH TURBINE INSTALLATIONS

When casualties occur to any part of a prupulsioi* installation or to vital auxiliaries, the officer of the watch should be notified as soon as possible; and he, in turn, should notify the officer of the deck, if the casualty is such that the engines will have to be slowed or stopped. Some of the casualties that may occur with turbine installations are described briefly below.

Vibration and Noise

Abnormal vibration or noise occurring while the turbine is in operation is generally caused by one or more of the following factors:

1. Water being carried over from the boilers

2. Bearing troubles

3. Bent or broken propeller blades

4. Broken or rubbing turbine blades

5. Rubbing of the carbon packing, labyrinth packing, or oil-seal rings.

If the trouble is due to water being carried over, slow down the main engines and find out what is causing the carry-over.

A rumbling sound heard from the turbine when it begins to vibrate is undoubtedly caused by water or other foreign matter in the turbine. If slowing down or correcting faulty boiler operation does not eliminate the trouble, the turbine must be shut down and inspected. A sharp metallic sound may indicate damaged blading.

Difficulty in Turning

Difficulty in turning the turbine after it has been jacked over or has been in operation is generally due to rubbing.

If the rotor does become distorted while standing by, the drains to the condenser should be closed and the pressure of the gland steam brought to between % and 2 psi (gage). The rotor is then spun alternately astern and

ahead, in order to avoid putting way on the ship. Instructions for performing this operation are given in chapter 41 of BuShips Manual.

Failure of Oil Supply

Maintaining the proper lubricating oil pressure to the main engines is highly essential, because extensive derangements of the propulsion machinery can result from the loss of lube oil pressure. The chief causes of oil supply failure are: (1) failure of the system itself, including the main lube oil pumps, and (2) failure of the power supply (steam or electric) to the main lube oil pumps, owing to an operational casualty or other damage to the boilers, steam lines, or electrical equipment.

If the oil supply to the bearings is interrupted or lost, the affected engine must be stopped immediately and steps must be taken to restore the oil circulation. These steps may include any one or a combination of the following :

1. Increasing the lube oil pressure by means of the pump in operation, or if necessary, by another pump

2. Checking and cleaning the lube oil strainers, if necessary

3. Checking the lube oil lines and repairing any leaks or breaks

4. Checking the level of the oil in the sump, and, if the level is low, striking down additional oil

5. Opening and inspecting turbine and gear bearings.

After a main engine lube oil supply failure, the engine should not be operated until an adequate inspection of bearings has been made.

Loss of Vacuum

Loss of vacuum in a plant that has been operating normally can usually be traced to one or more of the following causes:

1. Excessive air leakage into the vacuum system. —Air leakage may be caused by insufficient gland sealing steam; loss of water in the loop seal; an open vent valve on an idle condensate pump; broken vacuum piping (as, for example, a broken vacuum gage line) ; or an open bypass valve or a float which is stuck in the open position on the drain collectinR tank. Air leakage will also occur if you attempt to take on makeup feed from an empty reserve feed tank.

2. Loss OF steam pressure to the air ejectors. — Loss of steam pressure to the air ejectors might be caused by clogged steam strainers, a clogged nozzle, a defective reducing valve, or other defects in the air ejectors or in the steam lines leading to the air ejectors.

3. Insufficient flow of circulating water through THE main condenser. —This may be caused by improper operation of the main circulating pump, or by foreign matter clogging the entrance to the condenser tubes.

4. Flooded condenser. —Failure or improper operation of the condensate pump, caused by such things as leaky glands, governor failure, a partly closed pump discharge valve, etc., may cause the condenser to become flooded. Flooding of the condenser may also be caused by defects in the thermostatic recirculating valves or in the recirculating lines.

5- Insufficient flow of condensate through the AIR ejector condenser. —This may be caused by failure or improper operation of the condensate pump or by failure of the thermostatic recirculating valves.

The cause of any loss of vacuum should be found and the trouble should be corrected without delay.

Throttle Jamming

If an ahead or an astern throttle valve should jam, the officer of the deck should be notified immediately. If the ahead throttle valve jams open, the guarding valve or the main line stop valve should be used to control the engine; or, in an extreme emergency, the astern throttle valve may be opened. If the astern throttle valve jams open, the main line stop valve should be used to control the engine; or, in an extreme emergency, the ahead throttle may be opened.

Failure of Circulating Pump

If the main circulating pump fails while the ship is getting under way or steaming at slow speeds, steam on one engine until the pump that is out of order is repaired. This applies to conditions where immediate repairs can be made. Also, the turbines affected may be warmed up and placed in service without the circulating pump after the ship has attained a speed of more than 5 knots, using the other engine (s). This procedure applies to ships with scoop injection.

Low Steam

If, at any time, the steam pressure suddenly drops, the demand for steam must be cut down immediately. This means closing the throttle valve as far as necessary to prevent the steam pressure from falling to less than 85 percent of the authorized boiler operating pressure. The bridge must be notified of the casualty.

Hot Bearings

Hot bearings may be traced to one or more of the following causes:

1. Improper or insufficient lubrication
2. Grit or dirt in the lubricant
3. Bearings out of line
4. Bearings improperly fitted
5. Poor condition of the bearing or journal surfaces
6. Obstructions in the lube-oil lines.

If the temperature of a bearing reaches or goes above ISO"" F., or if it increases at a rate greater than 5"" per minute from a steady condition, the quality and quantity of the oil must be checked. The supply should be increased and the lubricant further cooled by increasing the flow of circulating water to the coolers. If these measures do not prove effective, the speed of the unit must be reduced, and, when the bearing has cooled sufficiently, the unit must be stopped. If the bearing has been slightly wiped, scraping the bearing surface may be sufficient to restore the bearing to a serviceable condition. If the bearing is severely damaged, a spare bearing must be fitted.

GENERAL CARE OF TURBINES

The maintenance of the turbine installations is as important as their proper operation. With proper attention given to maintenance, the development of abnormal conditions will be less frequent. Let's consider briefly some of the maintenance requirements.

Engines to Be Jacked Daily

When not under steam and not in drydock, the main engines should be turned at least 1^ revolutions every day, with the forced lubrication system in operation. Care should be taken not to bring the turbine to rest in the same position on any two consecutive days. An entry is made in the engineering log when this jacking over has been accomplished.

Inspection of Turbine Interiors

Inspection of interior of turbines is made quarterly, or more frequently if thought necessary. Accessible corrosion is removed from the rotors and casings, and other defects are remedied if practicable. If time permits, all turbines are inspected after each long run or after considerable cruising.

Lubrication of Idle Turbines

When propulsion machinery that is fitted with forced lubrication is not in use, oil should be circulated through the entire system for 15 minutes daily. The circulation of oil should be carried out prior to and while jacking the turbines. This test of the lubricating system is to be logged.

Care of Bearings

In handling the parts having bearing surfaces of babbitt or a journal finish, be careful to prevent other parts from striking or damaging the surface. Any injured spots on the thrust collar or the journal should be smoothed with a fine oil stone and kerosene to restore a smooth finish. Bearings that are to be disassembled for any length of time should be coated with a rust-preventive compound.

SAFETY PRECAUTIONS

The following safety precautions are prescribed by BuShips and must be observed when handling propulsion turbines:

1. Be certain that the lubrication system is in operation before turning over the main engines.

2. Turn the rotors at least every 5 minutes after disengaging the turning gear, while warming up the main engines.

3. Never fail to investigate any noise emanating from a turbine.

4. Do not put way on the ship when spinning the main turbines by steam during the warm-up period.

5. If a turbine vibrates, slow down, investigate, and endeavor to locate the cause.

6. Except in an emergency, do not admit steam to an astern turbine until the steam to the ahead turbine has been secured, and vice versa.

7. In getting under way, be sure that all steam lines are properly drained, in order to prevent water hammer.

8. When steam pressure drops, do not open the throttle to such an extent that the operating pressure of the steam is brought to a dangerously low point.

9. Stop the engines if the oil supply fails.

10. If the throttle valve sticks, close the bulkhead stop as soon as possible.

11. Exercise extreme care to prevent the entry of foreign matter into a turbine opened for inspection.

12. Close the turbine drains about 24 hours after securing, when the turbine is thoroughly

cooled.

REDUCTION GEARS

Reduction gears are classified as single reduction and double reduction. In gears of the single reduction type, a high-speed pinion, driven by the prime mover, engages directly with a low-speed gear that is mounted on the forward end of the propeller shaft, and the entire reduction is accomplished in one step.

Ships built since 1935 have double reduction gears. In this type of gear set, a high-speed pinion, connected to the turbine shaft, drives a large intermediate (first reduction) gear. This first reduction gear is connected either by a solid shaft or by a quill shaft to a small gear (slow-speed pinion) that drives a large (second reduction) gear mounted on the propeller shaft. Double reduction gears are of three'^types: (1) simple, or articulated; (2) nested; and (3) locked train.

The efficient lubrication of reduction gears is of utmost importance. It is essential that oil at the designated operating pressure and temperature be supplied to the gears at all times when they are being turned over, either with or without a load. The lubricating system must be kept clean; particles of lint or dirt, if allowed in the system, are likely to clog the oil spray nozzles. The lubricating oil must be kept free from all impurities and foreign matter in order to prevent damage to the reduction gears. If for any reason the supply of lubricating oil to the gears fails, the gears shall be stopped

and locked until it can be ascertained that the bearings are not damaged. The gears should not be operated again until the normal supply of oil has been restored. When churning or emulsification of the oil develops in the gear case, the gears must be slowed or stopped until the defect can be remedied.

Any unusual noises should be investigated at once and the gears operated with caution until the cause of the trouble is discovered and corrected.

QUIZ

1. What two modes of operation are possible on a vessel that has a cruising turbine?

2. Describe the flow of steam through the propulsion turbines, when the cruising combination is being used.

3. For what purposes is the maneuvering combination used?

4. When the maneuvering combination is being used, what steam passes through the cruising turbine?

5. By what means is a vacuum maintained in the cruising turbine, when the maneuvering combination is being used?

6. How would you describe the compounding of a turbine which consists of one Curtis stage followed by several Rateau stages?

7. Define a reaction stage.

8. How many pressure drops occur in one reaction stage?

9. When warming up propulsion turbines, which turbines should be spun first?

10. Should the turbine rotors be turned before or after sealing steam is admitted to the glands?

11. What air ejector is used to maintain vacuum on the main condenser under the conditions of '^standing by for getting under way"?

12. On a ship with scoop injection, the main circulating pump should be stopped when the ship reaches what speed?

13. What combination must be used when a ship is coming to anchor?

14. At what point in the securing procedure should the turbine drains be closed?

15. What is most likely to cause a rumbling sound in a turbine?

16. How frequently must idle propulsion turbines be turned over?

CHAPTER

ENGINEERING CASUALTY CONTROL

MISSION OF CASUALTY CONTROL

Engineering casualty control is concerned with the prevention, minimization, and correction of the effects of operational and battle casualties to the machinery, electrical, and piping installations^ Its mission is the maintenance of all engineering services in a state of maximum reliability under all conditions of operation.

The first objective under this mission is the effective maintenance of propulsion, auxiliary and electric power, lighting, interior and exterior communications, fire control, electronic services, ship control, firemain supply, and miscellaneous services, such as heating, air conditioning, and compressed air. Failure to provide all normal services will affect the ship's ability to function effectively as a fighting unit, either directly (by reducing its mobility, or its offensive and defensive power) or indirectly (by reducing habitability and thereby lowering personnel morale and efficiency).

The second objective is the minimization of personnel casualties and secondary damage to vital machinery, since minimizing these factors will contribute in a large degree to the successful and continued accomplishment of the first objective.

For more information, you should familiarize yourself with the Engineering Operation and Casualty Control

Manual. This book gives the organization and the procedures to be followed in case of engineering casualties or damage.

FACTORS INFLUENCING CASUALTY CONTROL

The basic factors influencing the effectiveness of engineering casualty control are much broader than the immediate actions or routines applied at the time of the casualty. Engineering casualty control reaches its peak efficiency by a combination of sound design, careful inspection, and thorough plant maintenance, including preventive maintenance and by effective personnel organizing and training. Casualty prevention is the most effective form of casualty control.

Influsnce of Design

Sound design influences the effectiveness of casualty control in two ways: (1) by the elimination of weaknesses which lead to material failure, and (2) by the installation of alternate or standby means for supplying vital services in event of a casualty to primary means.

Both of these factors are given maximum practicable consideration in the design of naval vessels. The second factor is provided in individual units by the installation of duplicate vital auxiliaries, by the use of loop systems and cross connections, and by the installation of complete propulsion plants designed to operate as completely isolated units (split-plant design).

Familiarity of Personnel with Plant Operation

Knowledge is the keystone of casualty control. Maximum knowledge of the details of the engineering installation, from the operating viewpoint, must be imparted to the greatest possible extent to all personnel concerned. Thorough instruction in the proper and normal operating procedures is the foundation upon which instruction in casualty procedures should be based. Complete familiarity with normal operation should be gained by all per-

ejonnel involved before any attempt is made to carry out simulated casualties. (From the design viewpoint, full mformation upon which to base improved designs should be contributed by the operating forces.)

Preventive Maintenance

Preventive inspection and maintenance are vital to successful casualty control, since these activities minimize the occurrence of casualties by material failures. Continuous detailed inspection procedures are necessary not only to discover partly damaged parts which may fail at a critical time, but also to eliminate the underlying conditions, such as maladjustment, improper lubrication, corrosion, erosion, and other enemies of machinery reliability which lead to early failure. Particular and continuous attention must be paid to the following external evidences of internal malfunctioning:

1. Unusual noises,
2. Vibrations,
3. Abnormal temperatures,
4. Abnormal pressures,
5. Abnormal operating speeds.

Operating personnel should thoroughly familiarize themselves with specific normal temperatures, pressures, and operating speeds of equipment corresponding to each normal operating condition, in order that departures from normal will be the more readily apparent. It must not be assumed that an abnormal reading on a thermometer, gage, or other instrument recording operating conditions of machinery is caused by an error in the gage. Each case should be investigated to establish fully the cause of the abnormal reading. The installation of a spare instrument, or a calibration test, will quickly determine if an instrument error exists. All other cases must be traced to their source if preventive maintenance is to be effective. Some specific examples of advance warning of ultimate failure are outlined in the following paragraphs.

Because of the safety factor commonly incorporated in pumps and similar equipment, considerable loss of capacity can occur before any external evidence is readily apparent. Changes in the operating speeds from normal for the existing load in the case of pressure governor controlled equipment should be viewed with suspicion. Variations in chest pressures, lubricating oil temperatures, and system pressures from normal are indicative of either inefficient operation or poor condition of machinery.

In some cases, it is necessary to start additional pumps, blowers, or other auxiliaries when the ship is operating above a certain speed. If past practice or reference to available design data indicates that these additional auxiliaries were previously not needed at this speed, tests should be conducted at the first opportunity to determine the underlying cause for their present need. This might be found either in the auxiliaries or main unit of machinery which they serve. Such departures from normal are indicative of internal wear or other depreciation of the equipment concerned and constitute an adequate reason for disassembly and inspection of the questionable unit at the first opportunity.

Capacity tests on pumps should be run at reasonable intervals where this can be done. With judicious planning these tests can be carried out in more cases than commonly realized.

The tests and inspection called for in various sections of this training course must be conscientiously performed, since they are based on the known requirements of preventive maintenance.

In all cases where a material failure occurs in any unit, a prompt inspection should be made of all similar units to determine if there is apparent danger of the same failure. Prompt inspection may eliminate a wave of repeated casualties.

Abnormal wear, fatigue, erosion, or corrosion of a particular part may be indicative of a failure to operate the equipment within its designed limits of loading, velocity.

and lubrication, or it may indicate a design or material deficiency. In any of the above cases, future inspections to detect repeated damage should be arranged for as a routine matter unless corrective action can be taken which will ensure that such repeated failures will not occur. The inspection interval should be considerably shorter than predictable failure life of the part concerned, based on past performance of the unit.

Strict attention must be paid at all times to the proper lubrication of all equipment, including frequent inspection and sampling to determine that the correct quantity of the proper lubricant is in the unit and that it is in good condition. It is a good practice to take samples of the lubricating oil in all auxiliaries at least daily. Such samples should be allowed to stand long enough for any water to settle.

Unusual quantities of fresh water in the oil normally indicate either poorly fitted or worn carbon packing on turbine-driven pumps. Salt water may enter the oil from salt-water pump glands, from salt-water-cooled oil coolers, or from salt water dripping or spraying on the unit. The presence of salt water in the oil can be detected by drawing off the settled water by means of a pipette and running a standard chloride test. A sample of suflicient size for test purposes can be obtained by adding distilled water to the oil sample and shaking vigorously, and then allowing it to settle before draining off the test sample.

Salt water in the lubricating oil is far more dangerous to a unit than is an equal quantity of fresh water, because of the corrosion effects. Salt water is particularly harmful to bearings. Where units are found to have been subject to salt-water contamination of the lubricating oil, it is essential to drain the oil as soon as possible, flush thoroughly, and refill with fresh oil. Early inspection of the bearings should be undertaken, since even minor corrosion leads to their rapid deterioration.

CASUALTY CONTROL TRAINING

In the preliminary phases of casualty control training, a so-called "dry run" is a useful device for imparting early knowledge of casualty control procedures without endangering ship's equipment by too realistic a simulation of a casualty before sufficient experience has been gained. Under this procedure, a casualty is announced and all individuals required to take action report as though action were taken, except for indication that action was only simulated. Definite corrective action motions can be made, and with careful supervision the timing of individual actions can be made very realistic. Regardless of the state of training, such dry runs should always be carried out before an actual attempt is made to simulate realistically any involved casualty. Similar rehearsal should precede relatively simple casualties whenever an appreciable proportion of men new to the ship are to take part and particularly after an interruption of regularly conducted casualty training such as is occasioned by periods of inacti v^ity during extensive overhaul period.

NEED FOR PROMPT CORRECTIVE ACTION

The speed with which corrective action is applied to an engineering casualty is frequently of paramount importance. This is particularly true in dealing with casualties which affect the main propulsion, steering, and electrical power generation and distribution. Casualties associated with these functions frequently become cumulative in nature, if not quickly corrected, and may lead to serious damage to the engineering installation. Damage of this type cannot be repaired without loss of the ship's operating ability. The commanding officer has the responsibility of deciding whether to continue operation of equipment under casualty conditions where possible risk of permanent damage exists, and such action can be justified only where the risk of even greater damage or loss of the ship may be incurred by immediately securing the

affected unit. In such cases, all possible steps must be taken to shorten the period of hazardous operation.

An example of this principle on a large scale could be the operation of the entire plant with abnormal salinity present, in order that your ship may steam clear of an area of possible enemy attack. A lesser case would be the operation of a main lubrication oil pump or of a main feed pump—although it shows evidences of overheating —during the time required either to secure the engine or boiler concerned or to establish standby service.

It is re-emphasized that whenever there is no probability of greater risk, the proper procedure is to secure the malfunctioning unit as quickly as possible even though considerable disturbance to the ship's operations may occur. Although speed in controlling a casualty is essential, action should never be undertaken without accurate information; otherwise the casualty may be mishandled, and irreparable damage and even loss of the ship may result. War experience has shown that the cross-connecting of intact plants with a partly damaged one must be delayed until assurance is gained that such action will not jeopardize the intact plant. Speed in the handling of casualties can be achieved only through a thorough knowledge of the equipment and associated systems and by thorough and repeated training in the routine required to handle specific predictable casualties.

PHASE NATURE OF CASUALTY CONTROL

The handling of any casualty can usually be divided into three phases—limitation of the effects of the damage, emergency restoration, and complete repair.

The first phase is concerned with the immediate control of the casualty to prevent further damage to the unit concerned and to prevent the casualty from spreading through secondary effects.

The second phase consists of restoration in so far as practicable of the services which were interrupted as a result of the casualty. In many cases, this phase when

completed eliminates all operational handicaps, except for the temporary loss of standby units—i. e., ability to withstand further failure. If no damage to, or failure of, machinery has occurred, this phase usually completes the operation.

The third phase consists of making repairs which will completely restore the installation to its original condition,

SPLIT-PLANT OPERATION

The fundamental of engineering damage control is SPLIT-PLANT operation. It follows the wisdom of the old adage of "not putting all your eggs in one basket,'* and its purpose is to minimize the damage that can be done by any one hit.

In most of our larger vessels built primarily as combat ships, there are two or MORE complete engineering plants. This is true of carriers, battleships, cruisers, destroyers, and destroyer escorts.

Suppose, for example, that you had four boilers and two turbine plants. The boilers would be grouped in pairs, each pair located in a watertight compartment of its own. Likewise, each turbine plant would be located in a watertight compartment of its own. But bear in mind that locating boilers and engines in different compartments has little or nothing to do with split-plant operation. If the engineering plant was so designed, you could still operate split-plant if all boilers and engines were located in the same compartment.

To place all boilers and engines into a single compartment, however, would result in a dangerously large compartment—dangerous because of the tremendous amount of BUOYANCY that would be lost if the area were flooded. Furthermore, such flooding would put out all boilers and all engines. To achieve buoyancy and stability and to protect machinery units against the grave effects of flooding, machinery spaces are usually subdivided as follows (working from forward to aft) :

Old Design Boilerroom Boilerroom Engineroom Engineroom
Present Design Boilerroom Engineroom Boilerroom Engineroom

Split-plant operation means dividing your boilers, engines, pumps, and other machinery so that you have two or more engineering plants, each complete in itself, and each operating its own fuel oil pumps and source of supply. The FORWARD pair of boilers would supply steam to the forward main engines, and the after pair of boilers would supply steam to the after main engines. Each turbine installation would be equipped with its own condenser, air ejector, lubricating oil pump, and other auxiliaries. Each engineering plant would operate its own propeller shaft. Then, if one pair of the boilers or one turbine or one propeller were put out of action by explosion, shellfire, or flooding, the other plant could probably continue in operation, driving the ship ahead at somewhat more than half speed.

Split-plant operation is not a good luck charm against devastating damage that would completely immobilize the entire engineering plant, but it will reduce the chances of such a casualty and prevent an injury to one plant from being transmitted to another or seriously affecting its operation. Thus, if you were not operating "split plant" and a shell ruptured the main steam line to the forward turbine, you would lose steam from all boilers. If you were operating split plant, with the cross-connecting valves closed, you would lose steam from only one set of boilers. Such operation is like having two automobiles clamped together side by side. If the engine of one car goes dead, the engine in the other car can provide motive power for both cars.

Split-plant operation has proved itself on more than one occasion. When a torpedo explosion wrecked one engineroom and destroyed one-half the boilers on a cruiser, the ship was able to steam away under its own power

by using the undamaged plant. Had the plant not been split, steam from the undamaged boilers would undoubtedly have escaped through the wrecked engineroom, resulting in the loss of propulsive power and the power necessary to control flooding, and possibly resulting in loss of the ship.

Failure on your part to understand the full meaning of split-plant operation may be the cause of unwarranted CASUALTIES. This lack of understanding caused heavy damage to one destroyer's main engine. The ship was operating with the plant split, when one boiler lost feed water. The Boilerman on watch, in the process of handling the low-water casualty, secured the bulkhead steam stops of both boilers. By so doing he cut off all steam to one engineroom, including the steam supply to the auxiliary machinery. The lubricating oil pump stopped, with the engine still turning over from the drag; as a result, the main engine bearings wiped, and the turbine blading was ruined. Steam should have been maintained to that engineroom by securing the stops only on the boiler that had low water, and the engineroom should have been notified to reduce speed.

In another case a cruiser was operating with the plant split when a shell penetrated a fuel oil service tank below the waterline. As a result contaminated fuel oil was fed to half the boilers, and the fires went out. But the ship was still able to make adequate speed on one-half her engines, and all would have been well had not someone opened the cross-connection leading to the tanks from which the other boilers were receiving their fuel oil. The result was that contaminated oil was fed to these boilers, the fires went out, and the ship was forced to stop.

Split-plant operation may have to be suspended because of a casualty. A destroyer steaming at high speed with all boilers in operation might be forced to secure one boiler because of a ruptured tube. The ship could continue steaming with two boilers supplying one engine, and one boiler supplying the other; but this would cause too great

a difference in shaft speeds. Cutting the three intact boilers into both main engines would remedy this condition, although it would render the plant more vulnerable.

The foregoing examples prove the value of operating ships with the machinery plant split, and they also show you the necessity of a complete understanding of this type of operation. The isolation or segregation of vital eng-i-neering systems may be considered as the initial step in splitting the plant. In the following sections we'll have more to say about that phase of engineering casualty control.

Cross-Connecting Valves

The main and auxiliary steam lines are provided with bulkhead and cross-connecting valves so that, by proper operation of these valves, any boiler or group of boilers, either forward or aft, may supply steam to one or all enginerooms. These most important valves should be kept in the best possible condition for quick emergency operation. When operating under battle conditions or in confined waters, the cross-connecting valves are closed, so that the engineering plants are completely isolated from each other, and are operating split plant.

Split-plant operation applies also to the electrical plant, to the boiler feed water system, and to the boiler fuel oil system.

Fuel Oil System

The fuel oil system is generally so arranged that by means of fuel oil booster and transfer pumps, suction can be taken from any fuel oil tank on your ship and the oil pumped to any other fuel oil tank. Fuel oil service pumps are then used to supply oil to the boilers from the service tanks. In split-plant operations the forward fuel OIL SERVICE PUMPS of your ship are lined up with the forward service tanks, and the after service pumps are lined up with the after service tanks. The cross-connection valves in the fuel oil transfer line must be closed except when oil is being transferred.

The main reason for securing the fuel oil cross-connecting lines is to prevent major casualties. For example, if all your boilers were being supplied fuel oil from the same set of tanks and one tank was ruptured by a torpedo, or a near miss bomb or shell, the results would be disastrous. The entire fuel oil system would become contaminated with salt water, the fires under your boilers would go out, propulsive power would be lost, and your ship would stop dead in the water.

Some ships are provided with sluicing lines that make it possible to sluice oil (or water) from one side of the ship to the other. The valves in these lines are normally kept closed to prevent the sluicing of liquid to the low side of the ship. On certain types of ships these lines may not be designated as sluice lines. Actually they may be common suction lines, but since they present all the dangers of the sluice lines, you'll have to observe the same precautionary measures with them as with the sluice lines.

Most fuel oil service tanks are provided with both high and LOW suction connections. Inasmuch as water is heavier than oil it tends to settle to the bottom of the tanks; and, as water is not uncommon in oil, the upper suction should be used at all times when supplying the boilers. Remember, however, that even though the tanks from which the suction is being taken may have a capacity of 10,000 gallons, this amount is not available for use if you are lined up with the upper suction connection. All that is available is the amount of oil above the upper suction level. You must be certain that you do not lose suction because you have overlooked this factor. Take steps to see that standby fuel oil service tanks are ready for immediate use. The standby tank should be filled with oil that has been tested for and found to be free of water. Be sure that the Boilerman in charge of the fireroom watch knows which tank has been designated for standby use, how much oil it contains, and how to shift suction to that tank.

Feed Water System

The same general principle of split-plant operation applies to your boiler feed water system, since it would be injurious, if not disastrous, to supply all boilers with sea water via a ruptured condenser, or to lose feed pressure entirely to all boilers because of a casualty to one de-aerating feed tank.

During split-plant operation the main feed system may be divided into two or more separate and complete systems. In an operation of this type the after boilers would be supplied from the after deaerating feed tank by means of the AFTER BOOSTER and MAIN FEED PUMPS; the same principle would apply to the forward system. In splitting the plant, use as little of the feed piping system as possible and isolate the rest of the system so as to reduce the area subject to casualties. Keep the emergency feed pumps warmed up and in a standby condition, ready to supply feed water to the boilers in the event of a casualty to the main feed and booster pumps. Emergency feed pumps are so arranged that they can take either a '*hot suction" from the booster pumps or a "cold suction" from the reserve feed tanks.

Main Steam System

The main steam system varies to a certain extent on the different types of ships. Certain types of ships may have the forward and the after systems CROSS-CON nected to I'orm a complete system throughout the enginerooms and urerooms, whereas other types have independent for-vard and after systems. The method of supplying steam 0 the propulsion turbines, however, remains essentially .he same. The forward boilers supply the forward main engines and the after boilers supply the after main engines. In ships having the cross-connection system, cut-out valves are provided in the cross-connection for the purpose of dividing the engineering plant for split-plant operation. Valves are also provided at strategic points, as at bulkheads and at the boilers, to permit effective isolation in case of damage.

FIREROOM CASUALTIES

Typical casualties and the general procedure recommended by BuShips for control of each are given in this section. It is desirable for the individual ship or class of ships to formulate special procedures applicable to their own plant, designating for each space its duties in controlling a particular casualty. With a well-trained crew many of the steps necessary for the control of a given casualty may be carried out almost simultaneously. Telephone talkers should be trained in the proper phraseology to be used in informing the engineroom of the casualties. When headsets are manned, it will be the talker's duty to inform the engineroom of the casualty and of the steps taken to control it.

It is to be borne in mind that any fireroom casualty that results in the securing of a boiler may also result in one or more associated or secondary casualties. Progressive casualties are of frequent occurrence. Casualties that are most frequently associated with the securing of a boiler are: (1) the loss of auxiliary steam pressure and steam-driven auxiliaries, (2) the loss of electric power and electric-driven auxiliaries, (3) the lowering or loss of the lube oil pressure—a serious casualty, (4) loss of feed booster pressure, (5) loss of feed pressure, (6) loss of vacuum, and (7) loss of auxiliary exhaust.

Casualties In The Control of Water Level

When the water level is too high, priming will occur. This is especially true when the steam demand is high and fluctuating rapidly. The proper water level to be maintained must be determined by experience with the installation in question.

Disappearance of the water level from the gages

MUST BE treated AS A CASUALTY REQUIRING IMMEDIATE

SECURING OF THE BOILER. In this connection, it is emphasized that differentiation between a full gage glass and an empty gage glass by mere observation is difficult, and that serious low-water casualties have occurred as a result of

a mistaken assumption of high water level (blowing down when a low water condition already exists). By careful observation it is sometimes possible to make such differentiation through presence or absence of condensate running into or down the inside of the gage glass. Presence of such condensate entering the glass is indicative of low water—i. e., an empty glass. However, the boiler must be secured whether the water is high or low. Then, if any doubt exists as to the location of the water level, the gage glass cut-outs and drain valve should be used to determine its location.

High water in boiler. —When the water level is above the 18-inch glass the following action should be taken :

1. Close throttles and stop shaft. Trip turbogenerator.
2. Secure feed check and stop valves.
3. Secure burners and air supply to boiler.
4. Close main, turbogenerator, and auxiliary steam stops.
5. Open cross-connection valves as directed.
6. Use the surface blow valve to blow the boiler down until the water is at the proper level.
7. Light off and cut in the boiler according to the light-ing-off procedure.

Low WATER.—This is one of the most serious and most frequent of fireroom emergencies. It is generally the

RESULT OF inattention ON THE PART OF THE BOILERMAN AND THE MAN TENDING THE CHECKS, OR OF THEIR ATTENTION HAVING BEEN DIVERTED TO SOME DUTIES OTHER THAN THE MAJOR ONE—THAT OF MAINTAINING THE PROPER LEVEL IN THE BOILER.

Low water in the boiler can result unless the following defects are corrected as soon as they are discovered: (1) failure of feed pumps, (2) the developing of leaks in the feed discharge Hne or elsewhere, (3) a defective check valve, (4) low water in feed tank, (5) defects causing incorrect water level to be shown.

Low STEAM PRESSURE DUE TO LOW WATER.—When the

water level in a boiler falls low enough to uncover por-

tions of the tubes, the heat-transfer surface is reduced, and, other conditions remaining the same, the steam pressure will drop. Ordinarily a drop in steam pressure is a result of an increase in the use of steam, and the natural tendency is to counteract it by cutting in more burners. This procedure is correct in such cases. If, however, the fall in pressure is due to low water, accelerating the combustion will result in serious damage to the boiler and possibly injury to the fireroom personnel. The possibility that a fall in pressure is an indication OF LOW WATER MUST ALWAYS BE BORNE IN MIND; WHENEVER THE STEAM PRESSURE TAKES A PRONOUNCED OR AN UNUSUAL DROP, THE REASON FOR WHICH IS NOT APPARENT, THE LEVEL OF WATER IN THE WATER GAGES MUST BE CHECKED BEFORE ADDITIONAL BURNERS ARE CUT IN.

Other effects of low water. —Maintaining a hot fire under a boiler in which there is insufficient water to absorb heat given off by the furnace is bound to cause warping of the boiler casing, distortion of the boiler-heating surface, destruction of the boiler brickwork, or serious steam and water leaks, with consequent danger of boiler explosion. In the event of low water, it is essential that no attempt be made to restore the normal water level by increasing the supply of feed. The boiler should be allowed to cool gradually, so that any parts that may have become overheated will be subjected to an annealing process, minimizing damage to the boiler pressure parts.

When the water level drops out of sight in the 18-inch glass the following action should be taken:

1. Cut off the fuel oil supply to the burners.
2. Close the feed check and stop valves.
3. Notify the engineroom.
4. Close the boiler steam stop valves and open the cross-connection valves as ordered.
5. Lift the safety valves by hand, to gradually relieve the pressure.
6. Let the boiler cool slowly and carry out the procedure as directed in chapter 51 of BuShips Manual. Loss OF FEED SUCTION.—If the feed suction is lost the following action should be taken:

1. Start the emergency feed pump on cold suction. (In a standby condition this pump should always be lined up on cold suction.)
2. Ring for booster and/or feed pressure.
3. Notify the engineroom.
4. If so ordered, open the main feed pump discharge cross-connection valve and take feed from this source if ordered.
5. Carry out the low-water procedure in the event that the water level cannot be maintained.
6. Keep the engineroom informed of the feed system in use, so that ship's speed can be held within the limits of boiler capacity.

When low feed pressure occurs the following procedure should be carried out:

1. Increase the feed supply by using the emergency feed pump.
2. Ring for more feed pressure.
3. Check the discharge pressure of the main feed pumps and booster pumps, the water level in the deaerating feed tank, the pressure of the deaerating feed tank, the auxiliary exhaust pressure, the feed check valves on idle boilers, and the operation of the constant pressure governor on the feed pump.
4. Notify the engineroom, and open the feed cross-connection valve if ordered.
5. Carry out low-water procedure, in the event that the water level cannot be maintained.
6. Keep the engineroom informed of how feed is being taken (main or emergency so that the ship's limiting speed may be estimated).

When a water gage glass on a boiler carries away, the following action should be taken:

1. Close the top and the bottom cut-out valves to the damaged glass.

2. Use the other glass to observe the water level in the boiler.

3. Open the drain on the water column of the damaged glass.

4. Renew the damaged glass.

5. Slowly open the bottom cut-out valve and then the top cut-out valve. Look for leaks. If there are no leaks, close the drain and compare the water level with that in the other glass.

Boiler Tube or Other Pressure Part Carries Away

When a boiler tube or other pressure part carries away the following action should be taken:

1. Secure the fuel oil burners.

2. Speed up the blowers to carry steam up the smoke-pipe.

3. Notify the engineroom.

4. Secure the boiler stops and open cross-connection valves as ordered.

5. Open the safety valves by hand to relieve steam pressure.

6. Continue to feed water to the boiler until it has cooled. Open the auxiliary check valve and start the emergency feed pump. The main feed supply must be shut off, if other boilers are being fed from it. Caution: Do NOT continue to feed water to the boiler if the casualty was caused by low water, or if the steam leak is so large that the water level cannot be maintained in the gage glasses.

7. Secure all air to the boiler as soon as the pressure has decreased and steam is no longer escaping into the fireroom. Allow the boiler to cool slowly.

Casualties In The Fuel Oil System

Loss OF SUCTION.—When the burners sputter, fires die out, or the fuel oil service pump suddenly starts racing, the fuel oil suction has been lost and the following action should be taken:

L Secure all burners, leaving at least one register on each side (superheater and saturated) open to expel any gases and to supply air for combustion of any oil that may have accumulated on the furnace decks. Keep forced-draft blowers running with approximately 2 inches air pressure in casing.

2. Start the standby pump on the standby service tank.

3. Notify the engineroom.

4. Open the cross-connection valves if directed by the engineroom.

5. Close the main, turbogenerator, and auxiliary steam stops.

6. Note the service pump discharge pressure. If there is water in the oil, the noise level of pump operation will very likely increase and there may be an increase in speed. If the pump is airbound, little or no pressure will be indicated and the pump operation will be extremely noisy; open the priming cock and the vent system. If there is water in the oil, run oil to overboard or, if the piping arrangements permit, to the contaminated oil tank.

7. Close the registers and slow down the forced-draft blowers after all oil has been burned from the furnace decks.

8. After it has been determined that good oil is available, light off and put the boiler on the line according to the lighting-off procedure.

9. Determine the cause of derangement.

Note. Modification of the above procedure will depend upon the steaming rate, in port or at sea, and split or open plant operation.

In the above procedure, the original casualty was isolated when the boiler stops were closed. Secondary casualties would have extended to all spaces where steam or electric auxiliaries were in use whose power was supplied by the secured boiler. The number of secondary casualties would be determined by the thoroughness and speed

with which the operating personnel handled the shift over at their stations.

Failure of fuel oil service pump. —In case the service pump fails the following action should be taken:

1. Cut in the standby pump. If that also fails, cut in the port-and-cruising pump, if installed. As a last resort, use a hand pump in order to maintain steam pressure until the difficulty can be located and corrected.

2. Notify the engineroom and open the cross-connecting valves if ordered.

3. Close the boiler stop valves if service cannot be reestablished immediately with the fuel oil service pump.

Note. If an electric (port-and-cruising) pump is placed in service, notify the engineroom so that the ship's speed can be held within the limited boiler capacity.

Ruptured fuel oil line. —When the fuel oil service line is ruptured on the suction side of the pump, the following procedure should be carried out:

1. Secure the fuel-oil service pump.

2. Notify the engineroom.

3. Secure the boiler (s).

4. Have fire-fighting equipment at the scene and in readiness.

5. Wipe up spilled oil, and/or flush the bilges as soon as possible.

6. Line up and use a different section of the service suction line.

When the fuel-oil service line is ruptured on the discharge side of the pump, the following action should be taken:

1. Secure the fuel-oil service pump, the discharge valve, and the quick-closing valve.

2. Notify the engineroom.

3. Secure the boiler (s).

4. Have fire-fighting equipment at the scene and in readiness.

5. Wipe up spilled oil and/or flush the bilges as soon as possible.

6. If it is necessary to operate temporarily with small leaks, smother the spray and catch the oil in a bucket or pan.

Oil in fuel oil heater drain. —When oil is sighted in the fuel oil heater drain, take the following action: 1- Shift drain to bilge.

2. Notify the engineroom.

3. Shift to another fuel oil heater.

4. When drain is clear of oil, shift the drains back to the drainage system.

5. Report the defective unit to the engineroom. Casualties that are frequently associated with the presence of oil in the fuel oil heater drain are:

1. Fuel oil in the deaerating feed tank;

2. Fuel oil in the boilers and in the feed system;

3. Contaminated drain line;

4. Fuel oil in the feed bottom.

Flarebacks

A flareback is likely to occur whenever the pressure in the furnace momentarily exceeds the pressure in the boiler air casing. Flarebacks are caused by an inadequate air supply for the amount of oil being supplied, or by a delay in lighting the mixture of air and oil.

Situations which commonly lead to flarebacks include: (1) attempting to light off or to relight burners from hot brickwork; (2) gunfire or bombing which creates a partial vacuum at the blower intake, thus reducing the air pressure supplied by the blowers; (3) forced draft blower failure; (4) accumulation of unburned fuel oil or combustible gases in furnaces, tube banks, uptakes, or air casings; and (5) any event which first extinguishes the burners and then allows unburned fuel oil to spray out into the hot furnace. An example of this last situation might be a temporary interruption of the fuel supply which would cause the burners to go out; when the fuel oil supply returned to normal, the heat of the furnace might not be sufficient to relight the burners immediately. In a few seconds, however, the fuel oil sprayed into the furnace would be vaporized, and a flareback or even an explosion might result.

To reduce the danger of flarebacks, the following PRECAUTIONS MUST BE OBSERVED:

1. Oil must not be allowed to accumulate in the furnace. All oil on the furnace floor must be wiped up and the furnace blown through with steam or air before lighting off the atomizers. The atomizer valves on secured burners must be kept tight at all times to prevent leakage of oil into the furnace.

2. Whenever atomizers are accidently extinguished, shut off the oil and blow through the furnace with steam or air before relighting the atomizers.

3. Never attempt to relight an atomizer from a HOT brick wall. USE A TORCH.

4. The man handling the torch must stand well clear to avoid injury in case a flareback occurs.

The following action should be taken for a major flareback which involves possible boiler damage and/or extinguishment of furnace fires:

1. Close the quick-closing master fuel oil valve to the burner manifold and the burner supply valves, and notify the engineroom.

2. Use fire-fighting equipment as necessary.

3. Secure the boiler main steam stop.

4. Adjust feed check valves or stop feed pumps as necessary to prevent high water.

5. Speed up forced-draft blowers to purge the furnace of unburned gases and burn any oil accumulations from the furnace floor.

6. Open the auxiliary steam cross-connection valves if necessary to maintain auxiliary steam service, and secure the auxiliary steam stop.

7. Inspect the boiler and light off, using normal procedure if conditions permit.

Minor flarebacks, or those which do not involve furnace tires being extinguished, do not necessarily require carrying out the above procedure. However, careful inspection of the boiler and surveillance of its operation should be carried out to ensure that fallen brickwork, etc., will not result in further damage if the boiler is continued in use.

High Salinity

When it is necessary to reduce the salinity in a steaming boiler the following steps should be taken:

1. Open the bottom blow guarding valves.

2. Open the bottom blow valve on the mud drum and blow down the boiler. Do not let the water level get below a safe level in the gage glass. Continue successive short blow-downs until the desired reduction in salinity has been achieved or until the quantity of make-up feed available for blowing down has been consumed.

3. Add Navy boiler compound to the boiler to bring up the boiler water to prescribed

alkalinity.

4. Do not blow down the division or water wall header while the fires are under the boiler. Reduction of boiler water salinity can be made by blowing down from the water drum.

Casualties to the Refractories

The following action should be taken when brick or plastic falls out of the furnace wall:

1. If practicable, secure all burners.

2. Notify the engineroom. Open the cross-connection valves as ordered.

3. If it is necessary to continue operating the boiler until another can be brought in on the line, cut out the burners adjacent to the damaged section, so as to avoid injury to the boiler casing.

Failure of Forced-Draft Blower

Should a forced-draft blower fail, the following action should be taken:

1. If one of two blowers in use should fail, speed up the other blower,

2. If one blower is in use, secure the burners at once to avoid a flareback.

3. Start the standby blower.

4. Notify the engineroom, because of possible need for limitation of or reduction in speed.

Fires

An oil fire may be caused by the ignition of the oil or oil vapor in any place where oil is allowed to collect by leakage from the system. Oil may accumulate in the double fronts from partially plugged atomizers, by continued drip when atomizers are secured and not removed from the burners, or from excessive deposition of carbon in registers or furnace opening rings. Such accumulations are potential sources of fire. Frequent observations through burner and furnace sight glasses and the correct settings of atomizers and register doors are helpful in preventing these occurrences.

Precautions. The following precautions must be taken:

1. Do not allow oil to accumulate in any place. Particular care must be taken to guard against this accumulation in drip pans under pumps, in bilges, in the furnaces, on the floor plates, and in the bottom of air-encased boilers. Should leakage from the oil system to the fireroom occur at any time, immediate action should be taken to shut off the oil supply by means of the quick-closing valves provided and to stop the oil pump.

2. Absolutely tight joints in all lines are essential to safety. Immediate steps must be taken to stop leaks whenever discovered.

3. No lights should be permitted in oil-burning fire-rooms, except electric lights (fitted with steamtight globes, or lenses, and wire guards), and permanently fitted smoke indicator and water gage lights. If work is being done in the vicinity of flammable vapors, or if rust-preventive compound or metal-conditioning compound is being used, all portable lights should be covered with steamtight globes as well as protected with wire or rubber guards.

4. During repair work in fireroom spaces, open-flame naked lights, hot work, and portable lights which are not protected by steamtight globes are permitted only after these spaces have been freed of flammable and explosive gases and liquids. The spaces should be inspected and when they are declared safe, a sign should be posted. The sign should not only state that the space is safe but should also limit the effective time.

Fire-fighting equipment. —The fireroom force must be thoroughly drilled to handle an oil fire promptly and efficiently, as it is but a matter of minutes before an oil fire assumes serious proportions. The Chief Boilerman should ensure that standard fire-fighting equipment is available and ready to use in all oil-burning firerooms. In most naval installations, permanently fitted steam-smothering lines are provided in the fireroom bilges and inside the casings beneath the boilers of air-encased boilers. These lines should be examined and tested by steam at least monthly to ensure satisfactory operating condition. Particular attention should be given to the condition of the line in casings beneath the boiler, because of its importance in the event of a casing fire.

Fuel oil fire. —When a major fire occurs in the fireroom and prevents operation of the boiler, the following action should be taken:

1. Close the quick-closing master oil valve and stop all fuel pumps.
2. Secure burners and close registers.
3. Stop forced-draft blowers.
4. Secure the ventilation supply and exhaust fans.
5. Use CO_2 hose and reel, foam, fog nozzle, and steam smothering, in that order.
6. Notify the engineroom, and open the cross-connection valves if ordered.
7. Secure the boilers.

Note. Both supply and exhaust ventilating fans should be secured except when ventilation might alleviate heat and gases interfering with fire-fighting personnel.

When a decision has been made to abandon a space, it may prove advantageous and even essential to continue to operate the fans, accepting the additional supply of oxygen supporting combustion in the interest of clearing the escape path of smoke. Whether both supply and exhaust fans, or exhaust fans only, need be started will depend on existing circumstances. For example, if the main access or escape trunk is open, it may be desirable to start only the exhaust fans, thereby clearing the access of smoke. On any occasion the decision concerning ventilation rests with the fire-fighting personnel.

Fire in boiler casing. —The following action should be taken when a fire occurs in the boiler casing:

1. Secure the burners and stop the fuel oil service pumps.
2. Stop the forced-draft blowers.
3. Turn on the steam smothering in the casing.
4. Notify the engineroom and open the cross-connection valves if ordered.
5. Secure the boiler.
6. After the fire has been extinguished, investigate, and remedy the cause. Report the amount of damage. Remove any remaining oil or vapors in casings or underneath boilers. Light off, using normal procedure if conditions permit.

Class C Combustibles. —The electrical equipment known as class C combustibles is widely distributed in naval vessels. This electrical equipment may ignite by short

circuits or by friction (static electricity), or class A or B fires might extend to and ignite it. When such equipment cannot be de-energized, the danger of electrical shock must be guarded against. While distilled water is not a conductor of electricity, fresh water is, because it usually contains minerals. The conductivity of sea water is many times greater than that of fresh water. Consequently a solid stream of water must not be used to extinguish class C fires. This hazard is greatly reduced, but not entirely removed, when the water applied is finely divided, as it is in fog.

Carbon dioxide does not conduct electricity, and it is, therefore, the indicated extinguishing agent for class C fires. Fog would be a second choice. Water or foam would cause damage to electrical equipment.

Class C fire in lighting or power panel. —When this type of fire occurs the following action should be taken:

1. Notify the engineroom and the distribution board to have the panel de-energized.

2. Fight the fire with carbon dioxide.

3. Have a fog hose in readiness in case the fire should spread to other combustibles.

4. When the fire is extinguished, have the electrician investigate, report, and repair the damage.

5. Have the circuit re-energized, and stand by with fire-fighting equipment until danger of re-flash is over.

QUIZ

1. What five general types of abnormalities give external evidence of the internal malfunctioning of machinery?

2. When material failure occurs in any unit, what should be done with respect to all similar units?

3. How frequently should samples of the lubricating oil be taken from all auxiliaries?

4. What kind of trouble is normally indicated by an unusual quantity of fresh water occurring in the lube oil of a turbine-driven pump?

5. Should realistic simulation of casualties be used for training purposes if the crew is very inexperienced?

6. Who has the responsibility for deciding whether to continue operation of equipment under serious casualty conditions?

7. What is the only possible justification for continuing to operate machinery under casualty conditions which involve possible risk of permanent damage?

8. What must be done immediately if there is any doubt whatever concerning the location of the water level in the boiler?

9. What is indicated by the presence of condensate running into or down the sides of the water-gage glass?

10. If steam pressure drops markedly for no apparent reason, what must you do before cutting in additional burners?

11. In the event of a low water casualty, why is it essential that you NOT increase the supply of feed in order to restore the normal water level?

12. What is the proper standby condition of the emergency feed pump?

13. If fuel oil suction is lost even though the oil level in the tank is above the suction line, what are the two most likely causes of the trouble?

14. If it is necessary to operate temporarily with small leaks in the fuel oil discharge line, what precautions should be observed?

15. What type of casualty is likely to occur when the pressure in the furnace exceeds the pressure in the air casing?

16. What is the first step to be taken when oil is sighted in the fuel oil heater drain?

17. If brick or plastic falls out of the furnace wall under conditions which necessitate the continued operation of the boiler, what action must be taken?

18. What danger exists in using a solid stream of water to extinguish a class C fire?

19. Is finely-divided water (fog) any safer than a solid stream of water, for use in fighting a class C fire?

20. What is the best extinguishing agent for a class C fire?

CHAPTER

RECORDS AND REPORTS

A MEANS TO AN END

Records and reports are necessary for the successful administration of the Engineering Department of any naval activity. In this chapter we discuss those which apply most directly to a ship. The basic data for all engineering records and reports aboard ship originate in the various machinery spaces. These data are required for all operations, repairs and alterations, casualties, material analysis, and various tests and inspections. Some of the records and reports will be prescribed by BuShips while others will be originated by the engineer officer of your ship. The responsibility for compiling the necessary data for records and reports will fall on the shoulders of the leading petty officers of your department. It is well to bear in mind that recordkeeping is not an end in itself but the means to an end. This chapter contains a general discussion and description of the more important records and reports.

MACHINERY INDEX

The Machinery Index is a comprehensive listing of all machinery and equipment, other than electronic equipment, installed on board each vessel. The data included in the index are required by the Bureau to provide adequate repair parts, battle damage components, and re-placement equipment to forces afloat. It is the basis for the maintenance of allowance lists, supply-demand reviews, and preparation of usage factors.

MATERIAL HISTORY

The material (machinery) history is a record of all repairs, alterations, inspections, derangements, measurements, renewal of parts, name plate data, length of service, and other pertinent data on each machinery unit.

If kept correct and up to date, the material history is the most valuable record found in the engineering department ; but if it is neglected it becomes a big headache to all personnel concerned. The material history is inspected during each administrative, material, and Board of Inspection and Survey inspection.

The following cards, of the loose-leaf binder type, are available for maintaining the material history:

Machinery History NavShips 527
Electrical Machinery History NavShips 527A
Unit Record Card NavShips 528
Repair Record Card NavShips 529
Alteration Record Card NavShips 530
Megger Test Record NavShips 531
Blank Utility Card NavShips 532
Bearing Record NavShips 533
Electronic Equipment History Card NavShips 536
Record of Field Changes NavShips 537
Tube Performance Record NavShips 538
Hull History Card NavShips 539

Of the above-listed forms, four (NavShips 527, 527A, 536, and 539) constitute the basis of the ship's material history: machinery history, electrical machinery history, electronic equipment history, and hull history.

Some of the cards listed here are of little interest to B division personnel.

Those cards which are of concern to the B division may be briefly described as follows:

Machinery history card. —The machinery history card, shown in figure 18-1, is the basic card used in setting up or revising the machinery history. The sequence of arrangement of the cards is based on the machinery index, which is in turn based on the Navy Filing ManiuxL A card is made out for each unit of machinery and for the major component parts.

The machinery history cards are placed in large binders which have a number of dividing fiber sheets with tabs. A group of cards is arranged for each dividing sheet of the binder. The individual cards are arranged from top to bottom in order of their filing numbers. This leaves the filing number and title of each card visible so that any card can be readily located. The tabs on the dividers in the binder are also numbered, so that each group of cards can be quickly found.

The original card is numbered 1, as shown in the top right corner of figure 18-1. When the space for remarks, machinery history, has been filled on this card it is continued on a blank utility card, which is numbered 2. This card is placed directly in back of the machinery history card. Other cards that are placed in back of the basic machinery history card are: Bearing record card,

UNIT record card, MEGGER TEST CARD, ALTERATION RECORD CARD, and the repair record card.

The alteration record and repair record cards are known as the CSMP cards, and have already been explained in chapter 13 ("Navy Repair Procedures") of this training course. They have an extended tab on the top left corner. When alterations and repairs are pending, these tabs are left showing above the machinery history card.

The following information is found on the machinery history card. (The numbers correspond to those encircled on the card illustrated in fig. 18-1.)

1. This is the machinery index number of the unit for which the card is prepared.

2. This entry gives the number and name of the unit.

I 2LCVI31. fO&CXD DHAUT

0-

#1

5-1 Stiu-board S14«, 1^. N DO-445-S5300-a '^^ 0 "5^53-1, 553-C. 553-3

Zm«%ruo(lo« look, BuSblp*

®

B-l Port Sid* fr, 3d4(MO. Jlook^M. J4fifi_

0

(n^^f '"* C^4J tf*«

S)0 ClU its P

tf*«tlQC)K>Ufl* Ilvctrlo Nfc.Co. Ublt Serial Mo. l-i^990 5-225 C^. (ralt^) 19300 CIM ITS P ?5.60 ina, vatar Gtp. (ovarload) 24«00 C7M at S P 38.00 ini. vater

(ratad) 5076 Powar ratla^ (ratad) 125 OT Po%ar ratlr^ (© ▼•rl<»«A) 22^ JHT Trpa^ _, .Propallar, OparatlD* Stag* praapura'"*"- 57& PSI. ^^-i-w

®

t-ls-iS S«a««*il ilihoi la lubt oil eooltr.

1-23-M Ttittd culoc r<ll<r tiIt* ud ut to lift Kt aS rSI. 3-I^&6 B»n«w«d btfarl&c, turtlaa .ad; el*a0«d flu«h«d lub* oil vltMl^

ioo«

19t3

I

Figura —Machinery hittory card.

3. The number placed in this space will be "1" on all -original cards in the set. Subsequent cards will be

numbered consecutively "2," "3," and so on.

4. This space was originally intended for the subject unit of the card. It is no longer used, as the subject now appears in space No. 1.

5. This entry gives the name and number of the compartment in which the unit is located.

6. In this space is entered the number of the page(s) on which the unit is listed in the BuShip's Allowance List.

7. This is the BuShip's plan number for the unit, or the first plan of a series of plans for a unit.

8. This space is for the alteration number if shown on BuShip's plans.

9. An entry here would represent the piece number. This space is normally left blank.

10. This is the manufacturer's drawing number (if given).

11. This is the spare parts box(es) number.

12. This entry gives the location of the spare parts boxes.

13. This space is for pertinent data such as the tile numbers of any letters concerning operation ana maintenance; repair parts list; manufacturer's instruction book number.

14. This entry is the name-plate data of the unit. (The data on the name plate should correspond with that in the machinery index.) When necessary this data can be recorded in the "Remarks" column.

15. Date entries made here refer to the data in the "Remarks" column.

16. The "Remarks" column is for recording tests, inspections, repairs, alterations, casualties, material analysis data, and the like.

17. These entries specify total hours the machinery unit had been in use at the time of the respective entry in the "Remarks" column.

Unit record card. —The unit record card, shown in figure 18-2, is used to record measurements of reciprocating machinery, such as internal combustion engines and reciprocating pumps. Unit record cards should be inserted behind the applicable machinery history card.

ill

m

Figure 18-2.—Unit record cord.

Bearing record card. —The bearing record card is provided to record bearing measurements of a particular unit. Cards should be inserted in the material history binder behind the applicable machinery history card.

BOILER RECORD SHEET

The boiler record sheet is a quarterly report that must be kept for each boiler. The record sheet must be maintained on a day-by-day basis. Boiler record sheets are retained on board ship, but transcripts of these records may be required by the Bureau of Ships.

The front of the boiler record sheet is shown in figure 18-3. For each day of each month of the quarter, it is necessary to record the following information:

1. The chloride content of the boiler water (epm) ;

2. The alkalinity of the boiler water (epm) ;

3. The number of pounds of boiler compound put into the boiler;

4. The hardness of the boiler water (epm) ;

5. The type of boiler blows given (if any). It is not necessary to record the number of boiler blows given on any day; merely enter the appropriate letter or letters to indicate what type or types of blows were given: B for water drum blows, H for header blows, and S for surface blows.

6. The number of hours under steam.

For each month of each quarter, it is necessary to record the following information:

1. The total number of pounds of boiler compound put into the boiler during the month;

2. The total number of hours under steam for the month;

3. The number of hours of steaming since the last cleaning of the watersides;

4. The number of hours of steaming since the last cleaning of the firesides.

CNClifimilt DATA Ul SURFACC VCSSflS

RICOKO or MllCi HO.

•OllCa ttCOOM SItCET rOi OUARTCfl CM) I He

^ •.8.S.

■ lOf

•OIUA

jiiAjio> iievs

I MSS Ul Cf« lOTE

IMU STEAM

HOUiS TO

OfCIMAL

CtlLO-IIM

Llim COM- AfSS

•OiU* ■LOVS

OMC 0€C IMAL

F0« MOiTN or

I 1

4

3.3. . O . Vif.J J j ,

^.J* 3./ o jVv^ 5.y 1.*-! .,.D4_ _i

^'S I'? ^tH jUa^.J I I

S.^ 3.S^S o i_T|_j, ^ ^ ^

- i

5*./ 3. A 5.3 3./ S,3 3.0

O

o o o o

o o

7,9 3.-?^ o t.o.s.s] o

7.5-:a.7,

7.7

i/5.7

j'-zj]—j—

3:s

3.V- Jo.

4^.0. 3.9 O

3.5* 7.d o B^^s ;^4-.o

a. 44^. 9

—i-

f.5 3.3 9-0 3 .^

O

/ /a3.7 Soo.H-

1206.7

O

/ 3 67.7

*orf wtrc «tv«l (fr a Mf. tntar n N «M/Of S.

VviMr of ft1«ft Mr My, Mt U M rt t ^r M .

Figur* 18-3.—Beilsr record th««t (front).

On the back of the boiler record sheet (shown in figure 18-4), you must keep an accurate, up-to-date record of the following:

1. The findings of all inspections and tests performed on the boiler during the quarter;

2. All repair and upkeep work done on the boiler during the quarter.

MCVAIIKS ftE:04«0 CONClSCtT THC FiNOlfVCS Of ALL tK;riCTlON5 TLSTS *5 *'*V i(Rl^UlDlO «V ■o^^'irS MAK^Al

A»rC ALL «0«R or nePAiR AM> U^KCC^

^ <J*Ji^ 'O&iM^ MU^*^. 4/^^^'"^^ ■f*^^*'*^*^ QjUoiA/^*jl- -i^t^-Wl*/!- - *.*>yaA^^4_ ill.£^L^€^u O^XA^L- XrU/t>XA^y

^a^*^ A'^:^

S4'.>.Mir» 114 mtv 1.441 «toAL«l

Figure 18-4.—BoiUr record shMt (r«v«rt« tid«).

BOILER TUBE DATA SHEETS

To assist in maintaining the record of defective tubes and of tube renewals, tube data sheets should be used. They should be kept for each boiler and filed with the applicable boiler record sheet. These sheets will show

SHIP NO. NAME PARTIALLY RETUBED BY

BOILER NO. DATE

THIS SM6ET APPLCS TO BSW EXPRESS TYPE BOILERS OF 00: 692-4, 69t-7«^ T2/>^ 72T-3t, 770, 775-«, aOO-S, mr-t5t 826, 629-53, 857-90.

DOWNCOMER TU8e6

1 T T T I 2 1 2

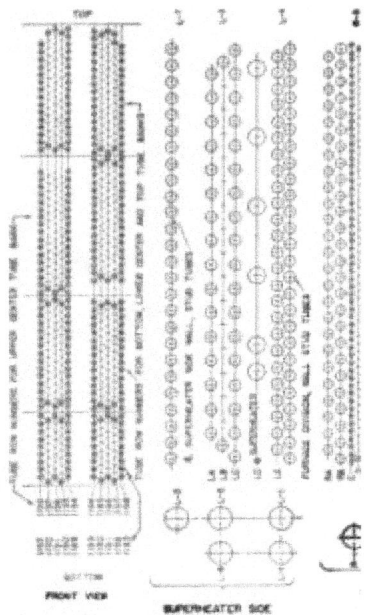

SATURATID \
8UPCPHEATCII
TUBES
FHONT OF BOILCII
80<L£R TUBES
m LK 80«XR8 NOS. 281 AS SHOWN, PQR RK BOlLCRS N06. l64 OPfOSfTC

Figure 18-5.—Tube data sh66t (B and W divided furnace superheat control boiUr).

ALTERATIONS
r34, 752-e7,
NOTE

I. WHEN REFERRING TO TUeES IN CORRESPONDENCE OR REPORTS OENTiFY TUBES AS FOLLOWS:

GENERATING TUBES-GIVE LETTER OF ROW AND igUMBER OF TUBE 00UNTVM3 FROW FRONT TO REAR OF SOM-ER

SUPERHEATER SlOE VUALL STVX) TUBES, FURNACE OfVlSON WALL STUD TUBES AND LC ROW STUD TUBES-G»Vt NAME OR LETTER OF ROW AND NUMBER OF TUBE COUNTING FROM FRONT TO REAR OF BOILER

SUPERHEATER TUBES-GIVE LOOP NUMBER, NAME OF TUBE BANK AND NUMBER OF TUBE COUNTMG FROM BOTTOM TO TOP OF INDMOUAL BUPERHEATCR TUBE BAfi<

ECONOMIZER ELEMENTS-GIVE LETTER ANO NUMBER OF TUBE AS iOC4TED ON CROSS SECTION OF ECONOMIZER.

INLET

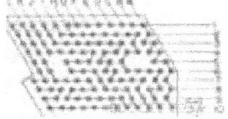

M oirnxT
BOTTOM
BOOMOMIZER SECBON LOOKING TOmPD REAR WALL

NAVY DIPAMTMIMT

WAftNINCTON. D C

BUREAU OF SNIPS

BOILER TUBE RENEWAL SHEET

FOR

B ANO W OIVIOED FURNACE, SUPERHEAT CONTROL BOILER WITH SINGLE UPTAKE

_FOft

CHtCF or ftUWCAU

OATt <7 JU^T 1949

BURCAU or ftMlPft NO

S5I0I 841402

SCALC- NONC

SMUT I or

Figurt 1^5.—Tub« data sheet (B and W divided furnace superheat control boiler) —Continued.

SHIP NO.

NAME

BOJLER NO.

FVXRTIALLY RETUeCD BY

DATE

THIS AARANG6MENT APPLIES TO B 6 W HEAOCR TYPE BOILERS OF A022 AND CVE 26-29

Alf^ATCR TUBES

CIRCULATWG TUBES

-0 0 00 Q> e€>^(M>e€>eG) 0 0

"S^ir^-^ o 0 0 0 -e>e (i>e <i><^e^>e G^ee

♦♦♦♦♦♦""♦^♦♦♦^ ^^^^^^ ♦♦♦^JMP^ ^^^^^^

GENERATING TUBES

SUPERHEATER

SUPERHEATER SUPPORT TUBES'

-^3 <]>—

MU) DRUM

(TiiTUUrihCk Af^fcjTWi^ilwTfc /Tttfytww%/fc itWlWWV%itl ffk/r-**W*rf7WTi ^mT. i V\ iTW% /?v«^*^/twm^/>^^

e<» ^< » ^ g » <!> 0<1<XK»<!> a<»3»0(3 » <XI > <»oa > MW»<t<K l> O^'iNie^ ^>0(»0^-0M

ee o<i><i><i> €Ki> eo €>e^ ee ee e e o<^-<r<i>-c«0- froxKBH^e-^©

0 CD-0 O (i)-€)-0<D QjCD 0^

^TERWALL RISE R TUB ES

■ WftTERWALL TUBES ■

WATERWALL CWWNCOMER TUBES

CROSS SCCTION VIEWED FROM BQLCR FROKT TOP

A^*fc^i*wwrw*v iTWv -^fc/r y*Vr^ ■/'v./'v /^^t'^^'^^ ^'^^^v^/^\ yVjrvy^fc^ ^^^^^^
AAtf^^^^^i^ i»wXV^Mr\j^v^ <m A ^yn^^^

•OTTOM

ARRANOEMCNT OF SUPCRHCATER TUKS ARRAM6MENT AS SHOWN FOR BLRS 164 TO OPPOSfTE HAMO FOR BLfS 2A9w

Figure 18-6.—Tube data sheet (B and W header-type boiler).

ALTERATIONS

NOTES

I WHEN REFERRING TO TUBES IN COR9ESP0NDC1>lCE OR REPORTS DENTIFY rU8E5 AS FOLLOWS

GE/€RAT1MG TU8ES-OVE HEADER NUMBCI? COUNn>i»G FROM LEFT TO RK3MT FACING FRONT OF BOILER. HANOHOLE FITTING NUMBER COUNTING FROM BOTTOM TO TOP OF HEADER AND LOCATE TUBE M T>C CLUSTER IN THE HANOHOLE RTTING CONCERNED.

SUPERHEATER TUBES-OVE WNER, •JTERMEDiATE, OR OUTER LOOP, AND NUMBER OF TUBE IN EAO-^ POW COUNTING FROM FRONT TO BACK OF BOILER

Fl/^NACE WATERFALL TUBES-GiVE NUMBER FROM BOTTOM TO TOP ANO SPECIFY LEFT OR RIGHT SIDE OF FURNACE FACING BOILER.

CIRCULATING TUBES, NIPPLES TO DO*/WTAKE ►CAOERS. ANO MUO DRUM MPPLES-GTVE HEADER NlMER CONCERNED COUNTING FROM LEFT TO RWHT FAONG FRONT OF BOILER

A«HEATER TUBES-GVE ROW NUMBER COUNTING FROM BOTTOM TO 7t3P FACING FRONT Of BOILEW ANO TUBE NUMBER W ROW COUNTING FROM LEFT TO RK5HT

WATERWALL DOWNCOMER ANO RISER TUBES-GIVE NUMBER COUNTWG FT»M LEFT TO RWKT FAONG FRONT OF BOILER

NAVY OC^ARTMCMT

WASMIMOTOM. O C

BUREAU OF SHIPS

Of riCUl tlONATUNC

1

BOILER TUBE RENEWAL SHEET

FOR

B ANO W HEAOER TYPE BOILER

row CHiir or ■uwtAO

S5I0I

scalc- none

841405

SMttr \ or I

Figure 1B-6.—Tube data sheet (B and W header-type boiler)—Continued.

all tube renewals, the date of renewal, the tubes that actually failed under operation, and the cause of such failures. A sample tube data sheet for a B and W divided furnace superheat control boiler is illustrated in figure 18-5. Figure 18-6 illustrates a tube data sheet for a B and W header-type boiler.

Tube data sheets are normally available for the different types of boilers. If none are available, a similar form can be prepared from the ship's boiler plans. Either a copy of the tube data sheet or a written summary covering all failures and renewals should accompany all requests for authority to retube or to cut exploring blocks, as described in chapter 9 of this training course.

The use of boiler tube data sheets helps to ensure an economical policy of tube renewals.

MATERIAL ANALYSIS DATA REPORT

The material analysis data report is intended to furnish information concerning the operating experiences of the fleet to the Bureau of Ships. When accurately filled out, this form gives factual information on the reliability of BuShips material; and thus in a very real way it may contribute to the development of better material.

All repairs to or replacements of hull, machinery, and equipment under the cognizance of the Bureau of Ships (except electronic equipment) must be reported on the material analysis data report. A repair is defined as work needed to keep a unit operating properly; however, such minor jobs as taking up on valve stem packing or pulling up on bolts are not considered as repairs and are not reported on this form.

A separate sheet must be made out for each item repaired. Instructions for making out the material analysis data report, together with coded CAUSE and remedy lists, are included in the pad of forms. Upon completion, the report is submitted directly to BuShips.

Figure 18-7 shows the front of the material analysis data report; figure 18-8 shows the reverse side.

MATERIAL ANALYSIS DATA This form bHoU uM«d t* rtpcrt utt fmaur^a from mn^ m»M whatever tm aU B^hip*

•urtHun «."<«> i-Mj material \'7bexeept KUt-r«0Nlcs). it kill be u$td to indieaU oam»4 and rata 0/ faUmrat a*

a biiJiit for tmprowod dteign.

MMAJIIU tM> MCOMMCaCMTOta

4pproxlMet«l7 16 hour* aftar # 1 forced draft blovar fead ba«a pat la sarrlca, •zeaa«iT« t«Bq9«r»itur* wma Bot«d oc turbln«-cn<*. bcarlrt^ und unit Tlbrated •scatslT«l7. Ublt vaa iBMdiat*!/ tacured. 8ub««<|u«nt Inapvctioa r«v«al«d vlpad bearing (turbine and). Renewed bearing. Cleared and fluehad lube oil ^jratea. Aaieabled and tsated unit. Operating cooaitloa oorual.

>. ti* • MM«aM fa# aadi ilaa r«»Mra4 aa^ Itol Ika aaltrial m rawatf tf^

a. HaWtl ^aayflf «a ra«» 1T|. Baraaa •# aM»*. I»«»artB«al a« tka Navy. Wm^ <- Il«a4 tmpt %m Tfr*. Cr*«». ar A4aMital(ali<t Cfiiiia IV r«^ir«4 k?

Figure 1S-7.—Moterial onalysis data report (face).

i-ier acow thk PAirre uero in tmc rctairom MAiMTENAHca or txa unit

Figure 18—8.—Materiol analysis data report (reverse side).

CHECK-OFF SHEETS

There are numerous tests and inspections that must be accomplished to ensure that the engineering plants and equipment are in the best operating condition. The engineer officer of your ship will undoubtedly prescribe the use of check-off lists when conducting tests and inspections. The use of these lists will ensure that no item of

U. 3. 3

rtrarooB Ho.

DAUT TES1S AMD IKPBCTICWS

VMk •Ddlng f?^*^, 19 -f^

Iteas vblcb abould be lUted on CSMP carda:

.^/^d4A 4^iuu^s^ y^/uAi?^. 7iu yd.^^ A^^t^

Reaarka

"B" DlTl»lon Offlcr

Figure 18-9.

—Daily check<off list.

equipment is overlooked and that no test is forgotten. The check-off lists illustrated in figures 18-9, 18-10, 18-11, and 18-12 are not standardized forms; however, similar lists are used on most naval vessels.

ENGINEERING LOG

The engineering log is an official midnight-to-midnighi daily journal of the ship's engineering department. It is

Itvos vbieh sbould b« •Dt.r.d on CSMP earda;

DlTlalOB Officer

Figurt 1t-10.—Weekly ch«ck-9ff

Mreroon lo. / NOIfreLT TESTS AHD IR3FBCTI0I6

Month •Dding //z::^ 19c /y

Items vhich should be entered on CSNF cards:

Beaarks

"B" Dirision Officer

Figure 18-11—Monthly check-off lilt.

an account of the operation of all engineering machinery and of all activities of importance that take place daily in the engineering spaces.

The ORIGINAL ROUGH LOG constitutes the official legal record. It is written in longhand, and no erasures in the entries are permitted. Any errors occurring should be overlined and, if practicable, the correct entry is made

U. S. S

FlrerooB t /—

QUAPTERLY TESTS AID WSPBCTIOIIS Co«i>let«d during quarter ending -^^ ^/diXcA 19

Inspect liquid-end valves, valve steinsf A valve springs on reciprocating punps; Inspect ateas valve gear; check setting of relief valvet. (BSM 1*7-35)

^-J- rire i, bilge

pump

/ eaiergency feed puap

Inspect all steaiB trap*.

(BSM 1*8-220)

Nike hydrostatic test of fuel oil tervlce systeo. (BSM 55-103)

KOrS: Enter results of tesU aod inspections in log, aod forward all Infon&ation for ■acblDtry blatory to log rocn.

0 gJ^Jtr Lr us^r

3l»t»loo Offlcr

Figur* 1S-12.—Quarterly ch«ck>eff list.

immediately following the entry struck out. While the ship is under way, the log is kept in the control engine-room ; in port, the log is kept in the live engineroom or in the engineering log room.

On smaller naval vessels the log is generally filled out by the machinist's mate of the watch or by the PO having the day's duty. Occasionally a Boilerman may be the

[■6iMCe»l»€ LOG - AIL SURFACf VESSCIS

SSTTT

TO TUmj: use BWOif ■!*«<•) Ml »llit« itful ' JPiC(i Ml'"fcoH! rtt MTI

—r Ti^ttr^ .

MIS CACM ULCBOM TUI

8

/a.3

93.3 /io. 0

12,0. O t 20,0

/ao, o

I Zo,o 10,0,0

! TOIil •fCM^TS

hf^QjIi^ 11.000

^S,3oS (,51

> TQtAl itfTRMO

X? a*? ■ /o- ai •■ fe - A?

A /V; 12.

U. lb.

n. lb. lb.

16.

u. lb.

(■STtVCTtMS

\» fA«in«rifif le^ Mr •ritttn ••»»» ^mc • I •» M*. "Ott CO*»^ni**t. 1^* 09IGIUI writing tkf LlOAl IKOtD atW aiitt to •r«»«rM«. It •• Mt M*cettarf to aak* • c»»7 •»c»H o»»» o* ■or* •««*• art »»»»t w«r •

• hip t*> CCM»I|»«01.

T«kit I Afw* (He tfiaicS ■Htl k« .ftlan •! IM tiat «v««t* o«c«r. Otl»«r t«kl«»
■«l b« oriltm b**ort i>oor th* loHotting 4al«.

TMf tH«li be t'^r^ l*f>'«*cr OM>c*r ol IM vr 0«r tofort fO)*« off

bAil#r» in u««: t«co<v«. wnqtnt c«*biAatiM ia tMftf. ■•jor t^M^ cl)aM«»t.

•« *o<»»-third,* *«Un«ar«.* 'fall*: to«rth. catMltiat to »*rM<M*«l or Bat*-rial BitHin or iii>»tr tN» co^n.iaitce of »i»fii>t«r 4«#«rtM«4tt: tkftk, tftcUl *Atria« r#^gir»d by lavr aaQulationt, Swroou of l^iM* MoMaal. iMttrifCl <•*• t0>4 l»ltrr» of t^r B«rrau of Sr>>»t.

airttATiOiS Ot (tASuifS lb TIM tfiaKIS UC lOT fCtMITTCO. ICCfSSMT COMfC-TIM^ SNAU «(laDl tT k ROTf lb TMC KMfflt.

0ISPO3ITIOB

Figur* 18-13.—Enginaoring log.

petty officer with the day's duty, and may thus be responsible for seeing that the engineering log is correctly filled out and kept up to date.

On larger ships, the engineering log is maintained by the engineering duty officer. Some

of the information required by this Jog must, of course, always be phoned in
to the control engineroom from the Boilermen in the firerooms.

The log is signed by the engineering officer of the watch, or by the engineering duty officer, before he goes off duty; by the engineer officer each day; and by the commanding officer at the end of each month.

Figures 18-13 and 18-14 show pages of the engineering log.

13^3 ff^^ Ju^ J^ti^^M^ ^>a«^»y<u»-^

Ais^-^J ^^^l. /S^^.jy.^^;

• 1>«AH«S l>»t I ' O a f ' HI I CQ^riNUtD OH ADDITIONAL »MttT OnAvSNifS « « 7A I Pnr,i sr, I ^^

WTl: AfftOm II SI«aCO 0«C(a MMTH on IM 0« UST IMT Ot Oa NTACIWCIT. CMIVfllM OTfiai WILL SIM 'MtttT' IT MM Of reiLOUlK MTC.

Figure 18-14.—Engineering log—Continued.

.555

The information required by the engineering log in eludes the following:

1. The name of the ship; the date; the place where the ship is moored or anchored, or the ship's route *

2. The all-shaft average rpm for each hour;

3. The average knots for each hour, as computed from the latest approved speed and rpm tables:

4. The major speed changes such as "one-third," "standard," and "full";

5. New standard speeds (if such are initiated) and the time the speed changes were made;

6. The engine combinations in use;

7. The boilers in use;

8. Fuel oil, fresh water, and lubricating oil on hand each midnight and the amounts received and ex pended during the preceding 24 hours;

9. The draft, displacement, miles steamed (from nav igator's data), engine miles steamed, and the number of days out of drydock;

10. Any special entries required by BuShips Mantiaf and by letters from BuShips;

11. Casualties to personnel or material in the engineering department and the remarks of the engineering officer of the watch (or, in port, the engineering officer of the day).

The remarks section of the log must be written up neatly, legibly, in complete statements, and with the use of standard phraseology. All watches are headed 00-04, 04-08, 08-12, or 12-16 (in port, watches may be 8 hours in length). The midwatch is required to enter complete information on the status of the engineering plant, including boilers in use and the fuel oil tanks supplying them, auxiliary machinery in use, generators in use, make-up feed tanks, and the standby fuel oil suction and standby make-up feed tanks. Any change in the status of any piece of machinery or tanks is also entered in the log.

EQUIPAGE CUSTODY RECORD

An Equipage Custody Record Card must be kept of all portable and semiportable equipment, certain individual spares, sets of spare parts, and special tools which are not subject to rapid wear and are relatively expensive items. These items are issued to ships as original outfits, replacements, or additions to allowances. You will be required to sign a record card as custodian of the equipage over which you will have immediate cognizance.

Equipage under the 12,000 series is inventoried annually, or when the person or persons having cognizance over the equipage is being relieved of his duties or is being detached or transferred from the ship. The inventory should be completed within a 30-day period. A copy of the custody record card is kept on file in the engineering office.

SURVEY OF EQUIPAGE

When material requires disposition and expenditure from the accounts of the ship, the heads of the departments submit requests for survey to the supply officer, stating the nature of the material and the conditions requiring the expenditure. The supply officer has the requested information and a description of the article (s) involved typed on an s. and A. Form 154. All papers are then submitted to the commanding officer for review and action. The commanding officer then designates the type of survey to be held; it may be either formal or informal.

The INFORMAL SURVEY is held if the circumstances surrounding the expenditure do not warrant disciplinary action by the commanding officer or if there are no peculiarities that would require a formal review. The department head requesting the survey is generally designated as the surveying officer.

The FORMAL SURVEY is held (1) when it appears that a person in the naval service will be held responsible for

the loss of material (that is, material lost through negligence, theft, or carelessness), or (2) when the circumstances otherwise indicate that a formal review of the case is required. In the formal survey, an officer not attached to the department where the request originated is designated as the surveying officer.

For surveys of equipage of value greater than $100 or when the commanding officer deems it desirable, a board of three officers is designated as the survey board. No member of the board will be from the department having control over the equipment under survey. After the survey has been completed an appropriate entry is made in the survey record book maintained by the supply officer.

Equipage Replacement

Equipage material should be replaced when it becomes apparent that it is no longer serviceable or cannot be accounted for. All items to be replaced must first be surveyed (as previously mentioned). When the request for survey is submitted to the supply officer, he considers it as an automatic request for replacement unless otherwise specified by the department head concerned. The supply officer uses the survey request as a guide in making out the necessary requisition s. and A. Form 43.

Material Replacement

Each ship is allotted a quarterly allowance of funds for operation, maintenance, and repair. This quarterly allowance is divided among the various departments of the ship by the commanding officer.

It is the responsibility of individuals—especially the leading petty officers or chief petty officers—within the department to ensure that material is not wasted or equipment carelessly handled or lost.

The replacement of consumable supplies can be requisitioned through the Supply Department by the use of the STUB REQUISITION (s. and A. Form 307). Each request should contain the following information:

1. Department requiring the material,

2. Date the stub was prepared,

3. Signature of the department head or of an officer authorized by the department head to sign in his stead,

4. Information properly identifying the material requested,

5. Quantity required,

6. Stock unit of material requested.

The Allowance Lists or the Standard Stock Catalogue should be consulted, where possible, w^hen filling out the stub requisition. This will make it possible to properly identify the material being requested and also to fill in the stock or class number. Only material in one stock or class should be entered on each stub.

If the material is not in stock, the stub will be canceled, and a new one must be submitted at such a time as the stock is replenished. If the material is not carried in stock aboard your ship, generally the supply officer will requisition the material from the nearest available source of supply.

REPAIR PARTS STOWAGE RECORDS

When the engineering department has custody of its own repair parts, as on many small ships, the leading petty officer should see that repair parts for his assigned space or machinery are properly stowed and inventoried, and that requisitions are submitted when stocks are below allowance. Too much stress cannot be put on the importance of maintaining a full allowance of repair parts, and the importance of their stowage. Laxity on this subject could result in a situation where a replacement part needed for an important repair and actually on board could not be located.

The engineering officer will usually assign a petty officer to the job of handling repair parts. Even though the task is not assigned to you, it is still to your advantage, as a Boilerman 1 or C, to give the repair parts man

all the cooperation and help that you possibly can. The Petty Officer assigned to the task will maintain a master list of the location and the contents of each box. However, leading petty officers in charge of spaces should maintain their own lists of the replacements for the machinery in their charge. There are various ways in which this can be efficiently accomplished. One method is to maintain a ledger in which the equipment listing is followed by a list of the allowance spares (copied from the allowance books), with the number and the location of the box containing them. For example, if a part was needed for a forced-draft blower, you could turn to the pages in the ledger headed '*F. D. Blowers," find the listing of the required part, its number, and the number and location of the storage box. When parts are removed you should see that requisitions for their replacement are initiated at once.

Frequent inspections should be made of all repair parts to determine their material condition and to see that they are properly preserved. Handling (during inventories or inspections, for example) sometimes causes the wrappings and preservatives to come loose, and the parts will be subject to corrosion. In such a case, preservatives should be reapplied, and the wrappings tied in place again.

If you are not already familiar with the BuShips allowance books, you should look them over carefully, especially groups 47 (pumps), 51 (boilers), 53 (F. D. Blowers), and 55 (F. 0. Equipment).

OPERATING RECORDS

Operating records, such as the fireroom log, are to be filled out completely and accurately by the men on watch in the engineering spaces. It is the responsibility of the petty officer of the watch or the petty officer having the day's duty to check all entries made, to ensure that they are correct and to assure himself that the plant is operating properly. He must also make a personal check of

the machinery spaces to see that all the required data are being accurately logged.

Inasmuch as operating records are the basis for a daily performance analysis of the engineering plant, the information given therein must be correct and concise.

Another important responsibility of the leading petty officer is the reviewing of the various check-off lists. No item contained in any of these lists should be overlooked or omitted, for such procedure may well lead to a casualty.

The numerous records and reports with which you will come in contact are necessary in promoting the over-all efficiency of the engineering department. The data given on these reports will be required whenever it becomes necessary to correct any casualty or to make any other repair or alteration. The reports will also be required when tests or inspections are to be made.

Some of the forms used for recordkeeping or reporting will be originated aboard ship, but many will be prescribed by BuShips. From time to time these latter forms will be subject to revision, but BuShips will send the revised forms to your ship.

QUIZ

1. What is the basic card used in setting up or revising the machinery history?

2. How are machinery history cards filed?

3. What is the purpose of the unit record card?

4. On what record or records should you report all repair and upkeep work done on a boiler?

5. What is done with boiler record sheets, after they have been filled out?

6. What is the purpose of the material analysis data report?

7. If an error is made in the rough engineering log, how should it be corrected?

CHAPTER

INSPECTIONS AND TRIALS

A naval ship must be inspected from time to time to ensure that her operation, administration, and material are maintained at a high standard of readiness for war. The frequency with which the various types of inspections are held is determined by CNO, fleet commander, and type commander. As far as the ship is concerned, the type commander usually designates the type of inspection and when it will be held.

Although a ship is frequently (but not always) notified some time in advance when an inspection will take place, it is a mistake to think that a poorly administered division or department can, by a sudden burst of energy, be made ready to meet the inspection. The only way to meet an inspection successfully is to keep up to date on such items as repair work, maintenance w^ork, operating procedures, training of personnel, engineering casualty control drills, material history, CSMP, operating records, and other records and reports. By using proper procedures and keeping up to date you will always be ready for an inspection.

Your ship may be required to furnish the inspecting party that will make an inspection on some other ship. When this occurs, a Petty Officer 1 or C may be assigned the duty as an assistant inspector. As a Boilerman 1 or C, therefore, you should know something about the different types of inspections and how they are carried out.

ADMINISTRATIVE INSPECTION

Administrative inspections cover administrative methods and procedures normally employed by the ship, and each inspection is divided into two general categories— general administration of the ship as a whole, and administration of each department. In this discussion we will consider the engineering department only.

The purpose of the administrative inspection is to determine (1) whether the department is being administered in an intelligent, sound, and efficient manner, and (2) whether the organizational and administrative methods and procedures are directed toward the objective of every naval ship—namely, being prepared to carry out her intended mission.

Inspecting Party

It is a routine procedure for one ship to conduct an inspection on another ship within the division. General instructions for the procedure of conducting the inspection are usually given by the division commander, but the selecting and organizing of the inspecting party is done aboard the ship that has been instructed to conduct the inspection.

The Chief Inspector, usually the commanding officer of the ship, will organize the assisting board. The organization of this board is in general conformance with the departmental organization of the ship. It is divided into appropriate groups, each headed by an inspector with assistant inspectors as necessary. Chief petty officers are usually selected as assistant inspectors, and on smaller ships such as destroyers, petty officers first class may also be assigned duties as assistant inspectors.

The engineering department inspecting group (or party) will be organized and supervised by the engineer officer. The thoroughness with which the inspection will be carried out will depend a great deal upon the knowledge and ability of the assistant inspectors.

General Inspection of the Ship

One of the two categories of the administrative inspection is that of the administration of the ship as a whole. Items of this inspection that will have a direct bearing on the engineering department, and for which the inspection report must indicate a grade, are as follows:

1. The appearance, bearing, and smartness of personnel.

2. The cleanliness, sanitation, smartness, and appearance of the ship as a whole.

3. The adequacy and the condition of clothing and equipment of personnel.

4. The general knowledge of personnel in regard to the ship's organization, ship's orders, and administrative procedures.

5. The dissemination of all necessary information among the personnel.

6. The indoctrination of newly reported personnel.

7. General educational facilities for individuals.

8. The comfort and conveniences of living spaces, including adequacy of light, heat, ventilation, and fresh water, with due regard for economy.

Engineering Department Inspection

The administrative inspection is primarily an inspection of the engineering department paper work, which includes numerous publications, bills, files, books, records, and logs. But the inspection will also include items with which the chief and first class will be more concerned— cleanliness and preservation of machinery and engineering spaces; training of personnel; assignment of personnel to watches and duties; proper posting of operating instructions and safety precautions; adequacy of warning signs and guards; the marking and labeling of lines and valves; and the proper maintenance of operating logs.

Administrative Inspection Check-Off Lists

Administrative inspection check-off lists are used as an aid to inspecting officers and chiefs to assist them in ensuring that no important item is overlooked. Check-off lists are usually furnished to ships by the type commander, but inspecting personnel are not to consider the lists as being all-inclusive. It is quite probable that, during inspection, it will develop that there are additional items.

The chief and first class petty officer should be familiar with the various check-off lists used for inspections. These lists will give you a good understanding of how to prepare for an inspection as well as for your daily supervisory duties. A knowledge of what is included in the check-off lists will be a great aid to personnel that are assigned duties as assistant inspectors. You will find it a good idea to obtain some copies of the various inspection check-off lists from the log room and to carefully look them over. They will give detailed information for your type of ship.

In order to give you a better understanding of the scope and purpose of administrative inspections, in comparison to other types of inspections, the following abbreviated sample of an engineering department check-off list is given:

1. Bills for Both Peace and War —

a. Inspect the following, among others, for completeness, correctness, and adequacy.

(1) Department Organization

(2) Watch, Quarter, and Station Bills

(3) Engineering Casualty Bill

(4) Fueling Bill

(5) Etc

2. Administration and Effectiveness of Training — a. Administration and effectiveness of training of

personnel for current and prospective duties:

(1) Are sufficient nonrated men in training to replace anticipated losses?

(2) BuPers training courses:

(a) Number of men enrolled

(b) Percentage of men in department enrolled

(c) Number of men whose courses are completed

(3) Are personnel concerned familiar with operating instructions and safety precautions? (Question personnel at random.)

(4) Are personnel concerned properly instructed and trained to handle casualties to machinery?

(5) Are personnel properly instructed and trained in damage control?

(6) Are training films available and used to maximum extent possible?

(7) Are training records of personnel adequate and properly maintained?

3. Dissemination of Information Within Department —

a- Is necessary information disseminated within the department and divisions?

b- Are the means of familiarizing new men with department routine orders and regulations considered satisfactory?

c. Etc

4. Assignment of Personnel to Stations and Watches —

a. Are personnel properly assigned to battle stations and watches?

b. Are sufficient personnel aboard at all times to get the ship under way?

c. Are men examined and qualified for important watches?

d. By random questioning, does it appear that men on watch have been properly instructed?

e. Etc

5. Operating Instructions, Safety Precautions, and Check-off Lists —

a. Inspect completeness of the following:

(1) Operating instructions posted at machinery

(2) Posting of necessary safety precautions

b. Are check-off lists (daily, weekly, monthly, etc.) properly maintained?

c. Are weekly tests and inspections of safety devices properly carried out?

d. Are responsible personnel familiar with current instructions regarding routine testing and inspections?

e. Are lighting-off and securing sheets properly used?

f. Etc

6. Procedures for Procurement, Accounting, Inventory, AND Economy in Use of Consumable Supplies, Repair Parts, and Equipage —

a. Is an adequate procedure in use for replacement of repair parts?

b. Are there adequate measures used to prevent excessive waste of consumable supplies?

c. Is there proper supervision in the proper supply of, care of, and accountability for hand tools?

d. Are inventories taken of repair parts which are in the custody of the engineering department?

e. How well are repair parts preserved and stowed?

f. What type of system is used to locate a repair part carried on board? (Have a PO first or chief explain to you how he would obtain a repair part for a certain piece of machinery.)

g. Are custody cards properly maintained for accountable tools and equipment?

h. Etc

7. Maintenance of Records and Logs —

a. Inspect the following for compliance with pertinent directives, completeness, and proper form:

(1) Engineering Log

(2) Bell Book

(3) Operating Records

(4) Machinery History

(5) Current Ship's Maintenance Project

(6) Alteration and Improvement Program

(7) Machinery Index

(8) Boiler Record Sheets

(9) Daily Oil and Water Records

(10) Engineering Reports

(11) Training Logs and Records

(12) Work Books for Engineering Spaces

(13) Etc

8. Availability and Correctness of Publications. Directives, and Technical Reference Material —

a. Engineering Blueprints

(1) Index of blueprints

(2)) Proper filing of blueprints

(3) Completeness and condition

b. Manufacturers' Instruction Books

(1) Proper indexing

(2) Completeness and condition

c. Type Commanders Material Letters

d. BuShips Mamial

e. General Information Book

f. Booklet Plans of Machinery

g. Etc

9. Cleanliness and Preservation —

a. Preservation and cleanliness of space (including bilges)

b. Preservation and cleanliness of machinery and equipment

c. Neatness of Stowage

d. Condition of Ventilation

e. Condition of Lighting

f. Compliance with Standard Painting Instructions

g. Etc

OPERATIONAL READINESS INSPECTION

The operational readiness inspection consists of a demonstration on the part of the ship of her readiness and ability to perform the operations which might be required of her in time of war.

The inspection will consist of the conduct of a battle problem and other operational exercises. A great deal of emphasis will be placed on AA and surface gunnery, damage control, engineering casualty control, and other appropriate exercises. Drills such as Man Overboard, Preparations to Abandon Ship, Fire, and Collision will be held and observed. The ship will be operated at full power for a brief period of time.

The over-all criteria of performance will be:

1. Can the ship as a whole carry out her operational functions ?

2. Is the ship's company well trained, well instructed, competent, skillful, and adept in all phases of the evolutions ?

3. Is the ship's company stationed in accordance with the ship's Battle Bill, and does the Battle Bill meet wartime requirements?

Observing Party

The observing party will consist of about the same personnel and organization as the inspecting party used for the administrative inspection. Usually more personnel are assigned to the operational readiness observing party. The added personnel are usually chiefs and first class petty officers, especially for smaller ships.

The observing party members should be briefed in advance of the scheduled exercises and drills that are to be conducted. The observers must have sufficient training and experience so that they can properly evaluate the exercises and drills that are to be held. Each observer will usually have an assigned station. He should be well qualified in the procedure of conducting drills and exer-

cises for that station. It is highly desirable that each observer be intimately familiar with the type of ship to be inspected.

Battle Problem

In this discussion we will take up the battle problem from the viewpoint of the observer. First you are given some general information on the requirements and duties of a member of the engineering department observing party. Then, knowing the viewpoint and duties of an observer, you can prepare yourself and your men for a battle problem and other appropriate exercises.

Preparation of a battle problem. —The degree of perfection achieved in any battle problem is a direct reflection of the skill and application of those who prepare it. A great deal depends upon the experience of officers and chief petty officers.

The primary purpose of a shipboard battle problem is to provide a medium for testing and evaluating the ability of all divisions of the engineering department to function together as a team in simulated combat operations, in order to accomplish the mission assigned by the problem.

Battle problems can be made the most profitable and significant of all peacetime training experience, since they demonstrate how i*eady a department is for combat. The degree of realism of this test governs its value: the more nearly it approximates actual battle conditions, the more valuable it is.

Conduct of a battle problem. —There is one element in conducting a battle problem which increases its value to the ship's company: the element of surprise. Of course, preparations for carrying out a problem can't be kept entirely a secret. The ship must be furnished in advance such information as:

1. Authority for conducting the inspection,
2. Time of boarding of the inspecting party,
3. Time ship is to get under way,
4. Time for setting condition one,
5. Time of conducting inspection for zero problem time conditions,
6. Zero problem time,
7. End of problem time,
8. Time of critique.

Observers should be proficient in the proper methods of introduction of information. Information delivered to ship's personnel should be verbal when practicable; and, in general, only that information which ship's personnel would logically determine from procedure and adequacy of investigation on search should be furnished by the observer. Should the inadequacy of procedure by ship's personnel result in the nondiscovery of a casualty imposed, observers may resort to coaching, but a notation should be made on the observers' form as to the time allowed before coaching and information was furnished. Special precautions should be taken to give the symptoms of casualty the same degree of realism that they would have if the casualty were actual rather than simulated.

It may be necessary, in order to provide the desired degree of realism, that valves be closed, switches opened, or machinery stopped. (Note: Care must be exercised in regard to the supply of lubricating oil to the main engines and the supply of feed water to the boilers.) In each case the observer should inform responsible ship's personnel of the action desired to provide the realism, and ship's personnel should operate the designated equipment. In the event that personnel injury or material damage may result from a lack of preparation, or from a lack of experience of personnel concerned, the casualty should be simulated or omitted.

An emergency procedure is set up, by the observing party with ship's company, to take proper action in case actual casualties—as distinguished from simulated or problem casualties—should occur.

During the conducting of a battle problem, the general announcing system (the IMC circuit) may be used by the ship, but observers normally will have priority in its use.

The problem time announcer will use the general announcing system to announce the start of the battle problem, and to announce the problem time at regular intervals, the conclusion of the problem, and the restoration of casualties. However, the general announcing system is kept available at all times for use in case of actual emergency. All other announcing system circuits and all other means of interior communications are reserved for the EXCLUSIVE use of the ship.

Engineering telephone circuits should be monitored by one or more observers. A check should be made for proper procedure and circuit discipline, and for handling of information or casualties.

An inspection should be made to see that the engineering plant is set up for battle conditions in accordance with current directives. Any fire hazards such as paint, rags, or oil should be noted. Check for missile hazards such as loose gear, loose floor plates, tool boxes, and repair parts boxes. The condition of fire fighting, damage control, and remote control gear should be carefully inspected.

Analysis of the battle problem. —The maximum benefit obtained from conducting a battle problem lies in the determination of existing weaknesses and deficiencies, and the resulting recommendations for improvement in organization and future training. Every effort should be made by observers to determine excellencies as well as deficiencies, since a knowledge of existing excellencies by ship's personnel will serve to heighten morale as well as to indicate those factors that presently, at least, may receive less emphasis in the shipboard training program.

Analysis of the battle problem affords an opportunity for the observers to present to the ship their opinion of her performance, and for the ship to comment on the observers' remarks; it also points up suggested improvements in doctrine or material. Analysis is conducted in two steps: the critique and the observers' reports.

A critique of the battle problem should be held on board

the observed ship before the observing party leaves in order that a review of the problem and the action taken may be made when both are fresh in the minds of all concerned. The critique is attended by all the ship's officers, appropriate chief and first class petty officers, the Chief Observer, and all Senior Observers. The various points of interest of the battle problem are discussed, and the Chief Observer comments on the over-all conduct of the problem after the Senior Observers have completed their analysis of the battle problem as developed from their observers' reports.

The observers' reports will be in the form prescribed by the type commander and with any additional instructions given by the Chief Observer. The reports of the observers are collected by the Senior Observer for each department. Senior Observers submit their reports to the Chief Observer. All observers' reports are reviewed by the Senior Observer for the respective department before the critique is held.

A second purpose of the reports is to furnish the inspected ship with all detailed observations which may not, because of time limitations, be brought out during the critique. The inspected ship receives a copy of all observers' reports, and each department is given an opportunity to view the detailed comments and to set up a training schedule to cover their weak points.

A brief example of an engineering observer's report form is given as follows:
Engineering Observer:
Location:
1. The engineering department's evaluation is based on: (a) extent of the department's preparation and fulfillment of the ordered conditions of readiness as appropriate to the problem, (b) extent of correct utilization of the engineering damage control features built into the ship, (c) extent to which proper engineering casualty control is accomplished, (d) extent to which on-station personnel take corrective

action for control of damage, (e) adequacy of reports and dissemination of information, and (f) the general handling of the plant in accordance with good engineering practice and the ability of the department to ensure maximum mobility and maneuverability of the ship and to supply all necessary services to other departments in fighting the ship.

2. Hit:

Exercise:

a. Preparation and status of the plant.

b. Fulfillment of proper condition of readiness.

c. Fire and missile hazards.

d. Condition of fire fighting and damage control g-ear.

e. Condition of personnel clothing and protection.

f. Stationing and readiness of personnel.

g. Investigation and interpretation of casualty.

h. Promptness and effectiveness in taking care of casualty.

i. Were proper doctrine and procedures used ?

j. Were prompting and additional information given

by observer? k. Were proper reports made? 1. Readiness of standby units, m. Readiness of alternate and emergency lighting

and power.

n. Were proper safety precautions observed?

o. Material deficiencies.

p. Coordination of personnel.

q. Coordination between engineering spaces.

r. Etc

3. Main Engine Control, Receipt of vital interior communications; origination and transmissions of required reports to Conn, Damage Control Central, and other stations.

4. Action taken by main engine control:

a. Correct action.

b. Sound judgment based on good practice.

c. Assurance.

d. Speed.

5. Recommendations.

The blank parts of the observers' report forms are filled in as applicable to the individual observer's station. Things that were not observed by him are either left blank or crossed out. Additional information, if required for a certain exercise or condition, may be written on the reverse of the form. A separate form or sheet is used for each exercise or drill. Remarks or statements made by the observer should be clear and legible.

MATERIAL INSPECTION

The purpose of material inspection is to find out the actual material condition of the ship in regard to the ability to perform all functions for which the items were separately and interrelatedly designed, and to recommend the repairs, alterations, changes, or developments which will ensure the material readiness of the ship to carry out her mission. In addition, the material inspection determines whether or not proper procedures have been carried out in the care and operation of machinery and equipment. Administrative procedures and material records which are inspected include such items as material history, CSMP, routine tests and inspections, boiler record sheets, quarterly readings for main turbine bearings, and weekly hull reports.

In brief, the prescribed requirements for material readiness are as follows:

1. Established routines for the conducting of inspections and tests, schedules for preventive maintenance, and a system which will ensure timely and effective repairs.

2. Adequate material maintenance records, that are kept in accordance with current directives and that will give the history and detailed condition of machinery and equipment.

3. The planned and effective utilization of the ship's facilities for preservation, maintenance, and repair.

A. The correct allocation of necessary work to the following categories: (1) the ship's force, (2) the tenders and repair ships, and (3) the naval shipyards or other shore repair activity.

The scope of the inspection will be similar to that of inspections made by the Board of Inspection and Survey. The inspection should be thorough and searching, and cover detailed maintenance and repair rather than general appearance. The distinction between administrative inspections and material inspections should be clearly recognized, and there should be as little duplication as possible. An examination of the material maintenance records and reports will be made to obtain data and material history for a proper understanding of the condition of machinery and equipment General administrative methods, general appearance, cleanliness of compartments, and cleanliness of machinery are not part of this inspection, except in cases where they have a direct bearing on material condition. Special painting should not be done solely in preparation for material inspection.

The inspecting party for the material inspection is similar to that of the administrative inspection party.

Preporation for the Inspection

At an appropriate time prior to the date of the inspection the Chief Inspector will furnish the ship with advance instructions, including:

1. List of machinery and major equipment to be opened for inspection. The limit that a unit of machinery or equipment should be opened is that which is necessary to reveal known or probable defects. Due regard should be had for the ship's operational and safety schedule. The units selected should be representative and, in a multiple-shaft ship, should not disable more than one-half of the main propelling units.

2. List of equipment to be operated. Auxiliary machinery such as the anchor windlass, winches, and steering gear are normally placed on this list.

3. Copies of the condition sheets. This is a form of check-off list which is used for the material inspection.

4. Any other instructions considered necessary by the type commander or other higher authority.

Each department will have to prepare WORK LISTS showing the items of the CSMP which have been assigned for accomplishment by naval shipyard, tender or repair ship, or ship's force, during an overhaul or upkeep period. This list is arranged according to priority and numbered according to current directives. A list of the outstanding alterations is also made up for the inspection. Work lists usually consist of cards (5"x8")» with one repair or alteration item on each card. The work list should include all maintenance and repair items, because material deficiencies found during the inspection will be checked against the work list, and if the item does not appear on the work list a discrepancy in maintaining the CSMP is noted by the inspector.

Condition sheets. —Condition sheets are made up in accordance with different material groups. The engineering department will be primarily concerned with the machinery, the electrical, and the hull conditions sheets. Condition sheets contain material in form of check-off sheets and material data sheets, and consist of a large number of pages. Items for data and check-off purposes are listed for all parts of the ship, and for all machinery and equipment on board ship.

In advance of inspection, the ship to be inspected must fill in a preliminary copy of the condition sheets. In order to accomplish this, detailed data must be obtained from the machinery index, machinery history, CSMP, and other records and reports.

An entry is made, in the proper place on the condition sheets, of any known fault or abnormal condition of ma-

chinery or equipment. Details and information are given, as necessary, to indicate the material condition to the inspecting party. If corrective work is required in connection with a unit or space, a reference is made to the work list item. Data and information requested in the condition sheets should be furnished whenever possible. The preliminary copy, if properly filled out, will represent the best estimate of the ship as to the existing material conditions.

When the condition sheets have been completed, they are turned over to the respective members of the inspecting party upon their arrival on board ship. During the inspection, the inspectors will fill in the various check-off sections of the condition sheets; the sheets are then used in making the final inspection report on the condition of the ship.

For more detailed information for your ship, you should obtain a copy of the applicable condition sheets from the engineer's log room.

Opening machinery for inspection. —The ship will open machinery as previously directed by the Chief Inspector, and as considered desirable, in order to obtain the inspector's opinion concerning known or probable defects. The information given in chapter 6 of BuShips Manual should be used as a guide in opening particular machinery units.

A list of machinery, tanks, and major equipment opened, and the extent of opening, should be supplied to the inspecting party on its arrival. Test reports on samples of lubricating oil should be furnished to the machinery inspector.

The following abbreviated example is for the purpose of giving you an idea of the preparation for inspection of material. It is not intended to set up any prescribed method of opening up machinery for inspection.

1. Main Turbines:

a. Lift at least one-half of the bearing caps.

b. Furnish a table of the latest bridge gage or depth

micrometer readings; both should be furnished if applicable. Any change from original readings should be noted, c. Remove the inspection plates on the casings and the exhaust trunks. Open casings should not be left unattended. Cover plates should be retained in place with 2 or more studs, until the arrival of the inspection party.

2. Auxiliary Turbines:

a. If any are known to be defective in operation, showing undue vibration, casing leaks, etc., open the casing and disassemble, if practicable.

b. Remove the inspection plates, if fitted.

c. Open half of the bearings, if practical.

3. Reduction Gears:

a. Remove inspection plates in the presence of the engineer officer and the inspecting party.

b. Furnish tables of bearing data, and of bridge gage, depth micrometer, or other types of readings, showing the last reading and any change from the original.

c. Have a sample of oil from the bottom of the lubricating oil drain tank ready for the inspecting party.

d. Lift the caps from one-half of the bearings, if this can be done without lifting the casing.

4. Condensers:

a. Remove inspection plates or manhole covers from the salt-water and steam sides, as far as practicable.

5. Deaerating Feed Tanks:

a. Remove the inspection plates.

6. Pumps, Reciprocating:

a. Open the steam and water cylinders and the valve chest of at least one of each type.

b. Furnish caliper measurements of cylinders, pistons, rods, valve chests, throat bushings, etc. (Cylinders are to be measured fore and aft and

athwartships, at top, middle, and bottom of piston travel.)

7. Pumps, Centrifugal:

a. Lift the casing of one pump of each type, if practicable.

b. Furnish measurements of internal clearances (wearing rings, etc.).

8. Pumps, Rotary (screw or gear type) :

a. Open one pump of each type, if practicable.

b. Furnish recent measurements of clearances (rotor, wearing plate, liner, etc., as set forth in chapter 47 of BuShips Manual).

9. Forced-Draft Blowers (Horizontal) :

a. Lift the bearing caps of one or more of those on board.

10. Forced-Draft Blowers (Vertical) :

a. Tag all blowers with operating oil temperatures to all bearings.

b. Dismantle all bearings on blowers which operate with lubricating oil temperature from bearings in excess of 175" F., or where the lubricating oil temperature through the bearings increases in excess of 50° F.

11. Boilers:

a. If practicable, open all boilers for inspection of firesides and watersides. In any case, open one half of the ship's boilers for inspection. Where applicable, open one boiler in each fireroom.

h. Internal fittings of one boiler should be removed to permit internal inspection of the fire-row tubes, the steam separator supports, the interior of steam separators, etc.

c. Boiler casings should be removed on one boiler to an extent that will allow an inspection of tube banks for soot deposits and an inspection of water drums for soot accumulations.

12- Uptakes:

a. Open for inspection all uptakes except those foi steaming boilers.

13. (Etc.)

a. Ship's company should have portable extension lights rigged up and in readiness for inspection of the units of machinery opened up.

b. The lighting of the space should be in good order.

c. The inspectors should be furnished flashlights, chipping hammers, file scrapers, and similar items.

d. Precision measuring instruments should be readily available.

Assembly of records and reports. —The material inspection also includes an inspection of various material records and reports. These documents are assembled so that they will be readily available for inspection. Records should be kept up to date at all times. It is good policy to check over all records to make sure they ARE up to date and that nothing has been overlooked. The individual records should be filled out and maintained in accordance with current directives. Where applicable, the chief petty officer who is in charge of an engineering space, or other assignment, should check on any records or reports that concern the material or the maintenance procedures of his space or assignment.

The following is a brief listing of some of the records and reports that are furnished to the inspectors, so that they may study and check them:

1. Condition Sheets, filled out for the inspection;

2. Copy of the latest Material Inspection Report by the Board of Inspection and Survey, or by Forces Afloat;

3. Weekly Hull Reports;

4. Latest copy of the Docking Report;

5. Material (Machinery) History;

6. CSMP;

7. Machinery Index;

8. Copy of the last Full Power Run;

9. Ship's Allowance Lists, by BuShips;

10. The Work List, prepared for the inspection;

11. The list of outstanding alterations;

12. The latest Ship Characteristics Card;

13. The Ship's Plan Index;

14. Boiler Record Sheets;

15. Operating Logs;

16. Etc

Conduct of the Inspection

The inspecting group for the engineering department should conduct a critical, and thorough, inspection of the machinery and equipment under the cognizance of the department. The condition sheets supplied by the type commander serve as a guide and a check-off list in making the inspection. Appropriate remarks, comments, and recommendations are entered on the condition sheets for each unit of machinery or equipment.

In conducting the inspection, the inspectors should cooperate with the ship's personnel. No attempt is made to follow a predetermined inspection schedule, but different units are inspected as they are made available by the ship's company. If the ship is prepared for the inspection there should be no delay between the inspection of the different units of machinery. It is not necessary that all machinery of one type be inspected simultaneously, nor is it necessary to complete the inspection of one space before going to another.

Some description of the main items of an inspection follow:

1. All opened machinery and equipment are carefully inspected, especially where repair work is needed as indicated on the work list.

2. An investigation should be made to locate any defects, in addition to those known to the ship, that may exist in material condition or design.

3. Operational tests of machinery and equipment, in accordance with the furnished list, are observed.

4. Ensure that electrical equipment is not endangered by salt water from hatches, doors, or ventilation outlets. Check for possible leaks in piping flanges

5. Ensure currently required fire fighting and damage control equipment in the engineering spaces is installed and properly maintained in accordance with current directives.

6. Inspect the supports and running gear of heavy equipment (boiler sliding feet, condenser saddles, and turbine supports).

7. Inspect holding-down bolts, plates, and other members of machinery foundations. Make free use of hammers for sounding, and of file scrapers for removing paint to investigate for metal corrosion.

8. The condition sheets should be checked to see that all the required information has

been filled in by the ship being inspected, and that all items have been checked off and filled in by the inspector.

9. Ensure that routine tests of mechanical and electrical safety devices are being conducted according to current directives.

10. Inspect the Material History and CSMP and make sure that they are being kept in accordance with prescribed procedures. A check should be made to see that all known repair requirements are listed in the CSMP.

Analysis and Reports

A critique should be held on board the inspected ship, at a convenient time after the completion of the material inspection, in order that the ship may derive the greatest benefit from the inspection. It should be attended by the ship's commanding and executive officers, heads of departments and such other personnel as may be designated from the inspected ship, the Chief Inspector, and Inspectors of each inspection group.

The inspectors, after receiving data from the assistant inspectors, submit reports of their inspections to the Chief

Inspector. These reports provide a means of furnishing the inspected ship with those observations that may not be fully discussed during the critique but are of interest to the ship's officers concerned. The inspector's reports should include his evaluation and any recommendations for the items inspected or observed. These reports can be used by the ship as a check-off list for corrective action and material improvement.

The Chief Inspector, after receiving the reports from the inspectors, will make up his report, evaluating and grading the inspection. The Chief Inspector's report should mention, with appropriate comment, the following:

1. Those conditions requiring remedial action which should be brought to the attention of the commanding officer of the ship inspected, and to higher authority.

2. Those conditions of such excellence that their dissemination will be of value in improvements to other ships.

3. Those suggestions or recommendations which merit consideration by higher authority.

The final smooth report is written up in a detailed procedure in accordance with the type commander's directives.

BOARD OF INSPECTION AND SURVEY INSPECTION

The (Main) Board of Inspection and Survey is under the administration of CNO. This board consists of a flag officer, as president, and of such other senior officers as may be required to assist him in carrying out the duties of the board. Regional boards and sub-boards are established, as necessary, to assist the Board of Inspection and Survey in the performance of its duties. In this discussion we are considering the shipboard inspections that are made by the sub-boards. These sub-boards consists of the Chief Inspector and about 10 or more members, depending upon the type of ship that is to be inspectedv

Material Inspections

The inspection made by the Board of Inspection and Survey is in many ways similar to the Material Inspection that has just been discussed. In fact, the Board of Inspection and Survey's inspection procedure, condition sheets, and reports are used as a guide in establishing directives for the Material Inspection. The primary difference, in view of routine shipboard material inspections, is that the Material Inspection is conducted by Forces Afloat, usually a sister ship, and the Board's inspection is conducted by a specially appointed board. Also, the Board of Inspection and Survey conducts other types of inspections which are of a different nature.

Inspections of ships are conducted by the Board of Inspection and Survey, when directed by CNO, to determine their material condition. This inspection usually takes place once in every 3 to 5 years. Whenever practicable, such inspections should be held sufficiently in advance of a regular overhaul of the ship to permit accomplishment, during such overhaul, of the authorized work resulting from the Board's recommendations. Upon the completion of its inspection the Board will report the general condition of the ship and its suitability for further naval service, together with a list of the repairs, alterations, and design changes which, in its opinion, should be made.

Acceptance Trials and Inspections

Trials and inspections are conducted by the Board of Inspection and Survey on all ships prior to final acceptance for naval service, to determine whether or not the contract and authorized changes thereto have been satisfactorily fulfilled. These inspections are usually conducted before a new ship is placed in commission. Similar inspections are made on ships that have been converted to other types. All material, performance, and design defects and deficiencies found to exist, either during the trials or as a result of examination on completion of trials, are reported by the

Board, together with its opinion as to the responsibility for correction of defects and deficiencies. The Board will recommend any changes in design which it believes should be made in the ship or in others of its type. Recommendations as to the acceptance or rejection of the ship are made to the Secretary of the Navy.

Except when the circumstances of war prevent, the preliminary acceptance trial takes place at sea over an established trial course. Tests include full-power runs ahead and astern, quick reverse, boiler overload, steering, and anchor engine. During the trial, the builder's personnel usually operate the ship and her machinery. Ship's personnel who are on board to observe the trial should carefully inspect the operation and material condition of machinery and equipment. Any defects or deficiencies should be noted and brought to the attention of division or engineer officer, so that the items can be discussed with the appropriate members of the Board of Inspection and Survey.

Survey of Vessels

Survey of a vessel is conducted by the Board of Inspection and Survey whenever a vessel is deemed by CNO to be unfit for further service because of material condition or obsolescence. The Board will, after a thorough inspection, render an opinion to the Secretary of the Navy as to whether the vessel is fit for further naval service, or can be made so without excessive cost.

If, in the opinion of the Board, the vessel is unfit for further naval service the Board will make appropriate recommendation as to its disposition.

FULL POWER AND ECONOMY TRIALS

Trials are necessary to test engineering readiness for war. Except while authorized to disable or partially disable, ships are expected to be able to conduct prescribed trials at any time. Ships normally should be allowed

approximately a 2-week period after tender overhaul, and a 1-month period after shipyard overhaul, to permit final checks, tests, and adjustments of machinery before being called upon to conduct a competitive trial.

Trials are also held from time to time to determine machinery efficiency under service conditions, the extent of repairs necessary, the sufficiency of repairs, and the most economical rate of performance under various conditions of service.

There are a number of different types of trials which are carried out under specified conditions. To afford a general idea of the different types of trials, a list comprising most of them is given here:

1. Builder's trials,
2. Preliminary acceptance trials,
3. Final acceptance trials,
4. Post repair trials,
5. Laying up or pre-overhaul trials,
6. Recommissioning trials,
7. Standardization trials,
8. Tactical trials,
9. Full power trials, 10. Economy trials.

The trials that are considered to be routine ship's trials are numbers 4, 9, and 10 of this list. The Post Repair Trial has been discussed in chapter 13 ("Navy Repair Procedures") of this training course. In this present chapter, we will discuss the full power and the economy trials. In case you should need or want information on the other types of trials, they are explained in chapter 8 of BuShips Manvud,

Inspections and Tests Prior to Trials

The full power and the economy trials, as discussed in this chapter, are considered in the nature of competitive trials. It is assumed that the ship has been in full operational status for sufficient time to be in a good material condition and to have a well-trained crew.

Prior to the full power trial, inspections and tests ot machinery and equipment should be made to ensure that no material item will interfere with the successful operation of the ship at full power. The extent of the inspections and the tests will depend a great deal upon the time available due to operational commitments, the recent performance of the ship at high speeds, and the material condition of the ship.

The inspection and tests of boilers, main engines, pumps, auxiliary machinery, safety devices, piping systems, and all equipment necessary for the proper operation of the engineering plant should be made: (1) as prescribed in those chapters of BuShips Manual which contain detailed instructions for the various units of the plant, or (2) in the absence of specific instructions, as the dictates of good engineering practice may require.

Not later than one day before a trial, the engineer officer should report to the commanding officer the condition of the machinery installation, stating whether or not it is in proper condition and fit to proceed with the trial, or wherein any part is, in his opinion, not in a safe and proper condition.

General Rules for Trials

During all full power trials, and during other machinery trials to which they may be applicable and consistent with the conditions imposed, the following general rules should be observed:

1. The speed of the engines should be gradually increased to the speed specified for the trial. Prior to commencing a power trial, the machinery should be thoroughly warmed up, as it would be by operating at a high fractional power.

2. The machinery should be operated economically, not exceeding designed pressures, temperatures, and number of revolutions.

3. The full power trial should not be conducted in SHALLOW WATER, which is conducive to excessive

vibration, loss of speed and overloading of the propulsion plant. Detailed information on the depth of water for your ship may be obtained from chapter 8 of BuShips Manual.

4. Should it be found desirable to continue a full power trial beyond the length originally specified, the observations should be continued until the trial is finished. The 4 hours, or any other predetermined time, of the trial should be continuous and without interruption. If a trial at constant rpm be discontinued for any reason, that trial should be considered unsatisfactory and a new start made. No major changes of the plant set-up or arrangement should be made during economy trials.

Trial Requirements

Trial requirements for each ship, covering the revolutions per minute for full power at various displacements and injection temperatures, are recommended by BuShips to CNO, who will furnish approved requirements to commanders and units concerned. The revolutions per minute for 15, 20, and 25 knots are also furnished for the appropriate ships.

Full power trials are of 4 hours' duration, as far as the report data are concerned. The usual procedure is to operate the ship at full power for a sufl^cient length of time until all readings are constant, before starting the oflRcial 4-hour trial period. The smoke prevention trial (Diesel-driven ships excepted) should be the last hour of the full power trial and should be run at the same speed. Economy trials are of 6 hours' duration; a different speed being run at each time a trial is made.

Trials once scheduled should be run unless prevented by circumstances such as:

1. Weather conditions which might cause damage to the ship;

2. Material trouble which forces the ship to discontinue the trial, or which might cause damage to the machinery if the trial were continued;

3. Any situation such that running or completing the trial would endanger human life.

If a trial performance is unsatisfactory the ship concerned will normally be required to hold a retrial of such character as the type commander may consider appropriate to demonstrate satisfactory engineering readiness.

The fact that a ship failed to make the required rpm for any hour during the trial, and the amount by which it failed, should be noted in the trial report. Similarly, the number of seconds smoke was observed during the smoke prevention trial should be noted.

Observation of Trials

When full power trials are scheduled, observing parties should be appointed from another ship whenever practicable. When a ship is scheduled to conduct a trial while proceeding independently between ports, or under other conditions where it is considered impractical to provide observers from another ship, the ship under trial may be directed to appoint the observers. For economy trials, observers may be appointed from the ship under trial.

The number of personnel assigned to the observing party will vary according to size and type of ship. The duties of the observing party are usually as follows:

1. The Chief Observer will organize, instruct, and station the observing party. He checks the ship's draft, either at the beginning of the trial or before leaving port; supervises the performance of the engineroom observers; checks the taking of counter readings; renders all decisions in accordance with current directives; and checks and signs the trial report.

2. The Assistant Chief Observer assists the Chief Observer as directed; supervises the performance of the fireroom observers; checks the taking of fuel oil soundings and meter readings; observes smoke as required; and makes out the trial report.

3. Assistant observers take fuel soundings and meter readings, counter readings, the ship's draft, and other data as may be required for the trial report. Some of the things that should be

accomplished or considered before starting the trial are as follows:

1. When requested by the observing party, the ship under trial should provide or designate a suitable signaling system so that fuel soundings and the readings of counters and meters may be taken simultaneously.

2. The ship under trial should furnish the Chief Observer with a written statement of the date of last undocking, and the authorized and actual settings of all main machinery safety devices and dates when last tested. The ship should have its draft, trim, and loading conform to trial requirements. In case a least draft is not specified, the liquid loading should equal at least 75 percent of the full load capacity.

3. The Chief Observer should determine draft and trim before and after the trial. He should verify the amount of fuel on board and correct this amount to the time of beginning the trial. He should determine the rpm required for the full power trial, at the displacement and injection temperature existing at the start of the trial.

4. The observing party should detect and correct promptly any errors in recording data, since it is very important that the required data be correct within the limits of accuracy of the shipboard instruments.

5. The Chief Observer should require members of the observing party to detect any violation of trial instructions, BuShips Manual, or of good engineering practice. The Chief Observer should verify any such report and then inform the commanding officer of the ship under trial. He should also include in the trial report a detailed report of any such occurrence.

Manner of Conducting Trials

Some of the requirements in regard to the manner of conducting full power and economy trials are as follows:

1. Unless otherwise ordered, a full power trial may be begun at any time on the date set, provided sufficient time remains so that the smoke-prevention run (if required) may be held during daylight.

2. The trial should be divided into hourly intervals, but readings should be taken and recorded every half hour. Data are submitted as hourly readings in the trial report.

3. Fuel expenditures for each hourly interval of the trial should be determined by the most accurate means practicable, normally by meter readings corrected for meter error and verified by soundings.

4. During the smoke-prevention run, the smoke pipes should be continuously observed by an officer. He should record in seconds the time during which smoke is observed.

5. The appropriate material condition of the ship should be set during the different trials.

6. During all trials the usual "housekeeping" and auxiliary loads should be maintained and the minimum services provided should include normal operation of the distilling plant, air compressor, laundry, galley, ventilation systems, elevators (if installed) and generators for light and power.

7. All ships fitted with indicators, torsionmeters, and other devices for measuring shaft or indicated horsepower should take at least two observations during the full power trial to determine the power being developed.

8. The Chief Observer should state in his report of the trial whether all rules for the trial have been complied with.

Some Hints in Regard to Full Power Trials

There are special forms used for full power and economy trial reports. Since illustrations of these forms are not given in this training course, it would be a good idea to obtain copies of these report forms from your log room, to get some idea of the data and readings that will be required for full power and economy trials.

Trial forms, and such items as tachometers, stop watches, and flashlights, should be available to the observing party and the personnel that take readings. Any gages or thermometers which are considered doubtful or defective should be replaced before trials are held. Usually a quartermaster will check and adjust all clocks in the engineering spaces and on the bridge before any trials are held.

A great deal of emphasis is placed on taking accurate fuel oil soundings during a trial run. Although it may be an easy thing to say, it may be a very difficult thing to accomplish. Take for example a destroyer running at full speed at sea, especially if the sea happens to be a little choppy. The difficulty in obtaining a fuel oil sounding can be readily understood. To overcome this difficulty in obtaining accurate soundings, a slightly different method can be used.

When the ship is tied to a pier or anchored in calm water the service tanks and standby service tanks are filled to 95 percent capacity. Accurate soundings are taken on the standby service tanks. When the ship proceeds out to sea and builds up to full speed the fuel oil suction is taken from the service tanks. At the moment the full power run commences, the suction is shifted to the standby service tanks for the official run. At the moment that the run has been completed, the fuel oil suction is shifted back to the service tank. When the ship has returned to the pier or to the anchorage in calm water, accurate fuel oil soundings are again taken on the standby service tanks. The amount of oil consumed is computed for each fireroom and compared against the total reading of the

fuel oil meter for the period of 4 hours during the trial. This method will give you the most accurate soundings possible, under practical shipboard condition, for the trial run.

It is a good idea to make careful inspections and tests of equipment and items of machinery that may cause difficulties during full power operation. It is possible that hidden defects or conditions will not be detected during operation at fractional powers, which is the normal operating condition of the ship most of the time. For example, sprayer plates should be carefully inspected. Any sprayer plate in doubtful condition should be replaced with a new one. Sprayer plates that look fairly good and give apparently satisfactory service during everyday operations may prove to be inadequate for full power performance.

It is usually a common procedure of many commanding officers, when making full power trials, to bring the ship up to a speed of one or more knots below the trial run speed of the ship. He then turns the control of the speed (except in cases of emergency nature) of the ship over to the engineer officer. The control engineroom, under the supervision of the engineer officer, will bring the speed up slowly, depending upon the conditions of the plant, until the specified speed has been reached.

In view of the fact that for most ships the designed boiler power is the first factor that establishes the maximum speed that a ship can attain, it is good policy to check boiler steaming conditions before ringing up additional turns. The boilers should not be loaded down faster than they are capable of taking care of the increased load. The steam pressure and temperature should be kept at full value for the appropriate steaming condition. In other words, the turbines must not get ahead of the boilers. The boilers should be the controlling factor and should be kept ahead of the turbines.

If the turbines are allowed to get ahead of the boilers, the main steam pressure and temperature will drop below

normal values for that particular steaming condition, or speed of the ship. Then in order to make up this bss in steam pressure and temperature and to meet additional increases of speed that may be rung up, the boilers must be fired at an extremely high rate. In some ships, this firing rate may exceed the full-load rating of the boiler and approach the maximum 120 percent overload capacity rating of the boiler.

The engineroom personnel should always make use of the acceleration curve or table, to prevent the overloading of the boilers; but the use of the acceleration curve is of particular importance when accelerating near full speed and full power.

QUIZ

1. What are the two general categories of an administrative inspection?

2. Who is responsible for organizing and supervising the engineering department inspecting group, for an administrative inspection?

3. What is primarily inspected in an administrative inspection of the engineering department?

4. Who usually furnishes administrative inspection check-off lists?

5. What must be demonstrated in an operational readiness inspection?

6. What type of inspection consists of the conduct of a battle problem and other operational exercises, with much emphasis on gunnery, damage control, engineering casualty control, etc.?

7. Which type of inspecting party is usually larger—an administrative inspection party, or an operational readiness observing party?

8. Of what value is realism in the conduct of battle problems?

9. Would the ship's company b? likely to profit more from a battle problem which contained some element of surprise, or from one in which all aspects of the problem were known beforehand?

10. By what method should an observer ordinarily communicate with ship's personnel, during conduct of a battle problem?

11. What should observers do if the ship's company fails to discover an imposed casualty?

12. If observers find it necessary to coach ship's personnel, what notation must be made on the observer's forms?

13. If it is necessary, during the conduct of a battle problem, to open switches, close valves, stop machinery, etc., should these actions be taken by the observers or by ship's personnel?

14. What provision is made for handling actual casualties which might occur during a battle problem?

16. Should a problem casualty be imposed if there is any danger that it might cause real injury to personnel or damage to machinery?

16. During the conduct of a battle problem, who will usually have priority in the use of the IMC circuit?

17. What announcing system circuits and other means of interior communication will be kept for the exclusive use of the ship, during the conduct of a battle problem?

18. What circuit is kept available for use in case of actual emergency, during the conduct of a battle problem?

19. Why is it important to indicate the ship's good points, as well as weaknesses and deficiencies, in the analysis of a battle problem?

20. What are the two steps of an analysis of a battle problem?

21. When and where is the critique of a battle problem held?

22. What four points form the basis upon which an observer in a battle problem judges the action taken by main engine control?

23. Under what conditions would general administrative methods, general appearance, and cleanliness of compartments and machinery be covered by a material inspection?

24. What is the limit to which a unit of machinery or equipment must be opened for a material inspection?

25. In a multiple-shaft ship, what proportion of the main propelling units may be disabled for a material inspection?

26. How frequently are ships inspected by the Board of Inspection and Survey, for the purpose of determining their material condition?

27. List three routine ship's trials.

28. Why should a full-power trial NOT be conducted in shallow water?

29. What is the duration of a full-power trial, as far as the report data are concerned?

30. For most ships, what is the first factor which establishes the maximum speed the ship can attain?

APPENDIX I

ANSWERS TO QUIZZES

CHAPTER 2

PUMPS AND FORCED DRAFT BLOWERS

1. The emergency feed pump.

2. High suction lift.

3. A broken valve in the water end.

4. Because these devices would tend to draw air into the feed water.

5. Tight packing.

6. Weekly.

7. The liquid end.

8. Allow slightly less than 0.001 inch for each inch of cylinder diameter.

9. Turn a groove abput inch deep and ⅜ inch wide around the middle of each ring, and drill three or four ⅜-inch holes through the grooved section of the ring to allow equalization of pressures.

10. By shims placed at the bottom or at the top of the stay rods.

11. Because the pump operates at maximum efficiency only when it is operating at rated capacity and discharge pressure.

12. The pump is likely to become overheated.

13. About once every 2 months.

14. Fit a fiber or asbestos washer between the end of the sleeve and the shaft shoulder, and fill all clearances with white or red lead.

15. When the clearance shown on the manufacturer's plans is exceeded by 100 percent.

16. Once a day.

17. Theoretical displacement.

18. Capacity is equal to the theoretical displacement times the speed (rpm), MINUS losses from slippage, suction lift, etc.

19. To allow time for the pump to fill.

20. They are positioned axially by jam screws which bear against the thrust plate on the top or bottom casing head; and circumferen tially by guide pins.

21. It ensures adequate lubrication of the bearings at low speeds; and it prevents the development of excessively high lube oil pressures when the blower is operating at high speeds.

22. Whenever the oil is renewed in the reservoir, and oftener if necessary.

23. Daily.

CHAPTER 3

AUXILIARY TURBINES AND ACCESSORIES

1. One set of nozzles and its succeeding row or rows of moving and fixed blades.

2. Only in the nozzles.

3. A simple impulse stage.

4. A single velocity-compounded stage.

5. Navy symbol number 3080.

6. Excessive bearing clearances; worn or damaged oil seals and deflectors; and clogged oil grooves and drain holes in the bearings.

7. To allow the bearings and the journal to cool gradually, so that the bearing metal will not freeze to the shaft.

8. The readings taken when the bearing was installed.

9. It should extend most of the way around the exposed half of the journal.

10. Move the shaft back and forth as far as it will go, and measure the amount of movement with a dial indicator.

11. About 0.008 to 0.012 inch.

12. The radial clearance between the nozzles and the blading, and the radial clearance between the blading and the redirecting or reversing buckets.

13. At least twice a year.

14. At least once a quarter; whenever a turbine is put back into service after prolonged idleness; and whenever any work has been done on the governor.

15. (1) The adjusting screw must be set to regulate the discharge pressure, and (2) the needle valve must be set to minimize hunting.

16. So that you can make allowances for any differences which may exist.

17. Once a week.

CHAPTER 4 FUEL OIL SERVICE EQUIPMENT

1. Only when the heaters can be set ashore.

2. It should not exceed 50 psi.

3. Trichloroethylene.

4. 187' F.

5. The temperature in the return generator.

6. About 10 gallons.

7. The liquid circulation method.

8. The size and form of the heater, and the degree of contamination.

9. Because efficient condensation cannot take place if the heater is too hot.

10. About 2 hours, varying according to the condition of the heater.

11. Orthodichlorobenzene.

12. Rub the affected parts with an animal or vegetable oil.

13. Death.

14. (1) The rate of oil flow through the heater, (2) the temperature to which the oil is heated, and (3) the nature of the oil.

15. The liquid circulation cleaning method.

16. Once every hour.

17. One heater at full load.

18. A temperature corresponding to a viscosity of 200 Seconds Saybolt Universal.

19. Corrosion is increased, and the life of the heater is shortened.

20. All parts of the system on the discharge side of the fuel oil service pump, including the fuel oil heaters in that fireroom.

21. Oil pressure equal to 150 percent of the designed operating pressure.

23. Cork washers and graphite.

24. At least once each day.

CHAPTER 5 BOILER FITTINGS AND INSTRUMENTS

1. After the initial lift of the valve, an additional area is presented for the steam pressure to act upon. Hence, the valve is held open until the pressure has dropped a specified amount.

2. Superheater safety valves.

3. None, provided the valves do not simmer at the boiler operating pressure.

4. Navy "fine" lapping and grinding compound.

5. Navy lapping and grinding compound with a grain of 600.

6. The distance from the top of the floating washer to the overlap shoulder.

7. A seat gage.

8. After each regular cleaning period or general overhaul of a boiler, before the boiler is put into service; and at any other time when you have reason to doubt their proper operation.

9. The pressure gage installed on the same boiler.

10. Make it handtight only.

11. Raise the adjusting ring.

12. The upper adjusting ring.

13. Raise it.

14. Screw the metering disks IN.

15. (1) Water level in the steam drum, and (2) steam flow from the boiler.

16. (1) What level in the steam drum, (2) steam flow from the boiler, and (3) feed water flow.

17. They must be cut in as soon as general quarters is sounded; and under normal conditions they must be used often enough V ensure their proper functioning under emergency conditions.

18. At all times.

600

19. Switch from automatic operation to remote manual operation, and operate the regulator manually from the control panel.

20. About 300 psi.

21. At least once a watch while under way, twice a day during in-port steaming, and just prior to securing a boiler.

22. At least once a year.

23. The pressure drop across the superheater.

CHAPTER 6
BOILER WATER TREATMENT AND FEED SYSTEMS

1. Parts per million (ppm) and equivalents per million (epm).

2. Parts per million.

3. Scale is formed by the deposit of crystals; baked sludge is formed when matter that

was suspended in the boiler water settles and bakes on the metal surfaces.

4. Calcium sulfate.

6. Because it can cause tube failures.

6. Low alkalinity.

7. Before the feed water leaves the deaerating feed tank.

8. Zero.

9. Dissolved oxygen.

10. By controlling the amount of dissolved and suspended solids in the boiler water.

11. Because its impurities are different from those of ordinary boiler feed water, and cannot be controlled by the usual water-treatment methods.

12. The chloride content.

13. By blow-down; or by securing, draining, and refilling the boiler.

14. Securing, draining, and refilling the boiler.

15. Preferably, not more than 30 minutes.

16. At least 0.02 ppm.

17. Weekly; and just before use.

18. Daily.

19. It is not as economical as recirculation from the condensate line ahead of the deaerating feed tank.

CHAPTER 7

BOILER OPERATIONS

1. 120 percent.

2. It is 103 percent of the steam-drum pressure.

3. Operating pressure.

4. The generating surface, the superheater surface, and the economizer surface.

B. One or more burners must be in operation on the saturated side before a burner is lighted off on the superheater side.

6. From 5000 to 9000 pounds per hour, depending on the type and size of the boiler.

7. The difference is referred to as "the degree of superheat."

8. 50** F every 5 minutes.

9. Approximately 600' F.

10. Radiant superheater.

11. Diesel oil.

12. The smallest capacity available.

13. The more rapid burning characteristics of the distillate fuel.

14. End points for water circulation, moisture carry-over, and combustion.

15. End point for combustion.

16. Checking on fuel oil consumption.

17. Oil pressures, and the number and size of sprayer plates in use.

18. Carbon formation.

19. The oil strikes the tubes or the brick before it has had time to burn.

CHAPTER 8 BOILER EFFICIENCY

1. In the fireroom.

2. Carbon and hydrogen.

3. Oxygen and nitrogen.

4. The combination of carbon and hydrogen in the fuel with the oxygen of the air, to form

carbon dioxide and water vapor.

5. Atmospheric air.

6. Oxygen, nitrogen, carbon dioxide, water vapor, and inert gases.

7. It absorbs heat during its passage through the furnace, and carries off a portion of the heat when leaving the boiler.

8. 20.91 percent.

9. It is the sum of the heat of combustion of each element in the fuel.

10. Carbon dioxide.

11. 11.52 lb.

12. More Btu's are given off.

13. Greater.

14. The difference between the heat input and the heat absorbed.

15. Loss due to heat carried away in the stack gases.

16. Causes it to decrease.

17. High excess air.

18. Centrifugal force, translation along the axis of the atomizer, and gravity.

19. Because they ensure better atomization.

20. Between 200 and 300 psi.

21. That required to reduce the fuel oil viscosity to 135 Seconds Saybolt Universal.

CHAPTER 9 BOILER RETUBING

1. The joints are likely to become strained, causing leaks and possible permanent damage to the boiler.

2. BuShips.

3. Dissolved oxygen in the feed water.

4. Interior scale.

5. U. S. Naval Engineering Experiment Station, Annapolis, Md.

6. At least 2 ounces.

7. 1-gallon sample.

8. Cut the defective tube flush with the water drum or header.

9. A safety ripping chisel.

10. A backing out tool.

11. This is done to prevent the tube from splitting when belled.

12. Cleaning the holes with a medium and then a fine grit emery cloth wrapped around a piece of hardwood.

13. At least 3/16 inch.

14. Just enough to secure tightness.

15. Until they are tight under hydrostatic tests; but they must not be rolled to the point where they become distorted or damaged.

16. The tube is belled to no more than inch and not less than inch.

17. Any past experience in doing exactly the type of job you wish to estimate.

18. Man-hours.

19. Never attempt to estimate any job that you are not sure you can finish.

20. The ^*time loss" factor is the percentage of time you estimate will be lost because of inexperience of men doing the work.

CHAPTER 10 BOILER REFRACTORIES

1. The same grade as that already installed in the remainder of the wall.

2. Slagging.

3. Because all fuel oils contain some slag-forming elements.

4. By causing spalling.

5. Because it is impossible to remove slag without also removing some firebrick.

6. Only when authorized by BuShips.

7. Grade C.

8. By the occasional removal and examination of one or more firebricks from the furnace floor.

9. From the refractory drawings.

10. One or two inches.

11. At least one anchor bolt per 100 square inches of surface area.

12. Cut it into small pieces and spread it out on wet rags or burlap bags; spray it with clean water; cover it with wet rags and allow it to stand until workable (about 4 or 5 hours).

13. The material must be pounded in a direction which will prevent the formation of vertical cleavage planes parallel to the front casing.

14. Severe cracking of the plastic; in some cases, this might occur rapidly enough to cause an explosion.

15. Air-drying.

16. 1800^ F.

17. As soon as possible—in any event, not more than 24 hours later.

18. Prolonged air-drying.

19. Within 30 minutes.

20. Because it is structurally weakened by this process.

21. The hardened part should be cut off and thrown away. Do not attempt to salvage this hardened material.

22. By stippling with a stiff wire brush.

CHAPTER 11 BOILER CLEANING AND MAINTENANCE

1. When the boiler is wet, its condition cannot be accurately determined.

2. In the boiler record sheet and in the engineering log.

3. As many times as necessary to ensure the removal of all scale, sediment, or rust which can be removed by this method.

4. Representative tubes in all parts of the boiler should be explored with a power wire brush.

5. Because it is often the same color as the tube.

6. Because such filling does not prevent further corrosion in the pits, but it does make it difficult to judge the condition of the metal on subsequent inspections of the boiler.

7. At least every 1800 or 2000 steaming hours.

8. At each alternate boiler cleaning period, unless unsatisfactory conditions in the superheater indicate the need for more frequent cleaning.

9. The soot may become wet and form sulfuric acid, which will attack the tubes and drums.

10. Overheating of the tube (because of dirty or scaled watersides), or lack of sufficient steam flow through the superheater.

11. At least 1 week, and preferably about 3 weeks.

12. Damage to the brickwork, and acid corrosion of tubes and drums.

13. The water lance method and the soot blower method.

14. The water lance method.

15. If the water lance method is ineffective, and if there is plenty of fresh water available.

16. Alternation of firing and idle periods.

17. So that the wet steam will not dry up immediately upon contact with the hot tubes.

18. The "sweat" method.

19. The sweat is formed by condensation of moisture from the air, when cooled water is circulated through the tubes.

20. If the process is interrupted, the slag may return to its original hardened condition.

21. (1) When it is newly completed, (2) after a major tube renewal, or (3) when scale is so bad that it cannot be removed by other methods.

22. (1) To prove the tightness of all parts; and (2) to prove the strength of the boiler and its parts.

23. After each general overahaul or repair, or at any time when it is considered necessary to test for leaks.

24. Every 5 years.

25. They should be kept full, except in emergencies which require that the water be maintained at steaming level.

26. Between 2.5 and 3.5 epm.

CHAPTER 12 FIREROOM MAINTENANCE

1. One type is flammable; the other type is poisonous.

2. Quarterly.

3. To load the bolts evenly, and to eliminate uneven stresses on the flange.

4. For temperatures up to 560" F.

5. Steel.

6. In terms of steam ratings.

7. Straight up.

8. In general, it is better to have the higher pressure above the disk; in some cases, however, it is essential to have the higher pressure below the disk.

9. Semiannually, or whenever they do not operate properly.

10. Adjust the adjusting spring.

11. To prevent steam from coming in contact with the diaphragm.

12. The dome pressure gage will read the same as the outlet pressure gage.

CHAPTER 13 NAVY REPAIR PROCEDURES

1. They should be recorded in the Current Ship's Maintenance Project.

2. NAVALT.

3. Alterations equivalent to repairs.

4. The repair record card.

5. (1) Repair record card, (2) alteration record card, and (8) record of field changes.

6. The material (machinery) history.

7. As soon as the defective material condition is discovered.

8. By checking on the **Alt" numbers.

9. By referring to the Ship's Plan Index.

10. S4700.

11. Restricted availability.

12. Technical availability.

13. No change in operating schedule is involved in voyage repairs.

14. Upkeep period. 16. The ship's force. 16. Two weeks.

17. Repair ships are concerned with maintenance of all types of vessels; tenders are concerned with repairs and maintenance of the specific types of ships to which they are assigned.

18. AR.

19. 30 days.

20. Representatives of the ship, of the repair department, and of the type commander.

21. After it has been approved by the repair activity and assigned a job order number.

22. By means of metal tags secured with wire.

23. As ship-to-shop jobs.

24. Appoint a chief or first class petty officer to act in that capacity.

25. Both the repair activity and the ship are responsible.

26. After a Naval Shipyard has made any major repair to propulsion machinery.

CHAPTER 14 THE OIL KING

1. A 50-percent level.

2. To retain maximum stability and maneuverability of the ship.

3. The officer of the deck.

4. One-half hour.

5. Phone circuits are tested, air hose is connected to the fueling connection, and thermometer and pressure gage are screwed into the fueling connection.

6. Request the tanker to lower the pumping rate, and direct the manifold men to throttle down so as to fill the tanks slowly, up to 95 percent of capacity.

7. It will be carried through overflow lines to an empty overflow tank.

8. Temperature correction and specific gravity correction.

9. 60° F.

10. By use of a hydrometer.

11. The oil, the hydrometer, and the thermometer must be at substantially the same temperature.

12. A temperature corresponding to 450 Seconds Saybolt Universal.

13. About 40 psi.

14. Once each week.

15. In the Damage Control Book (for the particular type of ship).

16. The weight of the volume of water displaced by the ship's hull.

17. The center of her underwater body.

18. Perpendicular distance from main deck to waterline.

19. List is the inclination of a ship around her longitudinal axis; trim is inclination around her transverse axis.

20. The deliberate taking on of sea water, to correct list or trim.

21. Counterflooding in two steps: (1) to correct for most of the list, and (2) to correct for trim and the remainder of the list.

CHAPTER 15 AUTOMATIC COMBUSTION CONTROL

1. To maintain the air input and the fuel oil input in accordance with the demand for steam.

2. Electrical, hydraulic, and pneumatic.

3. Pneumatic.

4. Measurement of steam pressure in the boiler steam drum or header.

5. Steam pressure.

6. Amount of fuel oil flow.

7. (1) As a direct source of power, and (2) as a controlling or balancing force.

8. (1) To register the steam pressure in the boiler, and (2) to operate the pilotrol, which sets up an air loading pressure to control the operation of the standatrol relay.

9. The standatrol relay.

10. The oil pressure in the line between the fuel line and the ratio controller.

11. By sending loading air pressure to the averaging relay.

12. (1) From the air control relay (relayed by the selector valve), and (2) from the ratio controller.

13. To operate the damper control drive.

14. 28 psi.

15. Loading air pressure and control air pressure.

16. They are interconnected by air pressures only.

17. 15 psi.

18. The desired—or standard—steam pressure required for the satisfactory operation of the boiler.

19. It allows speed in loading air pressure changes, together with stability in the control system operation.

20. It will increase to some value above 15 pounds.

21. From 5 to 25 psi.

22. The admission of air to, or the discharge of air from, chamber D

23. 15 psi.

24. Chamber C.

25. The setting of the bleeder valve.

26. Hand setting.

27. The ratio controller.

28. 15 pounds.

29. The bleeder valve.

30. Chamber A.

31. The air must be clean and dry.

32. The fouling of pilotrol ports with dirt, oil, or water.

33. From 120 psi to 300 psi.

34. The system will tend to hunt.

35. Steam pressure will vary beyond the acceptable limits.

36. The PILOTROL and the standatrol.

37. Brushing with soapsuds.

38. About y2 cubic foot per minute.

39. Once during each watch.

40. At least every 3 months.

41. Loading air pressure received from a master sender.

42. It should be set on manual.

CHAPTER 16

PROPULSION TURBINES

1. (1) The cruising combination, and (2) the maneuvering combination.

2. It passes through (1) all stages of the cruising turbine; (2) all stages of the high-pressure turbine; and (3) all stages of the low-pressure turbine. It is then exhausted to the main

condenser.

3. For high-speed operation, and for maneuvering at any speed.

4. Cooling steam taken from the first stage of the high-pressure turbine.

5. By connecting the cruising turbine exhaust to the main condenser.

6. It is a pressure-velocity-compounded turbine.

7. A reaction stage consists of one row of fixed blades and its succeeding row of moving blades.

8. Two.

9. The astern turbines.

10. Before.

11. The second-stage air ejector.

12. 5 knots.

13. The maneuvering combination.

14. After the turbine has completely cooled, and within 24 hours of the time of securing.

15. The presence of water or other foreign matter.

16. Daily.

CHAPTER 17 ENGINEERING CASUALTY CONTROL

1. Abnormal noises, vibrations, temperatures, pressures, and operating speeds.

2. They should all be inspected to determine whether there is danger of the same type of failure.

3. At least daily.

4. Poorly fitted or worn carbon packing.

6. The commanding officer.

?• That there is risk of even greater damage in immediately securing the affected unit.

8. The boiler must be secured.

9. Low water.

10. Check the water level in the water gages.

11. Because the boiler must be allowed to cool gradually.

12. It should always be lined up on cold suction.

13. Contaminated oil, or clogged suction line.

14. Smother the spray and catch the oil in a container such as a bucket.

15. Flareback.

16. Shift the drain to bilge.

17. The burners adjacent to the damaged section must be cut out.

18. The danger of electric shock.

19. Yes. The danger of shock is greatly reduced, although not entirely removed.

20. Carbon dioxide.

CHAPTER 18 RECORDS AND REPORTS

1. The machinery history card.

2. They are arranged in sequence according to the Navy Filing Manual^ and are kept in large binders.

3. It is used to record measurements of reciprocating machinery, such as internal combustion engines and reciprocating pumps.

4. The machinery history card and the boiler record sheet.

5. They are retained on board ship, but a transcript may be requested by the Bureau of Ships.

6. It is intended to provide BuShips with information concerning the reliability of BuShips material.

7. It should be overlined, but NOT erased; and, if practicable, the correct entry should be made immediately after the entry that has been struck out.

612

CHAPTER 19

INSPECTIONS AND TRIALS

1. (1) The general administration of the ship as a whole, and (2) the administration of each department.

2. The engineer officer.

3. Engineering department paper work.

4. The type commander.

5. The ship's readiness and ability to perform wartime operations.

6. Operational readiness inspection.

7. Operational readiness observing party.

8. The more nearly the test approximates actual battle conditions, the more valuable it will be as training experience.

9. The element of surprise increases the value of the exercise.

10. Verbally, whenever practicable.

11. Observers may coach ship's personnel, if necessary.

12. (1) The time allowed before coaching, and (2) the information furnished.

13. By ship's personnel.

14. An emergency procedure is set up by the observing party with the ship's company.

15. No. In such cases the casualty should be simulated or omitted.

16. The observers.

17. All announcing systems and all other means of interior communication, except the IMC circuit.

18. The IMC circuit.

19. (1) Because it will heighten morale, and (2) because it will indicate which factors will require less emphasis in the shipboard training program.

20. (1) The critique, and (2) the observers' reports.

21. On board the observed ship, before the observing party leaves.

22. (1) Correctness of the action, (2) soundness of judgment, (3) assurance, and (4) speed.

23. Only if these factors had a direct bearing on material condition.

24. Only as much as necessary to reveal known or probable defects.

26. Not more than one-half.

26. About once in every 3 to 5 years.

27. (1) Post repair trials, (2) full-power trials, and (3) economy trials.

28. Because shallow water may cause excessive vibration, loss of speed, and overloading of the propulsion plant.

29. 4 hours.

30. The designed boiler power.

APPENDIX II

QUALIFICATIONS FOR ADVANCEMENT IN RATING

BOILERMEN (BT) RATING CODE NO. 4000

General Service Rating

Boilermen operate all types of marine boilers and fireroom machinery; transfer, test, and take inventory of fuels and water; maintain and repair boilers, pumps, and associated machinery.

Emergency Service Ratings

Boilermen G (Shipboard Boilermen), Rating Code No. 4001 . BTG Operate and maintain all types of marine boilers and associated machinery; transfer, test, and take inventory of fuel and water.

Boilermen R (Boiler Repairmen), Rating Code No. 4002 BTR

Repair and overhaul marine boilers and associated equipment aboard repair ships or at ship repair activities.

Navy Job Clatslflcationt and Codes

For specific Navy job classifications included within this rating and the applicable job codes, see Manual of ErUiated ,Navy Job ClasBifica-twMy NavPers 16105 (Revised), codes BT-4500 U> BT-4599.

Oualificaiions for advancement in rating

100 PRACTICAL FACTORS

101 Operational

1. When standing a boiler check watch under
^^ay:

a. Regulate water level in a steaming boiler..

b. Blow down gage glasses _

c. Detect high- and low-water conditions by observing gage glass and using gage glass blow-down valves

d. Detect signs of oil on surface of water in gage glass

e. Test low-pressure feed water alarms

f. Detect signs of priming and foaming in steam drum

g. Detect and report abnormal feed water conditions to boilerman in charge of fire-room

2. Line up and start forced-draf tblowers, and check for normal operating conditions

3. Line up and start fuel oil service pump, and check for normal operating conditions

4. Line up fuel oil service pump and stand by tanks to fuel oil service pumps

6. Shift fuel oil suction and clear fuel oil system of water

6. Regulate forced-draft blowers for proper combustion of fuel oil by watching periscope and noting furnace conditions and changes in burner conbinations . -

7. Cut in and secure feed water regulator

8. Regulate fuel oil temperature

9. Light off burners ^\ith torch

10. Cut in, cut out, change, and properly assemble burners

11. Regulate oil pressure by use of a micrometer valve

12. Adjust openings of air registers for proper combustion

Applicable Rates

BT BTG BTR

3 3 3 3

3

3 3

3 3 3 3

3

3
3
3 3
3 3
3 3
3 3
3 3
3 3
3 3 3 3
3
3
3

Oualificalions for advancement in rating

101 Operational —Continued

13. Detect signs of oil in heater drain collection tank, shift drains to bilges, and shift heaters. _

14. Line up, start, and operate fire and/or bilge pumps to:

a. Pump bilges

b. Supply water to fire and cooling main

15. Line up, start, and operate emergency feed pump to:

a. Feed steaming boilers, using hot or cold feed water suction

b. Fill idle boilers w4th feed water

c. Add boiler compound to boilers

16. Shift superheater drains from low-pressure (funnel) drains to high-pressure drain main

17. Open root valve, drain, and operate soot blowers for blowing tubes, in proper sequence

18. Sound fuel oil tanks while refueling ship

19. Line up fuel oil system for recirculating and warming up of oil prior to lighting off boilers. _

20. Test boiler water for content of chloride, alkalinity, and hardness

21. Use calculating charts for determining boiler compound dosage

22. Test boiler feed water for dissolved oxygen content

23. Shift fuel oil service pump from manual operation to pressure regulator control

24. Line up and cut in fuel oil heaters and oil system to burner manifold

25. Use portable tachometer to check speed of fuel oil service pumps and forced-draft blowers. _

26. Line up, start, and operate fire and bilge pump to ballast and deballast fuel oil tanks

27. Line up, start, and operate emergency feed pump to transfer feed water

28. Line up, start, and operate fuel oil transfer pump for fuel oil transfer

Applicable Rates

BT BTG BTR

3 3

3 3 3 3

3 3 3

3 3

3
3
3
2
2
2
2
2
2
2
3 3 3
3 3
3
3
2
2
2
2
2
2
2
2

Qualifications for advancement in rating

101 Operational —Continued

29. Light off, operate, and secure superheaters on superheat control boilers

30. Split or cross-connect the following engineering systems:

a. Main steam

b. Auxiliary steam

c. Main feed

d. High- and low-pressure drain

e. Firemain

f. Cooling main

g. Fuel oil suction

h. Auxiliary exhaust main

31. Line up, start, and operate emergency feed pump to hydrostatically test boilers

32. Line up fuel oil system for fueling ship

33. Test fuel oil service, stand-by, and storage tanks for presence of water-

34. Empty fuel oil storage tanks and ballast with sea water, using proper sequence ^

102 Maintenance and/or Repair

L Clean watersides of a boiler, using power-driven tube cleaners

2« Clean boiler firesides, using scrapers, wire brushes, and lances

3, Clean boiler handhole seats and handhole plate seats

4. Spot in, repack, and renew Bonnet gaskets in high-pressure steam valves

6. Replace boiler gage glasses

6. Replace zincs in all fireroom auxiliary machinery equipped with lubricating oil coolers

7. Change lubricating oil in all fireroom auxiliary machinery

8. Spray fireside surfaces with metal conditioning compound

9. Clean and replace parts on smoke indicators. _ 10. Disassemble, clean, and assemble lubricating

oil pppj^r9 pn all fireroom auxiliary machines..

Applicable Rates

BT BTG BTR

3

3 3

3

3

3

3

3

3

3 2

3

3

3 3 3

Qualifications for advancement in rating

102 Maintenance and/or Repair —Continued

11. Locate principal isolation valves in engineering and adjacent spaces

12. Locate and grease boiler sliding feet

13. Clean oil burner atomizers

14. Inspect ana clean all strainers. _

15. Clean and test air register for proper operation

16. Renew weak or broken valve springs on pump end of reciprocating pumps

17. Maintain and repair boiler power-driven tube cleaners

18. Read and work from mechanical drawings

19. Dismantle and reassemble boiler internal steam drum fittings

20. Repair burner cone opening, using plastic firebrick

21. Replace smaH areas in furnace floor and walls, using insulating material and firebrick.

22. Make repairs and replacements to gaskets on boiler outer casing panels and doors

23. Repack and replace parts on soot blowers.

24. Replace lubricating oil pump parts on all fireroom auxiliary machinery

25. Spot in valve seats and disks on pump end of a reciprocating pump

26. Adjust tappets for proper piston stroke on reciprocating pumps

27. Repack pump end of a reciprocating pump..

28. Set all relief valves on fireroom auxiliaries

20. Clean flanges and replace gaskets in main and auxiliary steam lines.

30. Spot in slide valve on steam chest of reciprocating pumps

81. Repair and replace parts of forced-draft blower shutters and toggle gear

32. Repair insulation and lagging on steam lines_

33. Disassemble, repair, or replace parts in high-and low-pressure steam traps

Applicable RaUs

BT BTG BTR

Qualifications for advancement in rating

102 Maintenance and/or Repair —Continued

34. Inspect oil burner atomizers for wear and damage

35. Clean firesides of a boiler, using the two methods: Hot-water washing and wet-steam lancing

36. Install handhole plates and tighten or make adjustments as necessary to pass a hydrostatic test

37. Set and test bolier safety valves

38. Test boiler casings for airtightness

39. Replace thrust and shaft bearings on all fire-room auxiliary machinery, exclusive of electrical equipment

40. Check and set soot blowers for proper blowing arcs

41. Replace chrome ore on boiler stud tubes

42. Disassemble, clean, and replace parts on pump pressure regulators

43. Replace carbon packing rings and oil seal rings in all fireroom auxiliary machinery

44. Replace or plug defective boiler tubes

45. Dismantle, clean, and replace defective parts and reassemble G-fin type of fuel oil heaters

46. Clean fuel oil heaters with chemical equipment. _

47. Reface main and auxiliary steam line flanges. _ 48*. Fit piston rings to steam cylinder of reciprocating pumps

49. Gag safety valves

50. Replace power and idler rotors of positive displacement rotary (fuel oil) pumps

61. Adjust oil burner atomizers with reference to the diffuser plates

52. Disassemble boiler safety valves and inspect and replace defective parts

53. Renew boiler furnace brickwork and insulation

54. Boil out boilers

55. Insi>ect boiler uptakes and smoke pipes to determine their condition

Applicable Rates

BT BTG BTR

Oualtficationa for advancement in rating

102 Maintenance and/or Repair —Continued

56. Conduct hydrostatic tests on boilers, and make required inspection for tightness or strength

57. Make detailed inspection of an open boiler to determine material conditions

103 Administrative and/or Clerical

1. Make entries in fireroom operating logs

2. Take charge of fireroom watch in port under auxiliary steaming conditions

3. Make entries in fireroom lighting off and securing sheets _

4. Locate and use appropriate sections of Bu-Ships Manual, manufacturers' instruction books, mechanical drawings, and handbooks to obtain data when repairing boiler and fireroom machinery

5. Compute and record daily fuel oil and water receipts and exp>enditures

6. Take charge of a fireroom watch or boiler control station when under way

7. Prepare fireroom entries in daily, weekly, monthly, quarterly, semiannual, and annual

check-off lists

8. Prepare boiler record sheets

9. Organize and supervise the work and training of personnel in all phases of operation, maintenance, and repair of marine boilers and auxiliaries

10. Estimate time, labor, and material needed for repair of boiler and fireroom equipment

11. Prepare naval shipyard and tender work requests

12. Organize and assign personnel to fueling stations

13. Order fireroom supplies and materials, using standard stock catalog

14. Supervise fireroom personnel when performing full power and economy runs

Applicable iiateB

BT BTG BTR

C C 3

t

1 1 1

C

c

c c c c

3 2

2 ,,

C

c

c c

c c

c c

c c

c c

c c

c

c

Qualifications for advancement in rating

200 EXAMINATION SUBJECTS

201 Operational

1. Safety precautions involved in performing tasks appropriate to applicable rates listed under 100 Practical Factors

2. Procedures to be followed in determining and correcting the following casualties:

a. Loss of fuel oil suction

b. Loss of feed suction

c. Failure of fuel oil service pump

d. Loss of feed pressure

e. Water gage glass carries away on boiler

f. Low wat^r in boiler,.

g. High water in boiler

h. Class A, B, or C fires

3. Purpose of the economizer

4. Chloride and hardness limits and make-up feed water

6. Types and frequency of boiler and feed water tests.,^

6. Internal and external fittings for naval boilers and the function of each fitting

7. Sources of salt contamination in boiler feed water

8. Characteristics of Diesel, Navy Special Grade 2, and commercial fuel oiL Purpose of each fuel oil test

9. Prescribed rates for raising and lowering superheat temperature on superheat control boilers.

10, Conditions which require lifting superheater safety valves by hand _

11, Procedures to be followed when superheat thermal alarm sounds

12, Procedures to be followed in determining and correcting the following casualties:

a. Failure of emergency feed pump to take suction

b. Boiler tube or other pressure part carries away.- _

Applicable Rates

BT BTG BTR

a 2

Qualifications for advancement in rating

201 Operational —Continued

c. Major fuel oil leak

d. Oil in fuel oil heater drain

6. Water in fuel oil

f. Fire in boiler casing...

g. Superheater is lit off and thermometer does not register normal increase in temperature.

h. Forced-draft blower failure

i. Brick or plastic falls out of furnace wall... j. Reduction of high salinity in boilers while steaming

k. Lighting off without steam pressure

1. Steaming boiler op>erating pressure drops

below 85 percent

13. Various types of naval boilers

14. Procedures to be followed when using smoke-

making atomizers for laying a smoke screen. 16. Construction and operating principles of superheater flow indicators

16. Procedures to be followed when spraying

metal conditioning compound on boiler firesides

17. Procedures to be followed when superheater

or superheat control boilers are lit off and additional boilers are to be cut in

18. Lighting off, operating, and securing proce-

dures on all types of naval boilers

19. Construction and operating principles of all

types of boiler and superheater safety valves.

20. Inspections to be made preparatory to report-

ing fireroom ready to answ er all bells

202 Maintenance and/or Repair

1. Types of power-driven tube cleaners used for

cleaning watersides

2. Methods of removing soot from the firesides of

a boiler

3. Construction and use of globe, gate, and needle valves

Applicable Rates

BT BTG BTR

2 2 2 2

2 1 1

1 1

1 1

1

1

3 3 3

2 2 2 2

2 1 1

1 1

1 1

1

1

1 1

1 1

C C

c c

c c

3 3 3

3 3 3

Qualifications for advancement in rating

202 Maintenance and/or Repair —Continued

4. Purpose of zincs in salt waterside of lubricating oil coolers

5. Limits of chloride and alkalinity of water in a steaming boiler

6. Effects of low and excessive alkalinity on the watersides of a boiler

7. Selection and use of packing and gasket material

8. Types, purposes, and location of boiler furnace refractories

9. Effects of scale on the watersides of a boiler.-

10. Effects of oil on the watersides of a boiler

11. Methods and procedures to be followed when washing boiler firesides with hot water

12. Factors governing boiler and fireroom machinery efficiency, causes of poor performance, and appropriate remedies

13. Effects of dissolved oxygen on the watersides of a boiler

14. Purpose and procedure for boiler hydrostatic

tests, and when conducted

203 Administrative and/or Clerical

1. Use of allowance lists for determining spare parts, tools, and supplies kept on board

2. Procedures for obtaining replacement parts and supplies; maintenance of inventory

3. Reports covering boiler damage and failures that are submitted to BuShips

Applicable Rates

BT BTG BTR

c c

c c

c c

c c

c c

c c

Acid refractories, 260, 261

Administration of repairs, 375

Advancement in rating, requirements for, 1-6, 615

Air systems, automatic combustion control, 436-438

Air-flow control, Hagan automatic combustion control, 482

Alignment, auxiliary turbine, 80, 81

Alkalinity

excessive; effect of, 149 low; effect of, 149

All Hands, 6

Allowance list, ship's, 365 Alterations, 347, 348

and improvement program, 352

and repairs, 347-353

equivalent to repair, 348

procedure for accomplishing, 378

record card, CSMP, 348-352

request for, 352, 353 Arrival conference

naval shipyard, 378

repair, 369

Atomization pressure, fuel oil, 230-233

Atomizer barrel; positioning of, in distance piece, 215-217

Atomizing temperature, fuel oil, 231-233

Automatic combustion control, 430-485

Auxiliary turbines and accessories, 52-86

Availabilities, 363-365

regular overhaul, 363

restricted, 364

technical, 364

upkeep period, 364

voyage repairs, 364 Averaging relay, automatic combustion control system, 464, 465

maintenance, 475, 476

Baffle mix, 286-289 Baked sludge, 148 Ballast systems, fuel oil, 419 Ballasting, 416-428 Barrel, atomizer, 215-217 Basic refractories, 260 Basic Hand Tool Skills, 5 Basic Machines, 5 Battle problem, 570-575 Belling tubes, 249 Blowdown, 114, 128, 157, 158 Blowers, forced draft, 7, 41-50

failure, 531

lubrication, 43

maintenance and repair, 46

safety precautions, 48 Blueprint Reading and Sketching, 5 Blueprints, 361-363 Boilerman 3 & 2, 5, 7, 52, 110, 402 Boilers

boiling out, 311-315

casings

corrosion, 321 maintenance, 320-322

compound. Navy; nature of, 155, 156

625

Boilers—Continued

corrosion of tubes caused by sulfur content in fuel, 299-301

feed water, 128-132, 144, 145, 147, 150, 152-155, 165-177, 520

firesides, 297-301

fittings and instruments, 110-142

frequency of mechanical cleaning

and inspection, 294 full-power and overload ratings,

206

idle, to be kept full, 318-320

lighting oflf, 183, 187-191

loads; checking on, 207-211

on cold ship; lighting off, 188-191

operation, 191-198

power, reduced; steaming with,

211-213 record sheet, 541 refractories, 257-289 retubing, 236-256 securing, 198-200 sliding feet, 323 steaming rates, 204-206 superheat control, 181-183, 188,

192-197, 200 tubes corrosion, 299-301 replacement, 236-256 uncontrolled superheat, 181-183,

187, 199

water, 144-152, 155-158, 165, 168, 238, 239 alkalinity, 149

and scale; submitting samples

of, 238, 239 care of, 155-158 chemical treatment, 145-152 abnormal boiling conditions,

prevention of, 152 carry-over, nature of, 152 requirements, 168 saltfi in, 145 tests, 144, 145, 165 Boiling out boilers, 311-315 Brickwork, care of, 257-260

Bulging, tube, 238

Buoyancy, 515

Bureau of Ships Journal^ 5

Bureau of Ships Manual, 1, 3-6, 14, 31, 41, 52, 61, 63, 67, 80, 110, 165, 196, 208, 218, 236, 327, 337, 360, 362, 377, 393, 413, 428, 501, 524, 556, 568, 580, 587-589, 591

Carbon formations, 214, 215 Carry-over, nature of; boiler water

treatment, 152 Casings, boiler, 320-322

testing for tightness, 321 Castable hydraulic refractory, high-temperature, 277-279

Casualty control. See Engineering

casualty control Catalog of Enlisted Correspondence

Courses, 6 Centrifugal pumps, 24-32 Check-off lists or sheets, 550

administrative inspection, 565 Chloride

concentration, boiler water, 157 test, 158, 159 Chrome castable refractory, 285, 286 Cleaning and maintenance, boiler,

291-323 Combustion, 225 and air, 225-227 and heat losses, 227-229 control system or equipment,

automatic, 430-485 temperature and heat of, 225 Compressed air, automatic combustion control system supply, 471 systems, 437 Conference, arrival, 369, 378 Contamination, feed water; sources

of, 153 Corrosion, boiler firesides, 299-301 watersides, 148

Counterflooding, 426 Cross-connecting valves, 518 Current ship's maintenance project (CSMP), 359-361 alteration record card, 348-352 repair record card, 356-359

Damage Controlman 1 & C\ 428 Damper control drive, automatic combustion control system, 466 Deaerating feed tank, boiler feed water, 171-174 recirculation from, 175 warming up, 174 Deaeration, process of, 150, 151 Defective tubes, removing, 241-244 Discharging fuel, 411, 412 Dissolved oxygen effect of, 150 samples

collecting, 161, 162 fixing, 162, 163

interpretation of result of test, 165

sources of, 150

test; equipment and procedures,

159-165 titration, 163, 164 Dock trial, 389

Drum exteriors, inspection of, 322 Drydocking ship, 387-389

Economizers, care of, 201, 295 Efficiency, boiler, 222-234 Kngineering casualty control, 508-534

factors influencing, 509 phajse nature of, 514 training, 513 Engineering log, 551, 567, 568 Engineering Operation and Casualty

Control Manualj 508 Engineering, Operation and Maintenance, 5 Fquipage custody record, 557 Equivalents per million (EPM), 144, 145

Expanders, tube air driven, 248 ball-drift type, 246 care of, 333 roller tyiKj, 246

Feed water, boiler, 12»-132, 144, 145, 147, 150, 152-155, 165-177, 520

care of, 152-155

contamination; sources of, 153

deaemting feed tank, 171-174

deaeration, 150

dissolved oxygen content, 144, 145, 150

make-up, 152, 154-156, 167-169, 175-177

oil and grease in; prevention of,

154, 155 regulators, automatic, 128-132

emergency operation, 131

maintenance, 132 requirements, 167 split-plant operation, 520 systems

open, 168

pressure-closed, 169-177 semiclosed, 168, 169 vacuum-closed, 169 tests and mandatory requirements, 165-168 transfer of, 176 Firebrick, 265-267, 270-273, 276 Fireroom

casualties, 521 maintenance, 325-345 operations, 179-221 Fires, 531 equipment to fight, 532 fuel oil, 532 in boiler casing, 533 in lighting or power panel, 534 precautions against, 531 Firesides, boiler, 297-301

cleaning; frequency and procedures for, 297 examining, 298, 299

Firesides, boiler—Continued water-washing of, 303-307 wet steam method of cleaning, 307-310

Fittings and instruments, boiler,

110-142 Flange joints, bolted, 335-337 Flexible couplings, 73 Flareback, 528-530 Forced draft blowers, 7, 41-50

failure of, 531 Forms

Alteration record card, 348, 359,

537, 538 Bearing record, 537, 538, 541 Blank utility card, 537 Boiler record sheet, 541, 542 Electrical machinery history card,

537

Electronic equipment history

card, 537 Engineering log, 554-556 Hull history card, 537 Machinery history card, 537, 538 Material analysis data report, 548,

549

Megger test record, 537, 538 Record of field changes, 537 Repair record card, 356, 359, 537, 538

Tube performance record, 537 Unit record card, 537, 538, 540

Fouling, fuel oil heaters, 98

Fuel oil

atomization pressure, 230-233 atomizing temperature, 231-233 combustion of, 222-224 control valve, automatic combustion control, 459, 460 discharging, 411, 412 saving of, 222 service equipment, 86-108

heaters, 86, 90-93, 98 tests, 100, 101

meters, 102-107

operation, methods of, 99, 100

Fuel Oil—Continued systems, 392-401, 419, 518, 519, 525-528 casualties in, 525-528 oil king's job, 392-401 water in, 413-415 Fueling ship, 401-411 Full-power and overload ratings,

boiler, 206 Furnace

floor; construction of, 271 front wall; construction of, 269, 270

lining; installing, 263-277

Geared-turbine installations, 488-490

Gears, reduction, 73, 506

General Specifications for Building

Vessels of U. S. Navy, 327 General Training Course for Petty

Officers, 2 G-fin fuel oil heater, 95-98 Governors

constant-pressure pump, 76-80

speed-limiting, 74-76

sf)eed-regulating, 74 Gravity, specific, 407, 421, 424

Hagan automatic combustion control equipment, 482-485

Handholes, 315-317

Hardness, boiler water, 156

Heaters, fuel oil, 90-98

High salinity, 530

Hot bearings, cause, 503, 504
Huddling chamber steam drum safety valve, 110-112, 121
Hydrostatic tests, 317, 318
Idle boilers to be kept full, 318-320 Impulse turbines, 52-85, 490-507 In-port steaming, 200 Inspecting party, 563 Inspections, 562-588 administrative, 563 check-oflf lists, 565-568
Inspections—Continued
and tests prior to trials, 587
engineering department, 664
general, of ship, 564
material, 575-585
operational readiness, 569-576 Instruments and fittings, boiler, 110-142 Insulating
block, 264
brick, 265
Jet flow steam drum safety valve, 110,112-114 Jig for fitting carbon packing rings, 69
Jigs; use of, in instruction, 261
Leakage, valve; cause, remedy, 340, 344, 345 Lieaks, fuel oil heaters, 98 Lighting off boilers, 183, 187-191 Lining, furnace; installing, 263-277 Liquid circulation cleaning method,
fuel oil heater, 90-93 List of Training Publications^ 6 Loading air pressure, 437 Loads, boiler; checking on, 207-211 Log, engineering, 551 Low
exhaust pressure (high vacuum),
maintaining, 499 steam pressure, procedure in case
of, 213 Lubrication
forced draft blowers, 43-46 pump, 28, 36, 37, 41 systems, pressure, 59-63
Machinery
index, 536
opening for inspection, 578-581 Main steam system, 520 Maintenance
boiler, 291-295, 298-301,303-316, 320
blowers, forced draft, 46
Maintenance—Continued feed water regulators, automatic, 132
fireroom, 325-345 piping, 334
preventive, 366, 510-612 pump, 13-22, 27-31, 36-40 ship's force, 365-367 valves, 337, 340, 341, 479 Make-up feed water, 154, 155, 175 Manholes, 315-317 Manual of Qualifications for Ad'-
vancement in Rating^ 1 Master controller, automatic combustion control system, 438-448 Material (machinery) history, 537-540
MathematicSy 5
Metal-conditioning compound, 301-303
Meters, fuel oil, 102-107 Military requirements, 1
Haval Reservist^ 6 Naval Training Bulletin^ 6 Navy Filing Manual, 350, 362, 538 Navy Regulations, 363, 376, 387 Neutral refractories, 261, 263 Nozzle reaction steam drum safety valve, 110, 112
Observing party, 569 Oil
fuel. See Fuel

king, 392-428

Pollution Act, 415

seal rings, 71-73 Open feed water system, 168 Operating records, 560 Operation, boiler, 191-198 Operational readiness, 569-575 Organization

naval shipyard, 379-381

repair, shore-based, 374 Overloading, 203-213 Overspeed protection devices, 74-76 Oxygen pitting, 237

Oxygen-testing equipment, care of, 160, 161

Painting, 327-332

equipment; care of, 331

materials used in, 327 Piping; care, maintenance, 334-337 Plastic

chrome ore (PCO), 279-285 front, furnace; installing, 272-276

Pressure atomization, 230-233 lubrication systems, 59-63

Pressure-closed feed system, 169-177

Pressure-velocity-compounded impulse turbine, 491, 492

Preventive maintenance, 366, 510-512

Professional requirements, 2-4 Progressman, ship's, 383 Propulsion turbines, 488-507

care of, 504

casualties, 500

operation, 496

safety precautions, 505

types of, 490

Pump governors, constant-pressure,

76-80 Pumps, 7-41 centrifugal, 24-32

maintenance and repair, 27

safety precautions, 32

tests and inspections, 31 reciprocating, 7-23

maintenance, 14

repair, 15-22

safety precautions, 22 rotary, 33-41

maintenance and repair, 36

safety precautions, 41

tests and inspections, 40

Qualifications, advancement in rating, 1-€, 615

Quizzes, 50, 85, 108, 143, 178, 221, 235, 256, 290, 323, 346, 390, 429, 485, 507, 534, 561, 595 answers to, 597-614

Radial bearings, 63-67 Ratio controller, 460 Reaction turbines, 490 Reagents, preparation of; stock solution for, 165 Reciprocating pumps, 7-23 Recirculation, condensate, 175 Records, 536-548, 552-560 boiler, 541

CSMP, 348-352, 356-359 equipage custody, 557 maintenance of, 567, 568 material; assembly of, 581 oj)erating, 560

repair parts stowage, 559, 560 Reducing valves, 341-345

in automatic combustion control system, 480 Reduction gears, 73, 506 Refractories, 257-289

casualties to, 530 Regulators, automatic feed water,

128-132 Repairs

administration, 375

and alterations, 347-353

conference, arrival, 369

job, estimating, 251-256

list, procedure for submitting, 376-378

parts stowage record, 559, 560

procedures. Navy, 347-390

pumps, 15-22, 27, 36

record cards, CSMP, 356-36i'

ships and tenders, 367, 368

ship's force, 365-367

shore-based, 373, 374 Reports, 536, 548, 561 Retractable-type soot blower,

single-nozzle, 135 Retubing, boiler, 236-256 Rotary pumps, 33-41

Rotation in use of boilers, 198 Rust prevention, 326

Safe minimum steam flow, superheater, 192 Safety precautions boiler, 296

blower, forced draft, 48

fireroom, 217-220

fuel oil heater, 90, 93

pump, 22, 32, 41

turbine, 505, 506 Safety valves, boiler, 110-128

authorized popping and reseating pressures, 112, 117, 126

gagging, 126, 127

huddling chamber, 110-112, 121

jet flow, 110, 112-114

lapping, 119, 120

maintenance and repair, 117-126

nozzle reaction, 110, 112, 127

overlap, 123, 124

pressure pilot-operated superheater, 114, 115

spring-loaded superheater, 114, 115

steam drum, 110-114

testing and setting, 126-128 Salts

scale-forming, 147

in boiler water, 145 Scale formation

efifect of, 147

process of, 146 Scale-forming salts, 147 Seal rings, oil, 71-73 Sea-water distillate, nature
of, 152 Securing boilers, 198^200 Selector valve, 456-459 Semiclosed feed water systems, 168,
169

Shaft glands, 68-71 Ship-to-shop jobs, 371 Shipyards, naval; repair procedures at, 375-
390

Shops, naval shipyard, 381, 382

Shore wat€r, undesirability of, 152, 153

Shore-based epair activities, 373, 374

Sliding feet, boiler; examination of, 323

Smoke screen, laying, 217

Smoke-pipe covers and rain gutters, 322

Soot blowers, 132-137

maintenance, 133 Split-plant operation, 515-520 Stability, theory of, 421-428 Staging and compounding, turbine,

54-56, 490-496 Standatrol relay, 434, 436, 447, 449-

458, 464, 465, 468, 470-473 Steady steaming, 233 Steam

flow indicators, superheater, 137-142

pressure, low; procedure in case of, 213

system, main, 520 Steaming ratBs, boiler, 204-206 Stowage and care, tools, 332-334 Strainers, fuel oil, 107

Superheat control boilers, 181-183,

188, 192-197, 200 Superheaters

and economizers, cleaning of, 295

definitions, 181

outlet temperature; steadiness of,

while steaming, 194 safe minimum steam flow, 192 safety valves, 114-116 slagged; cleaning of, by "sweating'^ method, 310, 311 steam flow indicators, 137-142 Superintendent, ship's, 382, 383 Surface and bottom blows, use of,

157, 158 Survey, of equipage, 557-559

Temperature

atomizing, fuel oil, 231-233 superheater, raising and lowering, 193

Tender repair jobs, checking progress of, 372 Test data, boiler, 229 Tests

chloride, 158, 159 dissolved-oxygen, 159-165 feed water, boiler, 144, 145, 165-168

heaters, fuel oil, 100, 101

hydrostatic, 317, 318

pumps, 31, 40 Thrust bearings, 67 Titration, dissolved-oxygen, 163,164 Tools; care, stowage, 332-334 Training Courses and Publicalions

for General Service Ratings, 6 Transfer systems, fuel oil, 393-401 Trials, 585-595

acceptance, 585

dock, 389

full power and economy, 586-595

general rules, 588

inspection and tests prior to, 587

manner of conducting, 592

observation of, 590

post repair, 389, 390

requirements, 589 Troubles, fuel oil meters, 105-107 Tubes, boiler

belling, 249

corrosion of, caused by sulfur content in fuels, 299-301 data sheets, 543 defective; removing, 241-244 expanders

air driven, 248

ball-drift type, 246

care of, 333

roller type, 246 exploring blocks, cutting, 239, 240 failure, 237-240 plugging, 250

Tu bes—Con t i nued

removing, for inspection, 236 replacement, 236-266

cleaning, installing, 244-249 Turbines

auxiliary, and accessories, 52-85 alignment, 80, 81 casing joints, 80 classification, 52-59 by staging and compounding,

54-56, 490-496 by steam flow, 56, 491 maintenance and repair, 63-81,

504, 505 safety precautions, 84, 505, 506 tests and inspections, 82-84 vibration, 82, 500 impulse, 52-85, 490-496 propulsion, 488-507 care of, 504 casualties, 500 operation, 496 safety precautions, 505 types of, 490 reaction, 490

Uncontrolled superheat boilers, 181-

183, 187, 199 Use of other than Navy Special fuel

oil, 201-203 U, S. Navy Safety PrecaiUionSj 3

Vacuum, loss of; cause, 501, 502 Vacuum-closed feed water system,

169 Valves

cross-connecting, 518 installation and maintenance, 337-341

leakage; cause, remedy, 340, 344, 345

reducing, 341-345

in automatic combustion control system, 480

safety, 110-128

maintenance and repair, 117 testing and setting, 126

632

Vaporization cleaning method, fuel

oil heaters, 86-89 Vessels, survey of, 586 Vibration

centrifugal pump, 27

forced draft blower, 47

turbine, 82, 500

Water

analyses, units for reporting, 144, 145

boiler, 144-152,155-158,165,168, 238, 239

feed, 128-132, 144, 145, 147, 150, 152-155, 165--177, 520

Water—Continued in fuel oil, 413-415 level, casualties in control of, 521-525

shore, undcsirability of, 152, 153 treatment and feed systems, boiler, 144-177

Watersides, boiler, 291-296 care of, after cleaning, 295 inspection and cleaning, 292-294

Work requests and job orders, 370 preparation of, 354-356

Yarway superheater steam flow indicator, 140, 142
it U. S. GOVERNMENT PRINTING OFFICE: 1957 396661